D1068906

THE LOEB CLASSICAL LIBRARY
FOUNDED BY JAMES LOEB, LL.D.

EDITED BY
G. P. GOOLD, PH.D.

FORMER EDITORS
†T. E. PAGE, C.H., LITT.D. †E. CAPPS, PH.D., LL.D.
†W. H. D. ROUSE, LITT.D. L. A. POST, L.H.D.
E. H. WARMINGTON, M.A., F.R.HIST.SOC.

THEOPHRASTUS
DE CAUSIS PLANTARUM

I

471

© *The President and Fellows of Harvard College 1976*

American ISBN 0-674-99519-8
British ISBN 0-434-99471-5

GRACE LIBRARY, CARLOW COLLEGE
PITTSBURGH, PA. 15213

THEOPHRASTUS
DE CAUSIS PLANTARUM

English + Greek

IN THREE VOLUMES

I

WITH AN ENGLISH TRANSLATION BY
BENEDICT EINARSON
THE UNIVERSITY OF CHICAGO
AND
GEORGE K. K. LINK
THE UNIVERSITY OF CHICAGO

QK
41
T23
1976
v. 1

LONDON
WILLIAM HEINEMANN LTD
CAMBRIDGE, MASSACHUSETTS
HARVARD UNIVERSITY PRESS
MCMLXXVI

CATALOGUED

Printed in Great Britain

CONTENTS

INTRODUCTION[a]

The Author and the Work

Theophrastus of Eresus in Lesbos lived to be 85. His death occurred in the year 288–287 or 287–286 B.C., which would put his birth in 372–371 or 371–370. He travelled and worked with Aristotle in 347–344 and resided with him at Assos and Stagira. On the death of Aristotle in 322 Theophrastus took over his teaching and research, as he had already done when Aristotle fled to Chalcis several months earlier. During the régime of his pupil, Demetrius of Phalerum (317–307), Theophrastus was enabled to acquire property at Athens and set up a regular school.

His original name was Tyrtamos,[b] and his father was a fuller; when Theophrastus speaks of the removal of stains from clothing (*CP* 2 5. 4) he is

[a] All references are to the *CP* (*De Causis Plantarum*) unless otherwise indicated; *HP* is the *Historia Plantarum, O* the *De Odoribus, Pl* the pseudo-Aristotelian *De Plantis,* U* the earlier draft of *HP* 9 8. 1–9 20. 5, appended in MS U as Book X. I have followed the chapters and sections of Wimmer's edition of 1854; at 2 7. 2 and 2 11. 4, where the sections are unusually inconvenient, I have followed his edition of 1866. When the peculiarity or word or expression referred to occurs more than once in a section the letters a, b and c distinguish the first, second and third occurrences.

[b] The name occurs nowhere else, but that is not unparallelled at Eresus: *cf. Τερτίκων,* implied by the patronymic *Τερτικωνείω* (W. Dittenberger, *Orientis Graeci Inscriptiones Selectae* [Leipzig, 1903; photographic reprint 1960], vol. i, p. 26 [no. 8. 37]).

speaking of what he knows.[a] Presumably the father owned a fuller's business; philosophy was not a gainful occupation, and independent means were required for its pursuit.

The *HP* and *CP* are parts of a single course of lectures that included the work *On Odours*.[b] The latest dates and datable events that are mentioned or implied in Theophrastus' works are the following:

HP 4 8. 4 Antigonus used to make the tackle of his warships out of the papyrus of lake Huleh. From 315 until his death in 301 he was often at war with Ptolemy I, whose control of Egypt cut off the papyrus import. The passage was probably written after his death.

HP 5 8. 1 A ***hendekḗrēs*** of Cyprian cedar built for Demetrius. Demetrius gained control of Cyprus in 306 B.C. and retained it until 294.

HP 5 2. 4 Sack of Megara by Demetrius: 307 B.C.

HP 4 3. 2 Ophellas' march against Carthage: 308 B.C.

HP 6 6. 3 Archonship of Simonides: 310–309 B.C.

CP 1 9. 5 Archonship of Nicodorus: 314–313 B.C.

On Stones, chap. viii. 59 Archonship of Praxibulus: 315–314 B.C.

HP 4 14. 11 Archonship of Archippus: 321–320 B.C. (another Archippus was archon in 318–317 B.C.).

[a] So he speaks of odours retained by wool and clothes (*CP* 6 19. 4), of fuller's earth (2 4. 3) and of the use of urine to penetrate and open up the roots of certain trees (3 17. 5–6; *cf.* 3 9. 3). One may speculate that Theophrastus' interest in the arts and crafts began at home.

[b] *O* 1. 1 ("in what precedes") refers to *CP* 6 9. 2; *O* 2. 5 ("as was also said earlier") to 6 18. 8; *O* 3. 7 to the lost *CP* VII; *O* 3. 11 to 2 18. 4 or 6 19. 2.

Presumably Theophrastus kept working at the *HP* and *CP* for the rest of his life; like Aristotle, he frequently mentions that a point needs further investigation.[a] The work is not a text-book; it is research. It was probably read to a chosen few and corrected after the ensuing discussions.[b]

The Structure of the Work

Aristotle first classified animals and gathered information about them in the *History of Animals*; he then proceeded to explain certain common or distinctive characters in the treatises *On the Parts of Animals*, *On the Generation of Animals*, *On the Motion of Animals*, *On the Soul* and in the group of shorter treatises known as the *Parva Naturalia*. So Theophrastus first classifies and identifies plants and gathers information about them in the *Historia Plantarum*, then proceeds to account for certain common or distinctive characters in the *De Causis Plantarum*. Aristotle's model was no doubt the study of astronomy: for astronomy observations were first required (*cf. Prior Analytics*, i. 30 [46 a 17–27]), and the theory of movements had to rest on them or at least do them no violence.[c] The very word that was

[a] δεῖ . . . σκέψασθαι καὶ ἀνιστορῆσαι *CP* 1 5. 5; ἐπισκεπτέον 2 4. 5; 3 14. 6; 4 6. 9; 4 13. 1; σκεπτέον 2 3. 6; 4 5. 5; 4 6. 8; 4 7. 7; 4 8. 4; 4 15. 1 *bis*; 4 15. 2; 4 16. 2; 5 16. 4; 6 15. 2.

[b] *Cf.* the fragment of a letter to Phanias of Eresus cited by Diogenes Laertius v. 37: "For it is not easy to obtain even a small circle (συνέδριον), let alone a crowd, of the sort one wishes; and reading lectures involves correcting them; and one's years no longer allow postponing and neglecting everything."

[c] Greek astronomers were private persons and astrology was not yet a source of income. Eclipses, transits and the move-

used by the astronomers for matters of observation—*φαινόμενον*—is occasionally used by Aristotle of the facts (or views) on which a theory is to rest,[a] and it is likely that the opposition of sense to reason rests at least in part on the difference between observation and theory in astronomy.

The *CP* is carefully planned: additions were constantly being made, and one had to know where to enter them. The great division is between the works of nature and those of art. The scheme is as follows:

I. The works of the plant's distinctive nature (Book I)

1. Generation (1 1. 1–1 7. 5)
2. Sprouting (1 8. 1–1 13. 12)
3. (Flowering and) Fruiting (1 14. 5–1 21. 3); Fruiting is the goal of the plant.

   Transition (1 21. 3) to the excursus on heat and cold: the causes of the phenomena discussed are (1) the sun and air and (2) the plants' own distinctive natures, to which, among other qualities, belong heat and cold.
4. Excursus on heat and cold (1 21. 4–1 22. 7)

II. The plan (2 1. 1): Many points about sprouting and fruiting have been passed over. These are now to be discussed under two heads: (1) the

---

ments of the planets had not been faithfully recorded for centuries as in Babylonia and Egypt. Hence the interest in the Egyptian and Babylonian records: *cf. Epinomis*, 986 E 6–987 A 6 and Aristotle's injunction to Callisthenes to send him the Babylonian observations (Simplicius, *Commentary on Aristotle's* De Caelo, p. 506. 10–19, edited by Heiberg).

[a] *Cf.* the passages in Bonitz' *Index Aristotelicus*, 809 a 40–809 b 5.

effects of seasonal occurrences and (2) the effects of agriculture.

(1) Seasonal occurrences (that is, external nature) and their effects (Book II)

1. The "Things Above" (winters, rain, winds, climate) [2 1. 2–2 3. 8]
2. The "Things Below" (soils, surface waters, localities) [2 4. 1–2 7. 5]
3. Various effects explained from seasonal occurrences (2. 8. 1–2 16. 8)
4. Miscellaneous difficulties

   A. Effects of one plant on another (2 17. 1–2 18. 4)
   B. Movements of plants (2 19. 1–6)

5. Program for the solution of difficulties that are omitted (2 19. 6): all other points are to be studied on the same plan: one starts with trees and ascertains (1) the distinctive nature of the plant and (2) the nature of the country.

III. The plan (3 1. 1): there are two types of enquiry, each dealing with one of the two great divisions of the subject: (1) with spontaneous phenomena (that is, phenomena not brought about by human skill; here the starting-point belongs to the plant [Book I] or the country [Book II]), and (2) with the phenomena initiated by human art, which either helps the nature of the plant to achieve its goal (Book III) or goes beyond it (5 5. 1–5 7. 3).

(2) The works of human art (Books III–IV)

1. A transitional group: the plants that reject cultivation (3 1. 1–6)
2. The aims and procedures of agriculture (3 2. 1–5)
   - A. Procedures with trees (3 2. 6–3 18. 3)
     - a. Procedures common to all or most trees (3 2. 6–3 10. 8)
     - b. Procedures special to the vine (3 11. 1–3 16. 4)
     - c. Procedures special to certain other trees (3 17. 1–3 18. 3)
   - B. Procedures with undershrubs and vegetables (3 19. 1–3)
   - C. Procedures with seed-crops (3 20. 1–4 16. 4)

IV. The plan (5 1. 1): the two divisions of (1) spontaneous phenomena and (2) the effects of agriculture are each subdivided, the first into the (1a) natural and (1b) unnatural, the second into (2a) the effects of art that cooperates with the nature of the plant and (2b) the effects of art that aims at the unique and extraordinary. We have dealt with 1a (Books I–II) and 2a (Books III–IV); we now deal with 1b and 2b (5 1. 2–5 7. 3).

1. The spontaneously unnatural, real or apparent (5 1. 2–5 4. 7)
2. The extraordinary effects of art (5 5. 1–5 7. 3)
3. Disease and Death (5 8. 1–5 18. 4), a topic connected with IV 1 and IV 2 because the unnatural is involved.

A. Diseases (5 8. 1–5 10. 5)
 a. Of the tree (5 8. 2–5 9. 13)
 b. Of the fruit (5 10. 1–5 10. 5)
B. Death (5 11. 1–5 18. 4)
 a. Natural (5 11. 1–5 11. 3)
  aa. Through old age and accompanying weakness (5 11. 1)
  bb. Through over-production (5 11. 2); whether we call it natural or unnatural or intermediate makes no difference
  cc. Choking of the pine with torchwood (5 11. 3); natural rather than unnatural
 b. Unnatural (5 12. 1–5 14. 9)
  aa. From hot and cold weather (5 12. 1–5 14. 9)
  bb. From violence and man and all that does not come under the weather and the nature of the plants (5 18. 4)

V. Flavours and odours (Book VI and the lost Book VII)

 1. Definition of flavour and odour (6 1. 1)
 2. The species of flavours and odours (6 1. 2–6 13. 5)
  a. Natural flavours (6 3. 3–6 13. 5)
  b. Natural odours (6 14. 1–6 20. 4)
  c. Flavours prepared by human skill (in the lost Book VII)
  d. Odours prepared by human skill (in the treatise *On Odours*).

Within this scheme certain uniform arrangements are found:

(1) trees are discussed before the lesser plants (*cf.* 2 19. 6);
(2) phenomena common to all or most are discussed before phenomena common to few or restricted to one;
(3) the order of discussion is from generation to fruiting and the preservation of the seed.

Nature and Art

The general scheme rests on nature as opposed to art. Nature is in turn either the " distinctive nature " of the plant or that of the country (or environment). Nature has its starting-point in itself (although Theophrastus does not work out the theory for the country and the weather); in the case of art the starting-point is outside the plant, in man.

This internal connexion with the environment is characteristic of plants; so Theophrastus says (*HP* 1 4. 4):

> It is perhaps proper to include the regions where the various plants naturally grow and do not grow. For this too is an important distinction and not least appropriate to plants, because they are bound to the earth and not like animals free of it.

He speaks with some reluctance of " spontaneous " generation in the sense of generation from earth, and not from a pre-existing plant or part, since here there is no internal nature from which the process is initiated.

The threefold distinction of the nature of the tree, the nature of the country, and the operation of man, is based on the program laid down in the *Phaedrus* (259 E 1–272 B 2; especially 271 C 10–272 B 2) for a true art of rhetoric (and indeed for any art: *cf.* 271 B 8–C 1): we must take whatever nature it is that we deal with and see whether it is of a single kind or of many; if it is of many, we must distinguish and list the kinds, and since the aim of rhetoric is to guide the soul by speech, we must divide the types of speech as we divided the types of soul, and show how and why a given type of speech affects a given type of soul. So in the *Cratylus* (424 B 7–425 B 3) the elements of articulate speech must be distinguished on the one side, the types of reality on the other, and the sounds must be rightly assigned to the realities, if a true art of name-giving is to exist.

So with the various plants there is a certain fitness of each kind for a certain kind of country, and the theorist, who is here the possessor of the art of agriculture, sees to it that the right plant is put in the right country, this being what is most important for the well-being and good fruiting of a plant (*CP* 3 1. 6). This well-being (εὐθένεια) corresponds to the persuasion and virtue imparted by philosophical rhetoric (*Phaedrus*, 270 B 8–9) and the health and strength produced in the body by the drugs and food administered by the art of medicine. So too with music, according to the account (derived from Aristoxenus) in [Plutarch] *On Music*, chaps. xxxii–xxxvi: the various parts of music must be judged by their " appropriateness," and this depends on their *êthos* or moral character (1143 A).

Hence Theophrastus' stress on the appropriate

" country "[a] and his persistence in including both " air " (or climate or weather) and soil under the general designation " country." In effect the threefold distinction of (1) " distinctive natures," (2) the country and (3) the effects of human ingenuity is reduced to a twofold one, the old distinction between " nature " and " art."[b] Hence the stress on enumeration as well as on description (*cf. Phaedrus*, 270 D 6) that we find in *CP* 1 1. 1 and 6 1. 2. Again, theory cannot do everything; the student must discover in each particular situation what elements represent the various types that have been distinguished in theory. In this connexion Theophrastus, like Plato (*Phaedrus*, 271 E 1), speaks of " sense "[c] and experience (*CP* 1 5. 4; *cf.* Plato's *μελέτη*, *Phaedrus*, 269 B 5).

In accounting for the suitability of a plant to an environment the governing idea is that there should be no excess or deficiency in quantity or in quality in either. This rightness is expressed by the words *συμμετρία* (and its forms), which refers more to quantity and the world outside, and *κρᾶσις* (with its forms), which refers more to quality, and is used about equally of both the plant and its environment. *Σύμμετρος* comes from the equivalence of a quantity

[a] *Cf. CP* 1 9. 3 *οἰκείαν χώραν*; 1 16. 11 *οἱ οἰκεῖοι τόποι πρὸς ἕκαστον*; 2 3. 7 *ὁ δὲ οἰκεῖος* (*sc. ἀήρ*); 2 7. 1 *τοὺς δὲ τόπους . . . τοὺς οἰκείους*; 3 6. 7 *οἰκειότατον . . . τόπον*; 3 6. 6 *ἡ οἰκεία* (*sc. χώρα*).

[b] For the distinction *cf.* Plato, *Laws* x. 888 E 4–890 A 9, 892 B 3–C 7; Aristotle, *On the Generation of Animals*, ii. 1 (735 a 2–4); iii. 11 (762 a 14–18); iv. 6 (775 a 20–22).

[c] *CP* 2 4. 8 *αἰσθητικῆς . . . συνέσεως*: for *σύνεσις cf.* 6 11. 2 and Plato, *Euthydemus*, 277 E–278 A, cited below (p. xxv, note *d*).

of a commodity to the measure (μέτρον) used to mete it out. But like the word μέτριος, of the same origin, it is more often opposed to " too much " than to " too little "; perhaps it was as much a seller or lender's word as a borrower or buyer's. The word κρᾶσις, which comes from the blending of wine with water, is used of qualities much as we use " temperature " of heat and cold, to indicate their intensity or degree. It usually implies that the qualities are not extreme. The qualities in question may be any that vary in degree from one opposite to another: heat and cold, open and close texture, dryness and fluidity,[a] heaviness and lightness, or all or some of them combined.

In the plant this " tempering of qualities " is part of its distinctive nature. Theophrastus very seldom suggests, and never attempts, an enquiry into the origin of this " temper " or nature.[b]

Theophrastus does not have the taxonomic terms of modern botany, such as class, genus, species, variety. He uses γένος (" kind ") or διαφορά (" difference ") of the different kinds of vine and also of the great classes " terrestrial " and " aquatic." This is because the γένος is of the nature or essence of a plant; a different γένος or διαφορά means a different plant. The mutations that Theophrastus recognizes are usually from one variety to another, but he speaks of them as an " entire change of kind " (*CP* 1 9. 1; 1 16. 12; 1 18. 2). Usually a different name (among the same speakers) means a different

[a] The Greek ὑγρόν, often rendered " fluid " or " moist," is not easy to translate; it often indicates a pliant or yielding or soft character, opposed to rigid or brittle.

[b] So with the citron at *CP* 1 11. 1 and with aquatic and terrestrial plants at 2 3. 5.

plant (ἐρινεός "wild-fig," συκῆ "cultivated-fig;" κότινος "wild-olive," ἐλάα "cultivated-olive;" ἀχράς "wild-pear," ἄπιος "cultivated-pear"), but by no means always: thus among cultivated trees the name "vine" includes a number of different plants (*CP* 1 18. 3–4). A change in γένος is a change of nature or mutation (μεταβολή); change for the worse is sometimes a "departure from the nature," ἔκστασις τῆς φύσεως (*CP* 3 1. 6; *cf.* 5 3. 5; 4 4. 6; 7 8. 10; 4 5. 4; 5 9. 3), as opposed to ἀλλοίωσις "alteration of quality" (but not of identity), although the words are not always so distinguished: a given quality, after all, may be part of the nature.

Nature, like art, is purposive, and the goal of the plant's nature is to ripen the fruit and reproduce. Here the distinction drawn between concoction—an old term derived from cooking and used of ripening and digestion—of the pericarpion and "concoction" (here not the accepted word) of the fruit or seed is of capital importance. The pericarpion is for human use, and "fruit" (seed) for the plant's own propagation. Plato had said that plants were created to serve man as food (*Timaeus*, 77 A 3–B 1, 77 C 6–7), and the language in current use coincides with this view. So wild trees are said to fail to "concoct" or "ripen" their fruit, because they do not make it edible; and the seed of willow is "raw" (*CP* 4 4. 1). Theophrastus' distinction between the two concoctions gets rid of the notion that the goal of a plant is to feed man, although the notion passes uncontradicted when the "fruit" (a non-committal term which can mean the seed with the pericarpion or the one or the other alone) is made the goal of the plant. Aristotle shared Theophrastus' view here, although he does not go so far as to speak of " ripening " or " concocting " the seed: *cf. Meteorologica*, iv. 3 (380 a 11–15):

> Ripening is a kind of concoction; for the concoction of the food in pericarpia is called ripening. Since concoction is a perfecting, the ripening is then perfect when the seeds in the pericarpion are able to produce another individual like the plant itself . . .

Predecessors and Sources

Throughout the *HP* and *CP* Theophrastus has his predecessors in mind. Aristotle is the most prominent, and is often corrected, although never mentioned by name. Where Aristotle fails him, as in the discussion of what an art must do (*CP* 2 19. 6; 3 2. 2; 3 2. 3 with the notes) or in the description of the effects of flavours on the organ of taste (*CP* 6 1. 3), Theophrastus resorts to Plato.

Of the older philosophers Democritus receives the most attention (*CP* 1 8. 2; 2 11. 7–9; 6 1. 2; 6 1. 6–6 2. 4 [*cf.* 6 10. 3]; 6 17. 11); next Empedocles (*CP* 1 7. 1; 1 12. 5; 1 13. 2; 1 21. 3). Anaxagoras is mentioned for his theory of seeds once in the *HP* (3 1. 4), where he is coupled with Diogenes and Clidemus, and once in the *CP* (1 5. 2).

Other writers on agriculture are occasionally mentioned, but only when (like the philosophers) they were no longer living. Far more frequently views are ascribed to the Greek equivalents of " people " or " they."[a] It was good form not to mention a con-

[a] M. Wellmann, " Das älteste Kräuterbuch der Griechen " (in *Festgabe für Franz Susemihl* [Leipzig, 1898], pp. 1–31) has

temporary by name. So Phanias of Eresus, a contemporary and townsman, who wrote on plants and corresponded with Theophrastus, is not mentioned by him.

With the exception of Androsthenes, an admiral who described some of the flora observed on his voyage from the Indus to the Persian Gulf (*CP* 2 5. 5), the writers mentioned in the *HP* and *CP* are these:

(1) Menestor of Sybaris, a Pythagorean and one of " the old natural philosophers " (*CP* 6 3. 5).[a]

(2) Clidemus.[b]

(3) Leophanes; an Athenian or Ionian to judge by the form of his name. He is mentioned once by Aristotle (*On the Generation of Animals*, iii. 4 [765 a 25], which is cited in the *Placita*, v. 7. 5 [p. 420 a 29 Diels]), and once by Theophrastus (*CP* 2 4. 12). His language is archaic Attic: *cf.* δοχός, a feminine adjective meaning " receptive."

(4) Chartodras; otherwise unknown (the name is corrupt, perhaps a conflation of Chaereas and Androtion); cited at *HP* 2 7. 4 for a list of manures in the order of decreasing pungency.

(5) Androtion, mentioned twice in the *HP* (2 7. 2, 3), once in the *CP* (3 10. 4), all three times in connexion with olives, myrtles and pomegranates.[c]

made it likely (see especially pp. 22–31) that some of the pharmacological material in *HP* IX was taken from Diocles of Carystus.

[a] Diels-Kranz, *Die Fragmente der Vorsokratiker*, vol. i [10], pp. 375–376.

[b] *Ibid.*, vol. ii [8], p. 50.

[c] Perhaps Theophrastus' favouring the form μύρρινος (masculine at *HP* 1 3. 3, 3 13. 3, 9 11. 9 [μυρρινος U; οἱ μυρρινοι U*]; *CP* 2 7. 3, 2 8. 1, 3 9. 3, 3 17. 7, 6 14. 6 [τουμυρινου U],

Athenaeus mentions him four times, the citations probably all coming from the grammarian Tryphon's work *On Plants*:

iii. 7 (75 D): "Androtion or Philippus or Hegemon in the book on agriculture lists the varieties of fig-trees as follows: 'In the plain, then, one should plant the *chelidóneōs*, *eríneōs*, *leucerineōs* and *phibáleōs* figs; the *opōrobasilís* can be planted everywhere. For each variety has something useful. The docked and *phormýnioi* and twice-bearing and Megarian and Laconian trees are profitable if they have water."

iii. 14 (78 A–B): "Tryphon, speaking of the name fig in his second book of *Enquiries on Plants*, says that Androtion (so Kaibel; δωρίωνα) in his book on agriculture tells the story that Syceus, one of the Titans, was received by his mother Earth when he was being pursued by Zeus, and that she sent up the plant to entertain her son, and from this comes the city of Sycea in Cilicia."

iii. 23 (82 C): "Androtion says in his book on agriculture: 'They call the apple trees *phaúliai* (coarse) and *strúthiai* (for the sparrows), since the apple does not drop from the pedicel of the *strúthiai*. The spring apple-trees are either Laconian or of Sidus or downy."

xiv. 63 (650 E): "Androtion (so Kaibel; ἀντιφῶν) in

---

6 18. 10 *bis* [feminine at *CP* 1 13. 10]) over the Attic ἡ μυρρίνη (found in Aristotle, *History of Animals*, ix. 40 [627 a 8, b 18]; Theophrastus has it at *HP* 4 5. 3 [together with an anarthrous μύρρινος], 6 8. 5, *CP* 18. 4) can be traced to Androtion, who speaks (*CP* 3 10. 4) of the friendly entwining of ὁ μύρρινος with ἡ ἐλάα.

his book on agricultural matters says that the Phocian is a kind of pear."

Even here the parallels with Theophrastus are noticeable: the Laconian fig (*CP* 5 1. 8) and its fondness for water (*CP* 3 6. 6); the " sparrow apple " (*HP* 2 2. 5); the Phocian pear (*CP* 2 15. 2, 5); twice-bearing figs (*CP* 2 9. 13; 5 1. 6).

Theophrastus also consulted informants, especially about wild trees and plants. So we hear of one Satyrus in *HP* 3 12. 4:

> As the Arcadians say, it (*sc.* κέδρος) has three sets of fruit on it simultaneously, that of last year, not yet ripe; that of the year before, already ripe and edible; and thirdly it begins to show the new fruit. Satyrus said that the woodcutters brought both (*sc.* specimens, that of *kédros* and *árkeuthos*) to him when they had no flowers.

At *HP* 4 12. 2 we hear of a whole clump of rushes " being brought," presumably to Theophrastus or his agent. Here there was no occasion to mention an agent: the specimen was satisfactory.

Theophrastus evidently made a point of collecting information not readily available in writers on agriculture: about foreign plants, about wild trees, and about aquatic and marine plants. His chief sources of information about wild trees were the woodcutters of Arcadia, Mt. Ida and Macedonia. He apparently used Satyrus' information about Arcadia, but he was well acquainted himself with the woodcutters of Ida and doubtless those of Macedonia as well. In his search for information about aquatic plants he appears to have visited Lake Copais.

We may suppose that his close ties with Aristotle made him familiar with the Chalcidice and Chalcis: so he speaks of Olynthus (*HP* 8 11. 7) and its territory (*CP* 1 20. 4), of Stagira (*HP* 3 11. 1; 4 16. 3), Torone (*HP* 4 8. 8), Acanthus (*CP* 3 15. 5), of the vine of Aphytis (*CP* 3 15. 5). At *HP* 4 14. 11 and *CP* 5 12. 4 he speaks of Chalcis and its Olympias wind and the effects of the great frost of the winter of 321–320 [a] (*HP* 4 14. 12; *CP* 5 12. 8). He appears also to have known Philippi (*cf. HP* 2 2. 7; 4 14. 12; 4 16. 2–3; 6 6. 4; 8 8. 7; *CP* 5 4. 7; 5 12. 7; 5 14. 5–6). With Lesbos and the Troad he is naturally well acquainted (although some information may have reached him from Phanias): so he mentions Lesbos (*HP* 3 9. 5; 3 18. 13), Pyrrha (*HP* 9 18. 10; *CP* 2 6. 4) and the Pyrrhaean mountains (*HP* 3 9. 5), Mt. Ordynnos (*HP* 3 18. 3), the Troad (*HP* 3 11. 2), Ilium (*HP* 4 13. 2), Mt Ida (*HP* 3 3. 4; 3 5. 1; 3 8. 7; 3 9. 2; 3 9. 3; 3 9. 5; 3 10. 2; 3 11. 2; 3 11. 4; 3 12. 3; 3 14. 1; 3 15. 3; 3 17. 3, 4, 6; 4 5. 4, 5; 9 2. 5, 7), Antandrus (*HP* 2 2. 6; 4 16. 2; 5 6. 1; *CP* 5 4. 7).

The Language and Style

Originally called Tyrtamos, Theophrastus won his name by the " divinity " of his style.[b] The ety-

[a] In the archonship of Archippus (*HP* 4 14. 11). Another Archippus was archon in 318–317. Presumably the first is meant, and the account was written within thirty months of the event. If it had been written later the two homonymous archons would doubtless have been distinguished.

[b] Diogenes Laertius, v. 38: " Aristotle changed the name Tyrtamos to Theophrastus because of the divine character of

mology is dubious: θεόφραστος presumably means "indicated by a god." Conceivably, since the name is not uncommon at Athens (among the bearers is the archon of 340–339 B.C.) and not common elsewhere, it had for outsiders an Attic sound, and Aristotle did in fact use it to refer somewhat teasingly to his friend's predilection for Attic speech. Cicero[a] tells the story of how this Attic betrayed him as a foreigner. It would not be hard to cite forms and expressions in Theophrastus that do not meet the

his style (τὸ τῆς φράσεως θεσπέσιον)." *Cf.* Cicero, *Orator*, 19. 62: ". . . Theophrastus divinitate loquendi nomen invenit . . .;" Pliny, *N. H.*, *Praefatio* 29: . . . Theophrastum, hominem in eloquentia tantum, ut nomen divinum inde invenerit . . .

[a] *Brutus*, 46. 172: ". . . ut ego iam non miror illud Theophrasto accidisse quod dicitur cum percontaretur ex anicula quadam quanti aliquid venderet et respondisset illa atque addidisset 'hospes, non pote minoris,' tulisse eum moleste se non effugere hospitis speciem, cum aetatem ageret Athenis optimeque loqueretur omnium." ("So I am no longer surprised at the story told of Theophrastus: that he enquired of a little old woman the price she was asking for some article, and when she told him, adding 'Stranger, it cannot go for less,' he was mortified at still sounding like a foreigner, although he lived at Athens and was second to none in his mastery of language.") Quintilian (viii. 1. 2) adds some details: "multos enim quibus loquendi ratio non desit invenias quos curiose potius loqui dixeris quam Latine, quo modo et illa Attica anus Theophrastum, hominem alioque disertissimum, adnotata unius adfectatione verbi hospitem dixit, nec alio se id deprendisse interrogata respondit quam quod nimium Attice loqueretur." ("You will find many, well grounded in the theory of eloquence, whose language, you would say, is more studied than it is Latin, as in the story of the old woman of Athens who called Theophrastus, a man by the way of great eloquence, a foreigner, noticing the preciosity of a single word, and when asked how she knew replied that she found him out because his Attic was too pure.")

Isocratean standard.[a] But perhaps the criticism also turns on the use of so polished a style for a technical treatise: such a writer would have been capable of haggling in literary prose.

At all events Theophrastus is more precious[b] and at times more Attic[c] than Aristotle. So he prefers the Attic ἦρος, ἦρι and ἠρινός (which occur sixty times in the *HP* and *CP*) to Aristotle's uncontracted forms (which occur twenty-two times), and he uses the form ξυν- in eighteen different words when Aristotle uses it in but one.[d]

[a] *Cf.* such archaisms and ionicisms as ξυν-, ὀδμή, ἀτάρ (*HP* 9 20. 3), σὺν αὐτοῖς ταλάροις "baskets and all" (*CP* 5 6. 6; σὺν avoids hiatus), ἰθύτατα (*CP* 3 5. 1).

[b] In the pharmacological part of *HP* (9 8. 1—end) poetical words are numerous, perhaps because Theophrastus is citing the root-cutters and druggists, who used inflated speech to vend their wares: *HP* 9 8. 2 ἀμερθῶσιν; 9 8. 5 ἀλειψάμενον λίπα; 9 11. 7 ἀμᾶται; 9 11. 9 ἀμῶσιν; 9 15. 2 οἰστούς; 9 16. 3 ὀροδάμνων; 9 18. 3 ὀρεινόμου; 9 18. 3 σίνεσθαι; 9 18. 10 παιδογόνον; 9 18. 10 τεκνούσσας; 9 20. 4 σὺν ὕδατι.

[c] In his Göttingen dissertation (*De Theophrasti Dicendi Ratione.* Pars Prima. Observationes de Particularum Usu [Arnstadt, 1874], pp. 48–49; henceforth dissertation) W. Müller points out that Theophrastus returns to the Platonic usage of γε δή, which does not appear in Aristotle.

[d] The word is ξυνίημι, which is in effect a reference to Plato, *Euthydemus*, 277 E–278 A, a passage that contains the germ of the distinction between Aristotle's first and second entelechy:

> For first, as Prodicus says, one must learn about the correct usage of words. The two foreign gentlemen are pointing this out to you—that you did not know that people use the word 'learn' of the case when a person begins with no knowledge of a thing and then acquires the knowledge later, and also use the same word 'learn' of the case where a person already has the knowledge, and with this knowledge considers the same thing as it is performed or expressed. Now people tend to call this

The use of ξυν- illustrates Theophrastus' fondness for variation, concern with rhythm and striving for certain literary effects, and well deserves careful study.

Ξυν- occurs forty-two times in the *CP*, seventeen in the *HP* (another indication that the *CP* is the more polished work). In the minor writings *ξυνίησι* occurs twice in the fragment *On the Senses*, chap. xxv (p. 506. 21 Diels) in a paraphrase of Alcmaeon. In the list that follows all passages where *ξ-* lengthens the preceding syllable are daggered.

CP 1 1. 4; 1 3. 3; 1 4. 6; 1 15. 2; 1 16. 5; 1 18. 3†;
1 20. 4†; 1 20. 6a†; 1 20. 6b; 1 21. 3;
2 1. 5†; 2 1. 7; 2 2. 3; 2 4. 3a; 2 4. 3b†; 2 4. 5;
2 6. 2; 2 8. 1; 2 9. 15; 2 13. 5; 2 16. 6; 2 17. 7†;
2 17. 10†; 2 18. 1; 2 19. 3;
3 7. 7†;
4 3. 4; 4 6. 4†; 4 7. 3†; 4 12. 3†; 4 13. 2; 4 14. 1;
5 6. 9†; 5 6. 10; 5 10. 2†; 5 10. 5; 5 13. 4†;
5 14. 9;
6 3. 4; 6 8. 8†; 6 11. 7; 6 11. 8
HP 1 10. 1;
3 13. 3;
4 5. 7; 4 14. 1a†; 4 14. 1b; 4 14. 7†;
6 2. 6;

'catch on' (ξυνιέναι) rather than 'learn,' but they sometimes also call it 'learn' . . .

For Aristotle compare ξυνιέναι *Posterior Analytics*, i. 1 (71 a 13), i. 2 (71 b 32); *Topics*, ix. 4 (165 b 33); *Nicomachean Ethics*, v. 5 (1137 a 11); ξυνίησιν *History of Animals*, ix. 46 (630 b 20); ξυνίεσαν *Prior Analytics*, i. 31 (46 a 38); ξυνίεσθαι *Posterior Analytics*, i. 10 (76 b 37); ξυνείη *On the Soul*, iii. 8 (432 a 8); ξυνιέντα *Metaphysics* Γ 3 (1005 b 15).

8 6. 1a†; 8 6. 1b†; 8 6. 1c†; 8 6. 3; 8 6. 5†;
8 6. 6a†; 8 6. 6b; 8 7. 7;
9 9. 6 (U*); 9 14. 3† [a].

Twenty-three passages are daggered. In twenty of them the rhythm is improved (⏒ indicates the lengthened syllable):

CP 1 18. 3 ⏑⏑⏒⏑
1 20. 4 ⏒⏑⏑⏑
1 20. 6a ⏑–⏑–⏑–⏒–⏑–
2 1. 5 –⏑⏒⏑––⏑ (two trochees for three)
2 4. 3b ⏑–⏑–⏑⏒–⏑–⏑– (three iambs avoided at the end)
2 17. 7 –⏑⏒–––⏑ (the hexameter rhythm is broken; here incidentally we have in ἐπιξυνδεῖ the only internal -*ξυν*-)
2 17. 10 –⏒–⏑––
3 7. 7 –⏑⏒––– (an hexameter rhythm is broken)
5 6. 9 –––⏑⏒–––
5 10. 2 –⏑⏒––––
5 13. 4 –⏒–––⏑–⏑–⏑–⏑––
6 8. 8 –⏑⏑–⏒⏑–⏑⏑–––⏑⏑–
HP 4 15. 1a ––⏒––– (an iambic rhythm broken)
4 14. 7 ⏑⏑⏑⏒
8 6. 1a ––⏒–⏑–
8 6. 1b ⏑⏑⏑⏑⏒
8 6. 1c –⏒–⏑–

[a] -*ιν* (*ν* movable) *ξ*- U; -*ι σ*- U*. It is true that the -*ν* is here sufficient to produce length by position, and is strictly unnecessary. It is possible that an -*ν* may sometimes have been added by a scribe: so at *CP* 1 13. 2 it spoils the metre, and the two versions of *HP* 9 8. 1—9 20. 5 occasionally disagree, the one adding an -*ν* before a consonant, the other not.

8 6. 5 ⏑–⏒–⏑–
8 6. 6a ––⏑–⏒–⏑–
9 14. 3 ⏑–⏒––– (a long however yields a sequence of twelve longs)

Three passages remain in which the rhythm does not appear noticeably improved:

CP 4 6. 4 –⏑––⏒–––
4 7. 3 –⏑⏑––⏒––⏑
4 12. 3 –⏑–⏒⏑–⏑⏑⏑⏑⏑–⏑⏑⏑–––

They are accounted for by Theophrastus' love of variation. Certain of the commonest words in *συν-* are especially often written with *ξυν-*: if the word cannot be varied, the form can. So we have

ξυμβαίνω *CP* 1 15. 2; 1 20. 6b; 2 8. 1; 2 9. 15; 2 13. 5; 2 18. 1; 3 7. 7†; 4 6. 4†; 4 7. 3†; 5 6. 10; 5 10. 2†; 5 10. 5; 5 13. 4†; 5 14. 9
HP 1 10. 1; 4 5. 7; 4 14. 1a†; 8 6. 1b†; 9 4. 13†
ξυμφέρω *CP* 2 1. 7; 2 4. 3b†; 2 4. 5; 4 14. 1
HP 8 6. 1a†; 8 6. 1c†; 8 6. 3; 8 6. 5†; 8 6. 6a†; 8 6. 6b; 8 7. 7
ξυνίσταμαι *CP* 2 1. 5†; 2 2. 3; 2 6. 2; 4 3. 4; 4 12. 3†; 4 13. 2; 6 3. 4; 6 8. 8†; 6 11. 8
HP 6 2. 6

In this last verb the forms in *ξ-* may have appeared especially appropriate for suggesting a certain condensation: *cf. ξυνεστραμμένην* *CP* 1 3. 3 of the root of bay; *ξυνηθροισμένην* 1 4. 6 of moisture collected (and thickened); *ξυμφράττονται* 6 11. 7; *ξυγκείμενον*

HP 3 13. 3 of the flower of bird-cherry, resembling a honeycomb composed of small flowers.

Variation of the form with ξ- with a preceding form with σ- is evident in the following passages:

CP 1 1. 3 *τὸ σύμφυτον θερμόν* . . . 1 1. 4 *θερμότητα τὴν ξύμφυτον*
1 15. 3 *πλείω τοῦ συμμέτρου* . . . 1 16. 5 *πλείων* . . . *τοῦ ξυμμέτρου*
3 22. 1 *ἐπικύπτειν συμφέρει* . . . 4 14. 1 *ἐπικύπτειν ξυμφέρει*
HP 9 9. 4 *συλλέγεται* . . . 9 9. 6 *ξύλλεγεται* (U*)

At *CP* 2 19. 3 we have *συμμύει* . . . *ξυνιόντος καὶ οἷον πηγνυμένου τοῦ ὑγροῦ*. Here the ξ- not only hints at the thickening but is a variation on the preceding *συμμύει*. At *HP* 8 6. 1–6 the cluster of words in ξ- is notable: 8 6. 1 *ξυμφέρει—ξυμβαίνει—ξυμφέρει*; 8 6. 3 *ξυμφέρειν*; 8 6. 5 *ξυμφέρει*; 8 6. 6 *ξυμφέρειν—ξυμφέρει*. Theophrastus has come to the sowing of grain, and like Hesiod may have felt that his language should rise to the occasion.

Two passages remain, and perhaps in these too a certain dignity is aimed at: *CP* 1 21. 3 *ἔφαμεν ξυμπονεῖν* (of the art of husbandry) and 2 16. 6 *μᾶλλον ξυγκεχυμένην* of the nature of plants as opposed to that of animals.

Theophrastus writes the *Kunstprosa* or euphonic prose of the day, such as we see it in the *Constitution of Athens*. Such prose endeavours to borrow the graces of poetry without ceasing to be prose. It remains prose by avoiding all poetical or non-current words and all oddities of syntax or idiom, and it seeks to please the ear by avoiding hiatus and having a certain rhythm and balanced structure of its own.

At the beginning and end of sentences we find the following rhythms:

Opening Rhythms

Rhythm *CP*	I	II	III	IV	V	VI	Total	Per Cent
–⏑––	35	30	38	36	29	41	209	9·8
–⏑–⏑	41	38	37	26	24	36	202	9·6
–⏑⏑–	27	32	36	24	42	36	197	9·3
⏑⏑⏑–	27	33	23	29	33	33	178	8·4
⏑⏑–⏑	27	20	29	23	22	27	148	7·0
–––⏑	31	23	21	10	28	31	144	6·8
––⏑–	24	20	22	16	18	33	133	6·3
⏑––⏑	26	21	23	10	14	26	122	5·8
⏑⏑––	20	26	11	20	19	26	122	5·8
––––	18	18	20	19	23	24	122	5·8
⏑–⏑–	18	13	22	14	14	26	107	5·1
⏑–––	15	20	19	10	9	26	99	4·7
–⏑⏑⏑	15	16	14	9	14	23	91	4·3
⏑–⏑⏑	13	10	17	11	14	20	85	4·0
⏑⏑⏑⏑	20	14	11	19	8	8	80	3·8
––⏑⏑	14	14	15	11	6	16	76	3·1
							2115	

Closing Rhythms

Rhythm *CP*	I	II	III	IV	V	VI	Total	Per Cent
––––	65	63	79	46	72	71	396	18·7
––⏑–	65	64	74	54	53	76	386	18·3
⏑–––	42	44	45	25	38	55	249	11·8
–⏑⏑–	50	47	39	36	28	44	244	11·5
⏑–⏑–	36	43	40	38	32	55	244	11·5
–⏑––	45	30	39	28	39	50	231	10·9
⏑⏑––	37	29	24	32	29	37	188	8·9
⏑⏑⏑–	31	28	20	28	26	44	177	8·4

Among the clausulae it is strange to see the fourth paean (⏑⏑⏑–) in the last place, a rhythm praised by Aristotle and favoured by the later Plato. It is to be expected that the elegiac, trochaic, iambic and hexameter closes should not be sought. The fourth paean doubtless appears to be avoided because it is likely to have a short final or to be preceded by a short, and Theophrastus is reluctant to allow four or more short syllables in succession.

Hiatus is avoided by changing the order of words or the word or expression itself by shifting number or voice or resorting to a paraphrase. Sense is here in a way subordinated to sound, and certain shades of meaning are apt to be neglected. But the writer is also released from the need to attend to irrelevant distinctions.

The word order resulting from avoiding of hiatus is sometimes so unnatural that editors have mistakenly altered it:

HP 3 7. 1 τῇ ⟨δὲ Wimmer or his printer⟩ σκληρότητι [δ' Wimmer or his printer] ὑπερβάλλων[a]
HP 3 4. 6 αἱ μὲν οὖν τῶν καρπῶν ἀποβολαὶ καὶ πεπάνσεις ⟨τῶν ἀγρίων Schneider, working before it was known that Theophrastus avoided hiatus⟩ τοιαύτας ἔχουσι διαφορὰς οὐ μόνον πρὸς τὰ ἥμερα [τῶν ἀγρίων Schneider] ἀλλὰ καὶ πρὸς ἑαυτά.
HP 3 12. 4 ἔφη δὲ ⟨Σάτυρος a⟩ καὶ κομίσαι τοὺς ὀρεοτύπους αὐτῷ [Σάτυρος a] ἀνανθεῖς ἄμφω.

We pass to a selection of other devices. Before a vowel διὰ τί ποτε is often found for διὰ τί. When ἐστι(ν) would have come after a vowel we often find τυγχάνει. Where ἡ ὥρα would have caused hiatus we find τὸ τῆς ὥρας (1 7. 5; 1 14. 2); so we have τὸ

[a] For similar postponement of δέ to avoid hiatus *cf.*
CP 3 24. 3 τῷ σίτῳ δὲ ἀσύμφορον
HP 3 10. 2 τῷ χρώματι δ' ἐρυθρόν
HP 9 14. 4 ὡς ἐπὶ τὸ πολὺ δὲ αἱ.

τῆς ὀσμῆς (6 14. 2 *bis*; 6 14. 4), τὸ τῆς αὐξήσεως (1 12. 4), τὸ τῶν ῥιζῶν (1 12. 6; 2 17. 4; 5 6. 5), τὸ τῆς δάφνης (5 9. 4), τὸ δὲ τῆς ἐρυσίβης (4 14. 4; here the hiatus would have come later). Where a dative singular would be natural we often find a dative plural instead, as in ταῖς ὥραις, τοῖς χρόνοις, ταῖς ἱστορίαις. There are other sudden shifts from singular to plural. So at 1 22. 4 Theophrastus replies to Menestor's views; Menestor becomes " they " and hiatus is avoided: ὅταν . . . λέγωσιν, ὀψικαρπότερα κτλ. So with Democritus (6 2. 1): " we must ask them . . .," παρὰ τούτων, ὥστε κτλ. At 3 10. 7 a singular verb is succeeded by a plural one (Theophrastus is speaking of the apple tree and pomegranate): καὶ γὰρ οὐ πολύρριζα, καὶ τροφῆς ἐλαφρᾶς δεῖται, καὶ ταχὺ γηράσκουσιν, ὥστε κτλ. At 2 3. 7 we have a plural of " air " (τοὺς ἀέρας. οὐ), at 4 14. 3 a plural of " sun " (ἐὰν ἥλιοι συνεπιλάμψωσιν, ὡς κτλ),[a] at 2 8. 1 of " cold " (fluid plants require more intense " colds " for ripening, drier plants require ἐλαφροτέρων, ἀποξηραίνει γὰρ . . . τὸ ἄγαν). There is a shift from passive to active: the common phrase ὥσπερ (or καθάπερ) ἐλέχθη is not used before a vowel, and to avoid hiatus Theophrastus often writes ὥσπερ εἴπομεν instead.[b]

[a] At 3 24. 4 " suns " (which avoids four shorts) becomes " sun " for variation and to avoid hiatus: ἡλίους ἐπιλαμβάνειν . . . ἐπιλαβὼν ὁ ἥλιος ἐρυσιβώσῃ.

[b] Ὥσπερ [or καθάπερ] (. . .) εἴπομεν in the *CP* never comes before a consonant; it comes before a vowel at *CP* 2 3. 2; 2 3. 3; 2 4. 2; 2 7. 1; 2 9. 7; 2 14. 1; 3 14. 2 (ὅπερ [ὥσπερ Schneider] εἴπομεν); 4 12. 10; 4 13. 1; 5 6. 2; 6 16. 8; 6 19. 3. In the HP the phrase occurs six times before a consonant (*HP* 1 9. 2; 1 11. 3; 3 11. 1; 3 15. 2; 8 4. 2; 9 2. 2), twelve times before a vowel (*HP* 1 1. 4; 2 2. 10; 3 3. 8; 3 4. 1;

The most familiar of these devices is the substitution of a synonym. A necessarily brief and incomplete list of such synonymous pairs follows:

ὥσπερ—καθάπερ
ὅτι—διότι
ὅτι (" that," or with superlatives)—*ὡς*
ἐστι—τυγχάνει
ἥ—ἥτις (and the like)
ᾗ—ᾗπερ (and the like)
ἄνω—ἄνωθεν (so with *κάτω*, *ἔξω*)
ἀνοίγω—διοίγω (and so with other compounds)
ἄρτι—ἀρτίως
ἄχρι—μέχρι.

Such substitution, however, is not entirely mechanical; euphony, and even the meaning, sometimes make toleration of hiatus the lesser evil, as the following studies show.

ὥσπερ—καθάπερ

Καθάπερ (henceforth *κ*.) replaces *ὥσπερ* after a vowel; compare *HP* 9 9. 5, where of the two versions U has *δύναται κ*., U* *δύναται καθαίρειν ὥσπερ*.

Ὥσπερ comes after a vowel

(1) in the phrase *καὶ ὥσπερ*. *Καὶ κ*. does not occur (at *HP* 4 13. 3 it is corrupt) because of the cacophony (although *κ. καὶ* is frequent).
(2) in the following passages:
CP 1 17. 8 *ἕκαστα ὥσπερ* (avoids -*κα*- *κα*-)
 5 17. 7 *ᾧ φαίνεται ὥσπερ* (we must suppose eli-

3 9. 2; 6 6. 8; 7 1. 6; 7 2. 2; 7 13. 6; 8 6. 5; 8 8. 3; 9 2. 7; 9 15. 3 [U*]). At *HP* 1 2. 3 we find *ὥσπερ εἰρήκαμεν* for the common *ὥσπερ εἴρηται*; a vowel follows.

sion; *κ*. would yield an iambic rhythm: – – ⏑ – ⏑ ⏑ – ⏑ – ⏑ –)

5 18. 2 *διὰ τὸ ὥσπερ ἐν* (*κ*. would produce ⏑ ⏑ ⏑ ⏑ ⏑ ⏑)

6 1. 2 *ἢ ὥσπερ Δημόκριτος* (hiatus is tolerated after *ἤ*; here *κ*. would produce – ⏑ ⏑ – – ⏑ ⏑ –).

K. comes after a consonant in twenty-four passages of the *CP*.

In six of them the consonant is the movable *ν* (2 18. 2; 3 6. 6; 4 3. 7; 4 9. 2; 5 14. 2; 6 11. 14). At 2 18. 2 it avoids five shorts; at 3 6. 6, 4 9. 2, 5 14. 2 it avoids four; and at 4 3. 7 and 6 11. 6 it avoids three. Three shorts are not generally avoided, but before *κ*. there is a slight break, and Theophrastus likes a long syllable before such breaks. So at 5 3. 1 *ἀνάπαλιν* precedes, and by letting *κ*. follow he avoids four shorts.

In eleven of the remaining passages the syllable *ωσ* (*ὥσπερ*, *ὡς*, *ὡσαύτως*, *ὥστε*) has occurred a few words before: 1 3. 1 (here *κ*. also avoids ⏑ ⏑ ⏑ ⏑); 1 12. 4; 1 17. 2; 1 21. 5; 2 9. 2; 3 7. 9; 3 20. 8; 4 3. 7; 4 5. 5; 4 9. 2; 6 11. 2.

Six passages are left:

1 9. 2 (*ὥσπερ* precedes by eight lines, *ὡσαύτως* by seven, *ἴσως* by six, *ἁπλῶς* by four and again by one)

1 13. 1 (the text is easily emended. Incidentally four lines before (in *τοῦθ' οἷον περίοδος*) *οἷον* varies with *κ*. and *ὥσπερ*, thus avoiding the jingle -*περ περ*-)

2 2. 2 (*ὅλως* has occurred four lines earlier, *ὡς* eight)

2 14. 1 (the text is corrupt; in any case *διχῶς* occurs in the preceding line, *ὥσπερ* five lines earlier)

5 9. 6 (the vowel ω has occurred five times in the seven preceding words)
5 12. 6 (ὥσπερ occurs five lines before).

In the *HP* our MS (U) lets a consonant precede κ. twenty-six times. Four of the passages have been emended: 2 5. 4; 7 3. 1; 8 1. 4; 9 1. 4. In seven the preceding consonant is a movable ν:

1 19. 8 (ὡς comes two words before; -ν avoids ⏑⏑⏑⏑)
4 3. 6 (ἄλλως comes six words before; -ν prolongs a pause)
4 13.1a (ὡσὰν comes eight words before; -ν prolongs a pause)
4 14. 3 (ὥσπερ comes seven words before; -ν avoids ⏑⏑⏑⏑)
6 7. 5 (ὥσπερ comes ten lines before; κ. avoids –⏑⏑––; -ν avoids ⏑⏑⏑⏑⏑)
7 4. 12 (ὥσπερ has occurred ten lines before, in 7 4. 11, twice in 7 4. 10, once in 7 4. 8; the last preceding κ. was in 7 4. 6. K. avoids seven longs; -ν avoids –⏑⏑⏑––⏑⏑⏑–)
7 7. 3 (κ. avoids ––⏑–⏑–; -ν avoids ⏑⏑⏑⏑. In this chapter and the following there are six successive occurrences of κ., headed by οἷον ἀπάπης; οἷον avoids -απ- απαπ-).

(Only once does U miss an opportunity to add -ν before κ., at *HP* 3 3. 4 φασὶ καθάπερ.)

In the fifteen passages that remain, bad rhythms are avoided in four: 1 6. 3 (–⏑–⏑–––⏑–⏑––); 1 6. 6 (–––––––); 4 3. 3 (⏑⏑⏑⏑); 4 5. 1 (––––––). In the rest the sound ωσ precedes: 1 7 .1 (ὡς two lines earlier); 1 10. 6 (ὥσπερ—ὅλως—ὅλως precede); 3 1. 2 (ὡς four lines before); 3 14. 4 (ὅλως six words before); 4 14. 2 (ὡς in the preceding line; -------- is also avoided); 6 8. 2 (ὥσπερ—ὡσαύτως occur 4–5 lines before); 7 9. 3 (ὥσπερ comes four words before); 7 11. 2 (ὥσπερ—ὥσπερ come earlier in the sentence); 8 1. 6 (ὡς comes two lines before); 9 2. 2 (οὕτως comes two lines before); 9 9. 1 (ὥσπερ—ὥσπερ—οἷον [ὥσπερ U*] precede).

ὅτι—ὡς

The variation between ὅτι and ὡς is restricted to the meaning " that." In his dissertation (see note *c* on p. xxv above) W. Müller has collected (pp. 54–56) the relevant passages in the surviving works. Revising his figures we find ὡς forty-four times before a vowel, twelve before a consonant; ὅτι forty-three times before a consonant, eleven before a vowel.

Ὅτι comes before a vowel in the following passages:

HP 2 8. 2 σημεῖον δὲ λέγουσιν, ὅτι ἐπειδὰν (" because; " in any case ⏑–⏑–⏑– is avoided)
5 6. 1 σημεῖον δέ, ὅτι οὐδέποτε (" because ")
6 2. 8 λέγω δὲ παραλλάξ, ὅτι οὐκ (" because ")
6 3. 4 τοῦτο λέγουσιν, ὅτι εὐθὺς (the ὅτι clause, which is definitely substantival, comports better with τοῦτο than a clause with ὡς, since ὡς retains some of the sense " how ")
6 4. 6 ἴδιον δὲ ἔχει τὸ περὶ τὸ φύλλον, ὅτι ἀφαιρούμενον (the clause is substantival, " the fact that;" in any case ὡς ἁπλῶς occurs a few lines before)
8 4. 5 σημεῖον δὲ λέγουσιν, ὅτι οἱ (" because ")

8 7. 6 *λέγουσιν οὐ κακῶς, ὅτι ἔτος* (avoids -*ῶς ὡς*)
9 1. 5 *φανερόν, ὅτι ὤν* (avoids *ὡς ὤν*)
9 18. 9 *λέγειν . . . ὅτι ἑβδομήκοντά ποτε* (*οὕτως*—*ὥστε* occur a few lines before)
CP 2 15. 5 *δῆλον οὖν* (*ὡς ἁπλῶς εἰπεῖν*), *ὅτι ἀφαιρέσεως* (avoids *ὡς -ῶς*—*ὡς -ως*)
On Winds 3. 25 The passage is corrupt.

Ὡς comes before a consonant in the following passages:

HP 3 8. 5 *λέγουσιν ὡς καὶ* (Theophrastus is dubious about the report; the opening ⏑–⏑– is preferred to ⏑–⏑⏑)
3 18. 12 *δῆλον ὡς καὶ ἀκρόκαρπον* (avoids ⏑⏑⏑⏑⏑⏑)
4 13. 5 *λέγουσί τινες, ὡς, παραιρουμένων* (avoids ⏑⏑⏑⏑⏑⏑)
5 4. 1 *δῆλον ὡς τῇ πυκνότητι* (avoids ⏑⏑–⏑⏑–)
9 17. 2 *εἰπεῖν ὡς τῇ* (avoids ⏑⏑–⏑⏑–)
CP 1 12. 10 *δῆλον ὡς δυνάμει* (avoids ⏑⏑⏑⏑⏑)
3 3. 2 *δῆλον ὡς φυσικώτερον* (avoids ⏑⏑⏑⏑⏑)
3 17. 3 *δῆλον γὰρ ὡς τῇ* (*διότι*—*ὅτι*—*ὅτι* occur in the preceding ten lines; the opening rhythm ––⏑– is preferred to ––⏑⏑)
4 2. 2 *εἰπεῖν ὡς κατὰ τὰς* (avoids ⏑⏑⏑⏑)
4 3. 2 *δῆλον ὡς δι᾿ ἀσθένειαν* (avoids ⏑⏑⏑⏑)
On Fire 17 *δῆλον ὡς διὰ τὸ* (avoids ⏑⏑⏑⏑⏑⏑)
On Odours 2. 4 *δῆλον ὡς δι᾿ ἐναντίωσιν* (avoids ⏑⏑⏑⏑⏑).

ὅπως—*ἵνα*

Ὅπως with the subjunctive is far commoner than *ἵνα* with the subjunctive, which in the twelve instances of its occurrence in the *CP* is used for variation or euphony:

CP 1 19. 2 περικαρπίοις ἵνα (ὡς occurs four words before; avoids ⏑–⏑–⏑–)
1 20. 6 ἵνα μηδὲ ἐκ ταύτης ᾖ (avoids ––––––)
2 7. 4 τοὺς μυρρίνους, ἵνα συσκεπάζωσιν (avoids a seventh sibilant)
2 9. 5 (ὅπως occurs two lines before; ἵνα avoids ⏑–⏑–⏑–)
2 9. 9 (avoids a fourth sigma)
3 2. 1 πως ἵνα (avoids πως -πως)
3 6. 3 (ὅπως occurs five and ten lines earlier)
3 12. 1 ἵν' ὡς μάλιστα (here one might have expected ὅπως ὅτι μάλιστα; so we have ὅτι μάλιστα at 1 6. 8, 3 5. 2, 3 5. 5, 3 11. 6, 3 12. 1. But ὡς μάλιστα here varies with the preceding ὅτι μάλιστα in 3 11. 6 and the following one in 3 12. 1, not to mention the plain μάλισθ' three lines before. So ἵν' varies with ὅπως and the subjunctive, which occurs three and six lines earlier.[a]

[a] Ὡς (instead of ὅ τι [ὅτι]) is used with superlatives to avoid hiatus at 3 7. 12 (ὡς ἐλαφρότατα)—where it also avoids ⏑⏑⏑⏑⏑⏑⏑—and 3 11. 5 (ὡς ἐκ ψυχροτάτης).

In eight passages such a ὡς precedes a consonant:

HP 3 5. 1, *CP* 3 5. 1 ὡς πάχιστα (avoids ⏑⏑⏑⏑. Although Theophrastus knows the form παχύτατος (*cf. CP* 6 11. 8) he uses the archaic πάχιστος here to avoid a further accumulation of shorts)
HP 2 5. 1 ὡς πλείστου (avoids ––⏑⏑––⏑⏑––)
HP 8 7. 4 τὸ ὡς πλεῖστον (avoids ⏑⏑⏑; but perhaps hiatus between ο ω was preferable to that between ο ο)
CP 3 12. 1 ἵν' ὡς μάλιστα (avoids ⏑⏑⏑⏑⏑⏑)
CP 3 12. 1 ὡς βαθυτάτας (avoids ⏑⏑⏑⏑⏑)
CP 3 12. 1 ὡς βαθύτατα (avoids ⏑⏑⏑⏑⏑⏑)
CP 3 14. 7 ὡς μακρότατα (avoids ⏑⏑⏑⏑⏑⏑).

3 14. 8 (varies with ὅπως and the subjunctive preceding and with ὅπως and the future following)
3 20. 4 (avoids ⏑–⏑–––⏑–⏑)
3 20. 4 ὡσαύτως ἵνα (avoids a third ωσ)
5 5. 3 (avoids seven s's in five successive words).

Other Hiatus-Stoppers

The pregnant construction is very natural in Greek, and only a few examples of " whence " adverbs for " where " will be cited: κάτωθεν αὔξειν (1 19. 4), τὴν ἐκεῖθεν ἀρχήν (1 4. 6), ἄνωθεν ἐβλάστησεν (1 12. 9), οὐδαμόθεν ἐχόντων σκέπην (4 12. 8), πόρρωθεν ὄζει (6 14. 11), κύκλωθεν αὗται (*HP* 4 6. 10). So some " whither " expressions avoid hiatus: εἰς Λέσβον (*HP* 3 9. 5), πρὸς περιττότητος χώραν (*CP* 1 19. 4), ἐπὶ φαρμάκου λόγον (6 12. 7). But at 3 10. 3 a consonant follows and we have ἐν φαρμάκου μέρει.

Διοίγω is confined to the position after a vowel: *CP* 2 9. 5, 2 9. 6 (δίοιξις 2 19. 3); οὐ allows a pleasant variation: οὐκ ἀνοίγουσιν and οὐ διοίγεται both appear at 2 9. 9. *'Αν-* occurs after a vowel at 1 4. 4 (καὶ ἀνεῳγμένου; there apparently is no form *διέῳγμαι) and 2 9. 8 (ἡ ἄνοιξις). So with other pairs:

ἁμαρτανόμενα 3 2. 5—διαμαρτάνω 3 8. 4, 3 20. 5
ἀμφισβητέω 1 22. 6 (τοῦτ' precedes), 2 3. 5 (but ἡ ἀμφισβήτησις 1 22. 2)—διαμφισβητέω 1 21. 4, 2 7. 9 (but χρήσεως διαμφισβητοῦσιν 3 6. 1)
ἑλκούμενα 5 16. 1 (but αἱ ἑλκώσεις 5 17. 3)—διελκούμενα 5 9. 3
ἡλιόω 3 4. 1, 3 7. 2, 3 20. 7—διηλιωθέντα 4 12. 12.

Aristotle uses both ὡς ἐπίπαν and ὡς ἐπὶ τὸ πᾶν, always with εἰπεῖν: compare ὡς (. . .) ἐπίπαν εἰπεῖν *Meteorologica*, ii. 3 (358 b 15); iv. 9 (386 b 23–24);

History of Animals, ii. 15 (506 b 7); *On the Parts of Animals*, iii. 6 (669 b 3), iv. 2 (677 a 23–24); *ὡς ἐπὶ τὸ πᾶν εἰπεῖν On Length of Life*, chap. v (466 b 14–15); *History of Animals*, vi. 18 (573 a 29); *ὡς ἐπὶ τὸ πᾶν βλέψαντας εἰπεῖν On the Generation of Animals*, ii. 1 (732 a 20–21). With the last passage R. Eucken (*Ueber den Sprachgebrauch des Aristoteles; Beobachtungen über die Präpositionen* [Berlin, 1868], p. 58) compares *ὡς ἐπὶ τὸ πολὺ βλέψαντας εἰπεῖν* (*On the Parts of Animals*, iii. 2 [663 b 30–31]).

These last two passages suggest that *ἐπὶ* (*τὸ*) *πᾶν* is an equivalent for *ἐπὶ τὸ πολύ* with a better rhythm and an ending secure against hiatus. The equivalence is confirmed by the two versions of *HP* 9 8. 1 (U and U*): *ὡς ἐπίπαν* U; *ὡς ἐπὶ τὸ πολὺ* U*. Again at *CP* 2 10. 2 we find *ὡς ἐπίπαν μείζω* followed by *πάνθ' ὡς ἐπὶ τὸ πολὺ μείζω* (which avoids the cacophony *πάνθ'—πᾶν*; compare *HP* 1 8. 2 *ὡς ἐπὶ τὸ πολὺ πάντα*).

Theophrastus uses *ὡς* (. . .) *ἐπίπαν* ten times in the *CP*,[a] eight in the *HP*[b]; he uses *ὡς* (. . .) *ἐπὶ τὸ πᾶν* four times in the *CP* (1 17. 1; 1 22. 6; 3 11. 1; 6 14. 4), eight in the *HP* (1 3. 2; 1 10. 8; 3 2. 1; 3 12. 6; 7 8. 1; 8 1. 6; 8 6. 6; 9 8. 2 U*).

A similar variation to this between *ἐπίπαν* and *ἐπὶ τὸ πᾶν*, where hiatus is not involved, is that between *ἐρρήθην* (*ῥητέον*) and the more usual *ἐλέχθην* (*λεκτέον*). In three passages the forms with -*ρη*- avoid repeating -*λε*-: *ἐπὶ πλέον ῥητέον* 3 22. 6; *ἐπὶ πλέον ῥηθείη* 4 2. 2; *λέγουσιν, ἐπὶ πλέον ῥητέον*

[a] 1 9. 1; 1 10. 7; 1 22. 5; 2 3. 1; 2 10. 2; 3 7. 4; 4 1. 4; 6 9. 4; 6 14. 2; 6 14. 9.

[b] 1 7. 2; 3 2. 6; (*cf. τὸ μὲν ἐπίπαν* 3 18. 12); 6 6. 8; 8 4. 2; 8 6. 1 (*πάνθ'* U); 8 8. 6; 9 8. 1; 9 8. 2 (U).

4 3. 7. At 2 1. 1 ῥητέον is preceded by εἴρηται ten lines earlier; so too at 2 19. 5 we have εἰρήσθω πρὸς τὸ πρότερον ῥηθέν. The sixth and last instance in the *CP* is at 5 9. 5 διὰ τὰς ῥηθείσας αἰτίας. Here the form appears merely due to a desire for variation (ἐλέχθη occurred at 5 9. 2).[a] An isolated λελεγμένα occurs at 5 7. 3, where it avoids an hexameter rhythm: τὰ δ' ἐξ ἀρχῆς λελεγμένα μᾶλλον κατὰ φύσιν.

To these we may add the many variations of form in the names of plants and fruits: the fruit of the almond is ἀμυγδάλη or ἀμύγδαλον, of the pear ὁ ἄπιος (*CP* 6 14. 4; 6 16. 2) or τὸ ἄπιον; the myrtle tree is ὁ μύρρινος or ἡ μυρρίνη, marjoram is ἡ ὀρίγανος or τὸ ὀρίγανον, thyme is τὸ θύμον or ὁ θύμος. Here another purpose than that of pleasing the ear is perhaps more decisive, although hiatus and monotony are incidentally avoided. Both Aristotle and Theophrastus vary expressions to show that the various authorities have been consulted and that their differences in views and language are familiar and have been reconciled.

The effects of avoiding hiatus are manifest on every page: order, syntax and choice of words are constantly influenced. To ignore these effects is to neglect an indispensable means of establishing, understanding and enjoying the text.

[a] For a striking example of variation at a distance *cf.*

CP 4 11. 1 οὐκ ἂν ὑπομείνειεν διὰ τὴν ἀσθένειαν
4 11. 4 οὐκ ἂν ὑπομεῖναι διὰ τὴν ἀσθένειαν.
Cf. also *HP* 4 8. 8 τούτῳ σίτῳ χρῶνται
HP 4 8. 11 τούτῳ χρῶνται σιτίῳ
HP 4 11. 5 πρὸ τροπῶν μικρὸν
HP 4 14. 11 μικρὸν πρὸ τροπῶν

πᾶς—ἅπας

We pass to another form of variation, less immediately concerned with hiatus. Ἅπας is preferred to πᾶς, and ὑπέρ to περί (both with the genitive), after a consonant; after a vowel πᾶς and περί are used to avoid hiatus.

Πᾶς and its inflexions (henceforth πᾶς) occur 256 times in the *CP* (omitting 3 10. 7, where the text is corrupt), in all but forty-four of them after a vowel; ἅπας and its inflexions (henceforth ἅπας) occur 83 times, in all but three of them after a consonant.

Two of the three occurrences can be eliminated by conjecture: we read κἂν (καὶ U) ἅπασιν at 5 17. 7 and ἀποσημαίνει⟨ν⟩· ἅπαντα at 1 22. 3. At 3 7. 3 the hiatus is clearly intentional: κόμη ends a section and a topic, and the following topic begins with Ἅπαν δὲ φυτόν. Neither of the opening rhythms involved is a favourite; in fact πᾶν would have yielded an opening rhythm (–∪∪∪) that is very slightly preferred. But Theophrastus allows serious hiatus in transitions. So with the formulae of transition that close one subject and open the next (or else close a digression and return to the original subject): compare θεωρείσθω. ἐν δὲ (1 13. 3), εἰρήσθω followed by ἐπεί (3 3. 4—3 4. 1), by ὅσα δέ (4 12. 13—4 13. 1), by ᾗ δέ (5 14. 7–8), by ὅτι δέ (6 10. 7–8) and by οἱ δέ (6 19. 1–2). So too in less formal transitions, all meriting at least a new paragraph in English, as at 1 6. 3, 1 15. 3 (first paragraph), 1 15. 3 (third paragraph), sometimes with a preceding μέν, as at 5 12. 2, 5 13. 7, 5 14. 3 (last paragraph), 5 15. 2 (second paragraph), 5 15. 5–6, 6 10. 7, 6 14. 3 (third paragraph).[a]

[a] *Cf.* the ten instances (mentioned below, p. xlv) of περί (after a consonant) opening a new topic.

We turn to the forty-four occurrences of πᾶς after a consonant. In two the consonant is nu movable (3 18. 2; 5 9. 2). Meaning or idiom appears to use πάντες in the sense " all (men) " (1 1. 2; 3 13. 1; 6 5. 5a; 6 8. 7), and again in the partitive genitive with superlatives, as at 2 4. 9a, 4 14. 4, 6 5. 2.

In the thirty-five passages that remain πᾶς yields a better rhythm. Ἅπας would produce

(1) six shorts at 1 11. 1, five at 5 3. 1, four at 2 4. 8, 5 7. 1, 5 18. 4b;

(2) a less favoured opening rhythm at 1 10. 6, 2 8. 1 (a comic iambic dimeter), 2 18. 3, 3 22. 6, 5 9. 5, 6 6. 3, 6 6. 5, 6 7. 8b, 6 17. 7;

(3) a less favoured closing rhythm at 2 6. 1, 2 16. 7, 6 7. 8a;

(4) fragments of verse at 1 4. 1 (⏑–⏑–––⏑–), 1 10. 7 (––⏑––––––––⏑–, a limping trimeter), 1 16. 3b (⏑⏑⏑–––⏑–), 2 1. 2 (⏑–⏑–⏑––), 2 1. 6 (⏑–⏑–⏑⏑⏑⏑–), 2 2. 1 (––⏑–––⏑–), 3 17. 5 (––⏑–––⏑⏑–), 4 3. 7 (⏑–⏑–), 4 13. 3 (–––⏑⏑––), 5 6. 3 (⏑–⏑–⏑–), 5 6. 7 (–⏑–⏑–⏑–), 5 15. 3 (–⏑–⏑–⏑), 5 17. 4 (⏑⏑–⏑–⏑–), 5 18. 4 (–⏑–––⏑–⏑), 6 11. 11 (⏑–⏑–⏑⏑⏑⏑–––⏑–), 6 12. 5 (–⏑–––⏑––), 6 14. 9 (––⏑–––), 6 18. 2a (⏑–⏑–⏑––).

περί—ὑπέρ

In the sense of " concerning " περί with the genitive (henceforth περί) occurs 104 times in the *CP* and varies with ὑπέρ with the genitive (henceforth ὑπέρ), which occurs (excluding the local sense) 43 times.

In a few passages ὑπέρ retains some of its older meaning and implies support: 1 16. 8 πίστις ὑπὲρ τῆς θερμότητος; 1 21. 7 τὰ μὲν ὑπὲρ τῆς θερμότητος λεγόμενα; 2 14. 3b αἰτίαι· ὑπὲρ δὲ τῆς ἀπὸ τῶν ῥιζῶν μεταβολῆς . . . ἐκεῖνο δεῖ λαβεῖν, ὅτι; 4 4. 4 τὰ *γὰρ τοιαῦτα δίδωσίν τινα ἔννοιαν καὶ ὑπὲρ τοῦ μὴ ἔχειν ἔνια τροφὴν πρὸς ἑαυτοῖς* (of seeds); 5 14. 6 *αἱ μὲν οὖν αἰτίαι ὑπὲρ ἑκατέρου*. The hiatus with *ὑπέρ* in the last three citations is connected with its use in the older sense: *περί* would have eliminated the hiatus (or at 5 14. 6 replaced it with a milder one), but altered the meaning.

The only other instance of hiatus with *ὑπέρ* is at 1 4. 1 *καὶ ὑπὲρ ἧς*. Here *περί* would have eliminated one hiatus but introduced another.

There are eleven instances of hiatus with *περί*: 1 6. 3, 1 21. 4, 2 3. 5a, 2 11. 4, 2 11. 5, 4 3. 6, 5 6. 2 (*περὶ ὧν* ego; *ὥσπερ* U), 5 17. 1, 6 6. 1, 6 10. 10, 6 17. 13. In all but one (at 4 3. 6) replacement with *ὑπέρ* would have introduced a less desirable hiatus with the preceding word. The hiatus at 4 3. 6 is discussed below (p. xlvi).

In twenty-one passages *περί* is found after a consonant.

In seventeen of them *περί* begins a sentence (or clause) in a transition, closing one subject or opening another. The formula of dismissal *περὶ μὲν οὖν* is used thirteen times, five of them after a consonant (3 9. 4; 3 17. 8; 5 1. 2a; 5 10. 5; 6 20. 4); the equivalent formula *ὑπὲρ μὲν οὖν* is used six times (1 13. 12; 2 14. 3a; 5 12. 9; 6 5. 6; 6 6. 2; 6 18. 12). Conservatism and the preceding pause help to account for these occurrences of *περί* after a consonant. In one of them a nu movable precedes (5 1. 2a); in another *ὑπέρ* would have yielded the rhythm ⏑–⏑–⏑–⏑– (5 10. 5). In any case the opening rhythm ⏑⏑⏑– is preferred to ⏑–⏑–.

Another phrase of dismissal, *ἀλλὰ γὰρ περὶ μέν*, occurs twice (2 11. 11; 5 14. 7); the combination *γὰρ ὑπέρ*, no doubt regarded as cacophonous, is not found. The phrase is a livelier equivalent of *ἀλλὰ περὶ μέν* (found at 1 10. 7; 3 18. 1; 4 2. 2; 4 5. 7; 4 6. 9; 6 3. 5; 6 9. 4; 6 10. 10) and has a much preferred opening rhythm. It may appear odd that *ἀλλ' ὑπὲρ μέν*, with its better rhythm, does not replace *ἀλλὰ περὶ μέν* more often, occurring only twice (at 1 16. 8b and 6 7. 6); but Theophrastus is on the whole reluctant to elide.

A new topic is often opened with *περὶ* (*ὑπὲρ*) *δέ* and a noun or noun phrase designating the topic, the preposition and its object often having the loosest of grammatical connexions with the rest of the sentence and approaching the function of a chapter heading. Of these topical introductions twenty-two begin with *περί*, seven with *ὑπέρ*; of the twenty-two ten follow a consonant (1 17. 1; 3 8. 3; 3 9. 1; 4 3. 5; 4 4. 5; 4 9. 1; 4 13. 2; 4 15. 1; 5 17. 3; 6 1. 1). In all but one (*Περὶ δὲ κοπρίσεως* 3 9. 1) *ὑπέρ* would have introduced a less favoured rhythm; in this passage however *ὑπέρ* would have yielded ⏑–⏑⏑, which is favoured over ⏑⏑⏑⏑ as 4.0 to 3.8. One suspects that *ὑπέρ* was avoided because of the rough iambic rhythm: ⏑–⏑⏑⏑⏑–⏑⏑–––⏑⏑⏑. It is true that with *περί* an iambic scansion is still possible, but the additional resolution disguises it. At 4 4. 5 *περί* avoids another iambic opening (⏑–⏑–––⏑–); here too the rhythm with *περί* could be scanned as verse (⏑⏑⏑––– ⏑–), but the trochaics are disguised by the resolution.

Four of the twenty-one passages where *περί* comes after a consonant remain:

1 20. 6 *ὁ λόγος περὶ πάντων* (*ὑπὲρ ἁπάντων* would have yielded six shorts, *ὑπὲρ πάντων* four)

2 17. 9 *ἐν ταῖς ἱστορίαις ταῖς περὶ τούτων* (*sc.* animals) *εἴρηται* (Aristotle refers similarly to the *History of Animals*: *cf. On the Generation of Animals*, i. 3 [716 b 31–32]: *ἐν ταῖς ἱστορίαις ταῖς περὶ τῶν ζῴων*)

4 3. 6 *ἄλογον. Περὶ οὗ δὴ καὶ ἀντιλέγουσί τινες* (*ὑπέρ* retains enough of the sense of support to make it unsuitable for referring to a position attacked)

5 15. 1 *λοιπὸν δὲ δὴ εἰπεῖν περί τε τῶν βιαίων παθῶν* (*ὑπέρ* would yield ––⏑–––⏑–⏑–⏑–).

The Calendar of Theophrastus

In the *CP* Theophrastus does not date the time of year by the Attic months; in the *HP*, no doubt following his informants (and perhaps when necessary translating local months into the Attic equivalents), he uses them occasionally, but more frequently than Aristotle adds an astronomical date.

Theophrastus, like Aristotle, recognizes four seasons: spring, summer, autumn and winter. Astronomical dates are given by the morning risings of the Pleiades, of Sirius and of Arcturus, and the morning setting of the Pleiades, together with the equinoxes and solstices. So Aristotle tends to confine himself to these, adding on one occasion (*History of Animals*, viii. 15 [599 b 10–11]) the evening setting of Arcturus.

In the treatise *On Airs, Waters and Places*, chap. xi. 2 (vol. ii, pp. 50. 18–52. 6 Littré) Hippocrates mentions all but one of the astronomical turning-points in Theophrastus:

> One must be most careful of all of the greatest changes of the seasons, and neither administer

> any drug of one's own initiative nor apply cautery to the belly at all nor perform any surgery until ten or more days have passed. The greatest and most dangerous changes of season are both solstices, the summer solstice more than the winter, and both of the so-called equinoxes, the autumnal more than the vernal. One must also be careful of the risings of the stars, as especially of the dog-star, in the next place of Arcturus, and furthermore of the setting of the Pleiades . . .

Autumn began with the morning rising of Arcturus, winter with the morning setting of the Pleiades, and summer with their morning rising. The rising of the dog-star announced the time of greatest heat. But there was no commonly recognized astronomical date for the end of winter and beginning of spring, and Theophrastus merely speaks of the season (or the air or the day) as " breaking into a smile " (*CP* 1 12. 8; 2 1. 4; 4 5. 1; *HP* 8 2. 4; cited p. lvi below).

In the Hippocratic *De Regimine*, iii. 68. 2 (vol. vi, p. 594. 9–14 Littré) the author speaks of the seasons as follows:

> I divide the year into four parts, the ones most recognized by most men, winter, spring, summer and autumn: winter from the setting of the Pleiades to the vernal equinox; spring from the equinox to the rising of the Pleiades; summer from the Pleiades to the rising of Arcturus; and autumn from Arcturus to the setting of the Pleiades.

The author proceeds to go into detail (iii. 68. 7–11, 13–14; pp. 594–604 Littré), beginning with the healthier half of the year. Some forty-one days have been passed over.[a] His scheme is as follows [b]:

Winter: ⟨morning⟩ setting of the Pleiades (November 9)
44 days
winter solstice (December 23)
44 days
west winds (February 5)
15 days
⟨evening⟩ rising of Arcturus and appearance of the swallow (February 20)
32 days ($44 + 44 + 15 + 32 = 135$ days of winter)

Spring: vernal equinox (March 24)
$6 \times 8 = 48$ days of spring

Summer: ⟨morning⟩ rising of the Pleiades (May 11)
⟨41 days⟩
summer solstice (June 21)
93 days ($\langle 41 \rangle + 93 = 134$ days of summer)

Autumn: ⟨morning⟩ rising of Arcturus and autumnal equinox (September 22)
48 days of autumn
⟨morning⟩ setting of the Pleiades (November 9)

[a] Perhaps intentionally, since they take account of the fraction over 365 days in the year. Carl Fredrich (Hippokratische Untersuchungen [Philol. Unters. herausg. von A. Kiessling und U. v. Wilamowitz-Moellendorff, Heft 15, Berlin, 1899], pp. 224–225) actually adds 41/42 days. The fractional or extra day would come before the summer solstice, at the end of the Attic year.

[b] The modern analogues are obtained by setting the summer solstice at June 21. The Julian year and the precession of the equinoxes are ignored, since our concern is with the weather, not with chronology.

A calendar transmitted in the manuscripts of Geminus [a] cites (among others) two contemporaries of Theophrastus: Eudoxus (395?–340?) and Callippus (second half of the fourth century). According to Ptolemy (*Apparitiones*, p. 67. 5–11 Heiberg) Callippus (of Cyzicus) made his observations at the Hellespont, Meton and Euctemon (Athenians of the Attic empire) made theirs at Athens and in the Cyclades and in Macedonia and Thrace, Eudoxus (of Cnidus) made his in Asia and Sicily and Italy and Democritus (of Abdera) in Macedonia and Thrace. Extracts follow:

Cancer: 31 days [June 21–July 21]

1st [June 21]: Summer solstice (Callippus)
25th [July 15]: Sirius rises in the morning (Meton)
27th [July 17]: Sirius rises in the morning: the Etesians blow for 55 days [July 17–September 10] (Eudoxus)
30th [July 20]: The south wind blows; visible rising of Sirius in the morning (Callippus)
31st [July 21]: The south wind blows (Eudoxus)

Leo: 31 days [July 22–August 21]

1st [July 22]: Sirius visible; oppressive heat follows; indications of a change of weather (Euctemon)
14th [August 4]: Greatest heat (Callippus)
17th [August 7]: The Lyre sets; rain follows; and the Etesians cease (Euctemon)

[a] It is published in C. [Karl] Manitius, *Gemini Elementa Astronomiae* (Lipsiae, 1898), pp. 210–233 (with a German translation) and in C. Wachsmuth, *Ioannis Laurentii Lydi Liber De Ostentis et Calendaria Graeca Omnia* (Lipsiae, 1897), pp. 181–195.

Virgo: 30 days [August 22–September 20]

5th [August 26]: The Etesians cease (Callippus)

10th [August 31]: Rising of Arcturus; storm at sea; south wind (Euctemon)

17th [September 7]: Indications of a change of weather; visible rising of Arcturus (Callippus)

19th [September 9]: Arcturus rises in the morning; winds blow for the following seven days; mostly clear weather; at the end of the period there is wind from the sunrise (Eudoxus)

20th [September 10]: Arcturus conspicuous; beginning of autumn; Capella rises and indications of a change of weather follow; a storm at sea (Euctemon)

24th [September 14]: Rain (Callippus)

Libra: 30 days [September 21–October 20]

1st [September 21]: Autumnal equinox (Euctemon, Callippus)

5th [September 25]: Pleiades seen at evening (Euctemon)

8th [September 28]: Pleiades rise ⟨at evening⟩ (Eudoxus)

Scorpio: 30 days [October 21–November 19]

4th [October 24]: Pleiades set in the morning (Democritus)

5th [October 25]: Arcturus sets in the evening (Euctemon)

8th [October 28]: Arcturus sets in the evening (Eudoxus)

15th [November 4]: Pleiades set (Euctemon)

16th [November 5]: visible setting of the Pleiades (Callippus)
19th [November 8]: Pleiades set in the morning (Eudoxus)

Sagittarius: 29 days [November 20–December 18]

Capricorn: 29 days [December 19–January 16]

1st [December 19]: Winter solstice (Euctemon)
4th [December 22]: Winter solstice (Eudoxus)
14th [January 1]: Middle of winter (Euctemon)

Aquarius: 30 days [January 17–February 15]

14th [January 30]: Clear weather; sometimes the west wind blows (Eudoxus)
16th [February 1]: The west wind begins, lasting 43 days from the solstice (Democritus)
17th [February 2]: Season for the west wind to blow (Euctemon); the west wind blows (Callippus)

Pisces: 30 days [February 16–March 17]

2nd [February 17]: Time for the swallow to appear (Euctemon); the swallow appears (Callippus)
4th [February 19]: Variable days, called the halcyon days (Democritus)
12th [February 27]: Arcturus rises in the evening; there is rain and the swallow appears (Eudoxus). (If Euctemon let winter begin with the setting of the Pleiades on the 15th day of Scorpio, and set mid-winter on the 14th day of Capricorn, winter has 115 days and the 12th of Pisces is the beginning of spring)

Aries: 31 days [March 18–April 17]

1st [March 18]: Vernal equinox (Callippus, Euctemon)
6th [March 23]: Vernal equinox (Eudoxus)
10th [March 27]: Pleiades set (Euctemon)
13th [March 30]: Pleiades set in the evening (Eudoxus); Pleiades set in the evening and remain invisible for 40 nights [March 30–May 8] (Democritus)

Taurus: 32 days [April 18–May 19]

13th [April 30]: Pleiades rise; beginning of summer (Euctemon)
22nd [May 9]: Pleiades rise ⟨in the morning⟩ (Eudoxus)
32nd [May 19]: Arcturus sets (Euctemon)

Gemini: 32 days [May 20–June 20]

13th [June 1]: Arcturus sets in the morning (Eudoxus)

The question when the year began might make some difference in the understanding of the words "early" and "late." The Attic year, which began about mid-summer, would make May late and July early. Theophrastus refers to difficulties about "early" and "late" in *CP* 1 10. 5, and in *HP* 7 10. 3–4 he indicates that the time after the vernal equinox is on the whole the one at which most plants could be said to be brought into being.[a]

[a] At *HP* 7 10. 3 U is corrupt: *πλην εἴ τις ὑποθοῖτο τοῦ ἔτους τὴν ἀρχην τινα προς τῆι ἴναξει.* If we read *πρὸς τῇ γενέσει* the connexion with the words that follow will be brought out.

Theophrastus' farmer's calendar would then run as follows:

[June 21 Callippus] summer solstice

CP 2 19. 1 ταῖς τροπαῖς ταῖς θεριναῖς
HP 7 15. 1 ἅμα γὰρ ταῖς τροπαῖς (-φαῖς U)
CP 3 4. 1 ἀπὸ γὰρ τροπῶν θερινῶν
HP 1 10. 1 μετὰ τροπὰς θερινάς; *HP* 7 1. 2 μετὰ τροπὰς θερινὰς (χειμερινὰς U), τοῦ δὲ Μεταγειτνιῶνος μηνός (August); *HP* 7 4. 11 μικρὸν πρὸ τροπῶν ἢ μετὰ τροπὰς
HP 1 10. 1 ὅτι γεγένηνται τροπαί
CP 1 13. 3 περὶ (επι U) τροπὰς; *CP* 6 10. 9 περὶ τροπὰς καὶ περὶ τὸ ἄστρον ἐνιαχοῦ καὶ ὅλως τοῦ θέρους; *HP* 1 9. 7, *HP* 6 2. 3 περὶ τροπὰς . . . θερινάς; *HP* 3 4. 3 περὶ τροπὰς θερινάς; *HP* 6 4. 7 περὶ τροπάς
HP 4 11. 5 τοῦ Σκιροφοριῶνος καὶ Ἑκατομβαιῶνος (June and July) ὡσπερεὶ πρὸ τροπῶν μικρὸν ἢ ὑπὸ τροπάς; *HP* 7 4. 11 μικρὸν πρὸ τροπῶν ἢ μετὰ τροπὰς (cited above)
HP 4 11. 5 ὡσπερεὶ πρὸ τροπῶν μικρὸν ἢ ὑπὸ τροπάς (cited above); *HP* 7 10. 4 ὑπὸ τροπάς

[July 20 Callippus] visible rising of Sirius; south wind. For the south wind *cf.* *CP* 1 13. 5 νότια πνεῖ; for the rising of Sirius *cf.*

CP 1 6. 3, *CP* 1 13. 3 Κυνὸς ἐπιτολή
CP 1 13. 5 ἐν τῇ τοῦ Κυνὸς ἐπιτολῇ
CP 6 7. 6 ἐπὶ τῆς τοῦ ἄστρου ἐπιτολῆς
CP 1 6. 6, *HP* 3 5. 4 ἐπὶ Κυνί
HP 1 9. 5, *HP* 4 2. 4, *HP* 6 3. 4 μετὰ Κύνα
CP 3 16. 2 μέχρι που (τοῦ U) Κυνὸς ἐπιτολῆς; *HP* 4 2. 4 μέχρι Κυνός

CP 3 3. 3 *περὶ Κύνα καὶ τοὺς ἐτησίας*; *CP* 2 17. 3, *CP* 6 10. 9, *HP* 2 6. 4 *περὶ τὸ ἄστρον*
CP 1 13. 3, *CP* 5 12. 1, *HP* 7 10. 4, *HP* 9 1. 6 *ὑπὸ Κύνα*; *CP* 1 12. 1, *CP* 1 13. 4, *CP* 3 3. 4, *CP* 5 9. 1 (⟨*τὸ*⟩), *HP* 9 6. 2 *ὑπὸ τὸ ἄστρον*; *CP* 1 13. 5 *ὑπὸ γὰρ αὐτὸ τὸ ἄστρον*; *CP* 5 9. 2, *HP* 7 5. 4 *ὑπὸ δὲ τὸ ἄστρον*

[August 26 Callippus] the Etesians cease

CP 3 3. 3 *περὶ Κύνα καὶ τοὺς ἐτησίας*
HP 2 7. 5 *ὅταν οἱ ἐτησίαι πνεύσωσιν*
HP 4 2. 5 *ὑπὸ τοὺς ἐτησίας*

[September 7 Callippus] visible rising of Arcturus

CP 6 8. 1 *τῇ τοῦ ἄστρου* (*sc.* *ὥρᾳ*)
CP 1 19. 3 *ἀπ' Ἀρκτούρου*
CP 1 13. 7 *ἅμα δὲ ἐν τῇ τοῦ ἄστρου* (*sc.* *ὥρᾳ*)
CP 1 6. 3, *CP* 5 10. 1, *HP* 3 5. 4 (first occurrence), *HP* 4 14. 10 (*ὑπ'* U) *ἐπ' Ἀρκτούρῳ*; *HP* 3 5. 4 (second occurrence) *ἐπὶ Κυνὶ καὶ Ἀρκτούρῳ*
CP 1 10. 5, *CP* 1 10. 6, *CP* 1 13. 3, *CP* 1 13. 5, *CP* 6 8. 1, *CP* 6 8. 5, *HP* 1 9. 7, *HP* 1 14. 1, *HP* 6 2. 6, *HP* 7 4. 10 *μετ' Ἀρκτοῦρον*; *HP* 9 8. 2 *τοῦ μετοπώρου μετ' Ἀρκτοῦρον*; *HP* 5 1. 2 *μετὰ τρυγητὸν καὶ Ἀρκτοῦρον*; *HP* 6 6. 9 *μετὰ γὰρ Ἀρκτοῦρον . . . καὶ περὶ ἰσημερίαν*; *HP* 7 10. 4 *μετ' Ἀρκτοῦρον καὶ ἰσημερίαν μετοπωρινήν*
CP 3 4. 1 *μέχρι Ἀρκτούρου*
HP 4 2. 4 *περὶ Ἀρκτοῦρον καὶ ἰσημερίαν*; *HP* 6 2. 6 *περὶ Ἀρκτοῦρον*
CP 6 8. 5 *πρὸς Ἀρκτοῦρον*
HP 4 11. 4 *ὑπ' Ἀρκτοῦρον Βοηδρομιῶνος μηνός* (September)

[September 21 Callippus] autumnal equinox

HP 6 2. 2, *HP* 7 10. 4 μετ' ἰσημερίαν μετοπωρινήν; *HP* 6 4. 2 μετὰ ἰσημερίαν φθινοπωρινήν; *HP* 7 7. 3 ἅμα τοῖς πρώτοις ὑετοῖς ἐστιν μετ' ἰσημερίαν; *HP* 7 10. 4 μετ' Ἀρκτοῦρον καὶ ἰσημερίαν μετοπωρινήν

HP 4 2. 4 περὶ Ἀρκτοῦρον καὶ ἰσημερίαν; *HP* 6 6. 9 μετὰ γὰρ Ἀρκτοῦρον . . . καὶ περὶ ἰσημερίαν

[October 28 Eudoxus] Arcturus sets in the evening

HP 3 17. 2 ἅμα Ἀρκτούρῳ δυομένῳ

[November 5 Callippus] visible setting of the Pleiades

CP 3 23. 1 ἅμα Πλειάσι δυομέναις; *HP* 6 5. 1 ἅμα Πλειάδι καὶ τοῖς πρώτοις ἀρότ[ρ]οις

HP 6 5. 2 ἄχρι Πλειάδος

CP 3 4. 1 (μετὰ Πλειάδος δύσιν δεξάμενοι) τὸ ἐπὶ τῷ ἄστρῳ ὕδωρ

CP 3 4. 1, *CP* 3 13. 2, *HP* 3 4. 4 μετὰ Πλειάδος δύσιν; *HP* 3 4. 5 μετὰ δύσιν Πλειάδος; *CP* 3 7. 10, *HP* 3 4. 5, *HP* 7 7. 3, *HP* 7 11. 3 μετὰ Πλειάδα.

HP 3 4. 4, *HP* 4 4. 10, *HP* 8 1. 2 περὶ Πλειάδος δύσιν; *CP* 3 7. 10 περὶ Πλειάδα, *HP* 6 6. 10 (⟨περὶ⟩)

CP 3 23. 1 πρὸ Πλειάδος

[December 22 Eudoxus] winter solstice

HP 4 14. 13 μετὰ τροπὰς μετὰ (περὶ τὰς Schneider) τετταράκοντα (for the forty days see the note on *CP* 5 12. 4); *HP* 4 14. 11 μικρὸν πρὸ τροπῶν ἢ μετὰ τροπὰς χειμερινάς; *HP* 7 1. 1 μεθ' ἡλίου τροπὰς χειμερινάς; *HP* 7 1. 2 μεθ' ἡλίου τροπὰς τοῦ Γαμηλιῶνος μηνός (January); *HP* 8 1. 2 ἀρχομένου τοῦ ἦρος μετὰ τὰς τροπὰς τοῦ χειμῶνος; *CP* 3 11. 6 μεθ' ἡλίου τροπάς; *CP* 3 13. 2 μεθ' ἡλίου δὲ τροπὰς

CP 3 23. 2 περὶ τροπάς; *CP* 5 12. 4 περὶ τροπὰς ὑπὸ τὰς τετταράκοντα
HP 3 4. 4 μικρὸν πρὸ ἡλίου τροπῶν; *HP* 4 14. 11 μικρὸν πρὸ τροπῶν ἢ μετὰ τροπὰς χειμερινὰς
CP 3 23. 2 ὑπὸ τροπάς

[February 20 Democritus] halcyon days (but the time is rather before and after the winter solstice: see the note on the passage)
CP 1 7. 5 περὶ τὰς ἁλκυονίδας

[February 2 Callippus] west wind
CP 3 13. 2 μεθ' ἡλίου δὲ τροπὰς καὶ μετὰ ζεφύρου πνοάς
HP 3 4. 2 πρὸ ζεφύρου καὶ μετὰ πνοὰς εὐθὺ ζεφύρου· πρὸ ζεφύρου μὲν . . . μετὰ ζέφυρον δὲ . . .

[February 18 Callippus] the swallow appears
HP 7 15. 1 ἅμα τῇ χελιδόνι (of the flowering of the celandine or " swallow-flower ")
[weather smiles]
CP 1 12. 8 ἅμα δὲ τῇ ὥρᾳ διαγελώσῃ; *CP* 2 1. 4 ἅμα τῇ ἡμέρᾳ διαγελώσῃ; *CP* 4 5. 1 πρὸς τὸ ἔαρ καὶ διαγελῶντος ἤδη τοῦ ἀέρος; *HP* 8 2. 4 διαγελώσης δὲ τῆς ὥρας

[March 18 Callippus] vernal equinox
CP 1 6. 3 ἐπ' ἰσημερίαις ἔτι κυόντων ; *HP* 3 4. 2 πρὸ ἰσημερίας δὲ μικρόν

[May 9 Eudoxus] (morning) rising of the Pleiades
CP 5 9. 12, *HP* 4 14. 5 ἐπὶ Πλειάδι.[a]

[a] Hiatus is avoided
(1) by the isolated plurals (*CP* 1 6. 3 ἰσημερίαις ἔτι; *CP* 3 23. 1 Πλειάσι δυομέναις ὥσπερ);

To these dates may be added those taken from the darkening of the grape [a] or from the vintage [b] or from the harvesting of wheat.[c]

Varro discusses the farmer's year in *R. R.*, i. 28–36. The four seasons each begin on the 23rd day in which the sun is in a sign of the zodiac:

23 Aquarius (a. d. VII id. Feb. = February 7); setting the summer solstice at June 21 we obtain [Feb. 4] as the modern equivalent for the weather. Beginning of spring, which lasts 91 days.
23 Taurus (a. d. VII id. Mai. = May 9 [May 6]); beginning of summer, which lasts 94 days.
23 Leo (a. d. III [tertium Jucundus; VII the MSS.] id. Sextil. = August 11 [August 8]); beginning of autumn, which lasts 91 days.
23 Scorpio (a. d. IV id. Nov. = November 10 [November 7]); beginning of winter, which lasts 89 days.

He can thus let the equinoxes and solstices split the seasons into approximate halves; into these he fits

(2) by using τοῦ ἄστρου for Ἀρκτούρου (*CP* 1 13. 7; *CP* 6 8. 1);
(3) by using τὸ ἄστρον for Κύνα (except at *CP* 5 9. 1 and 5 9. 2, where τὸ ἄστρον points to the etymology of ἀστροβολία; at *CP* 6 7. 6 the phrase ἐπὶ τῆς τοῦ ἄστρου ἐπιτολῆς conceivably refers to some similar name for this change in wine);
(4) by using περί for ὑπό (except at *HP* 4 11. 5 ἢ ὑπὸ τροπάς; *CP* 1 13. 3 καὶ ὑπὸ Κύνα; *CP* 5 9. 1 καὶ ὑπὸ τὸ ἄστρον).

[a] *HP* 3 4. 4 ἅμα τῷ βότρυϊ περκάζοντι; *HP* 9 11. 7 ὅταν ἄρτι περκάζῃ σταφυλή.

[b] *HP* 5 1. 2 μετὰ τρυγητόν; *HP* 9 11. 8 ἅμα τρυγητῷ.

[c] *HP* 3 4. 4 περὶ πυροῦ ἀμητόν (read πυραμητὸν); *HP* 7 6. 2 ὑπὸ πυραμητὸν; *HP* 9 9. 2 περὶ πυραμητόν; *HP* 5 1. 2 μετὰ πυροτομίαν; *HP* 9 8. 2 ὑπὸ πυροτομίαν (-ας UU*); *HP* 9 11. 11 περὶ πυροτομίαν (-ας UU*).

the west wind, the rising of the Pleiades, the dog-star and the setting of the Pleiades:

(1) the west wind (a. d. VII id. Feb. = February 7 [February 4]; this is 23 Aquarius, the beginning of spring).
    45 days (so Schneider; 40 the MSS.)
(2) vernal equinox
    44 days
    rising of the Pleiades
    2 days
(3) 23 Taurus, the beginning of summer. 45 + 44 + 2 = 45 + 46 = 91 days of spring.
    46 days
(4) summer solstice
    27 days
    dog-star
    21 days
(5) 23 Leo (a. d. III id. Sextil. = August 11 [Aug. 8]); the beginning of autumn. 46 + 27 + 21 = 46 + 48 = 94 days of summer.
    46 days
(6) autumnal equinox
    32 days
    setting of Pleiades
    13 days
(7) 23 Scorpio (a. d. IV id. Nov. = November 10 [Nov. 7]). The beginning of winter. 46 + 32 + 13 = 46 + 45 = 91 days of autumn.
    44 days
(8) winter solstice
    45 days
(1) 23 Aquarius; the west wind. 44 + 45 = 89 days of winter.

Using our own analogues of the dates, we obtain the following calendar:

Feb. 4: west wind; beginning of spring.
March 21: vernal equinox.
May 4: rising of Pleiades.
May 6: beginning of summer.
June 21: summer solstice.
July 18: dog-star.
August 8: beginning of autumn.
September 23: autumnal equinox.
October 25: setting of Pleiades.
November 7: beginning of winter.
December 21: winter solstice.

The Manuscripts

The eight Greek manuscripts of the *CP* (U Nv MC H PB), the Aldine (a) and Gaza's translation are related as follows:

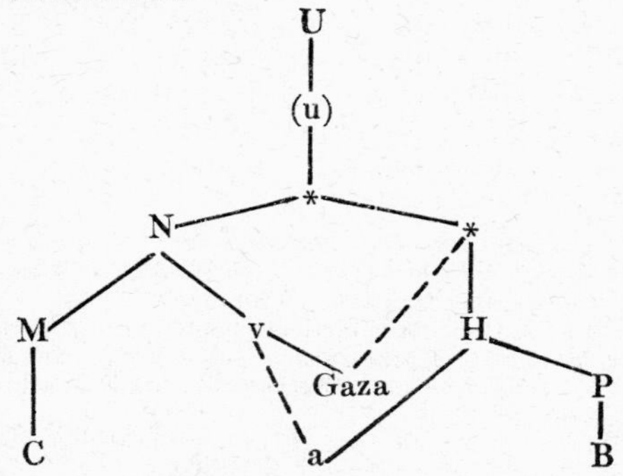

U Vatican City, Urbinas graecus 61; 11th century; *HP CP*.
U[d] The diorthotes of U.

u Correctors (more probably an Italian corrector) of the 15th century. The corrections of u are in *CP* apparently all conjectures.

N Florence, Laurentian Library, desk 85, 22; 15th century; *CP Pl* (= [Aristotle] *De Plantis*) *HP*.[a]

v Venice, Library of St. Mark 274; dated January 3, 1443; *HP CP*. A copy of N.

Gaza The Latin translation, completed in 1451, of Theodorus Gaza. In addition to the editio princeps of 1483, which is full of misprints, seven manuscripts survive.[b]

M Florence, Laurentian Library, desk 85, 3; 15th century; *HP CP Pl*. An emended copy (or descendant) of N.[c]

C Oxford, Corpus Christi College 113; 15th century; *inter alia* it contains *HP CP Pl*. In *HP CP* it is a copy of M.[d]

H Harvard College Library 17; 15th century. When the Aldine was printed from it this manuscript (one of several bound together) was intact. It now contains, misbound in the wrong order, the following fragments:[e]

[a] Dieter Harlfinger, *Die Textgeschichte der pseudo-aristotelischen Schrift ΠΕΡΙ ΑΤΟΜΩΝ ΓΡΑΜΜΩΝ* (Amsterdam 1971) p. 410, ascribes the title to Camillus Venetus.

[b] Listed in C. B. Schmitt, article Theophrastus in *Catalogus Translationum et Commentariorum: Mediaeval and Renaissance Latin Translation and Commentaries*, vol. ii (Washington, 1971), p. 273.

[c] D. Harlfinger, *op. cit.*, pp. 417f, identifies the scribe with that of 13 other MSS or portions of MSS.

[d] D. Harlfinger, *op. cit.*, p. 412, identifies the scribe as Petros Hypsilas.

[e] For the scribe *cf.* Harlfinger, *op. cit.*, p. 415.

HP 9 16. 8 |λ ἐκ σούσων—*CP* 2 6. 1 μάλιστα,|
CP 2 9. 6 |ἄλλων—*CP* 3 2. 7 ῥιγοῦν|
HP 1 8. 3 |μεῖζον—*HP* 3 11. 1 κοινῷ|
HP 3 12. 1 |δὲ· τὸ μέν, ἄρρεν—*HP* 4 6. 5 ἕως|

a The fourth volume of the Aldine Aristotle; dated Calendis Iunii M IIID; *inter alia HP CP*. Printed from H, then intact; readings have also been taken from v and Gaza.

P Paris, National Library 2069; 15th century; *HP CP Pl.*[a]

B Vatican City, Vaticanus graecus 1305; 15th century. One of the two manuscripts bound together in this volume contains *HP CP Pl* and is a copy of P.[b]

U is the only source of the manuscript tradition; the rest need only be cited for conjectures or when U has become illegible. It is a very incorrect manuscript: accents and breathings are often omitted or interchanged and the Byzantine homophones (including single and double consonants) are constantly confused. To lighten the apparatus errors in spelling are regularly omitted and many inconsistencies have been allowed to remain, some of them conceivably due to the author or his amanuenses, such as variation between α and αι in such words as ἀεί, ἐλάα, between ο and οι in ῥόα and πόα, and between ττ and σσ in θάλαττα and the like.

The following superior numbers and letters are used, illustrated here with the siglum U:

U^c^ indicates a correction by the first hand

[a] For the scribe *cf.* Harlfinger, *op. cit.*, p. 413.
[b] For the scribe *cf.* Harlfinger, *op. cit.*, p. 418.

U^{cc} such a correction made in the course of writing
U^{ac} the reading before correction by the first hand
U^{r} a reading due to erasure
U^{ar} the reading before erasure
U^{m} a reading or note in the margin by the first hand
U^{t} a reading in the text
U^{ss} a superscription
U^{1} a reading by the first hand.

The two scribes of U were calligraphers; when an erasure is clumsy we therefore ascribe it at times to u, not to U^{r}. Again, u in making his corrections often avoids erasure, and simply adds the correct accent without deleting the accent of U.

Editions, Translations and Commentaries

1483 Theodorus Gaza, Latin translation of the *HP CP*. Colophon: IMPRESSVM TARVISII PER BARTHOLOMAEVM CONFALONERIVM DE SALODIO. ANNO DOMINI. M.CCCC.LXXXIII. DIE XX. FEBRVARI. Gaza completed his translation in 1451.

1497 The fourth volume of the Aldine Aristotle; dated Calendis Iunii M IIID; *HP CP inter alia.*

Itali Conjectures in the margins and between the lines of a copy of the first Aldine now in the Royal Library at Copenhagen (Inc. 4338). They come from two sources, " v. c. 1 " and " v. c. 2 " (vetus codex primus and vetus codex secundus), the first the marginalia in an Aldine, the second and fuller source a manuscript bearing the names of Hermolaus Barbarus (1454–95) and Scipio Carteromachus

(Forteguerra; 1460–1515). Both sources were apparently written in the same hand. Substantially the same corrections are also found in a manuscript in the former Ducal Library at Weimar, in the form of corrigenda to the Aldine, the number of the folio (*ante* or *a tergo*) and the verse being given; the same corrections are also found in Vat. Ottobonianus 316 (pp. 41–56). Most were suggested by Gaza's Latin. A small selection is also found in Isaac Vossius' (1618–89) copy of Heinsius' edition (1613) in the University Library at Leyden (758 A 12), ascribed to " v. c."

1541 Theophrasti . . . Opera . . . omnia . . . Basileae; *inter alia HP CP*. The preface (in my copy by Joachim Camerarius) is dated XI. Cal. Septembris . . . M. D. XLI. The Greek colophon gives the date as the second of September (*μαιμακτηριῶνος*), 1541, the publisher as Ioannes Oporinos (Johann Herbster).

1550 THEOPHRASTI DE CAVSIS Plantarum liber primus. PARISIIS Apud Vascosanum . . . M. D. L. The copy in the University Library at Leyden (758 C 2 ^{2}/) is bound with the Greek text (without the title page or translation or notes) of Adrianus Turnebus' edition of Theophrastus *De Odoribus* (Lutetiae, apud Michaëlem Vascosanum, M. D. LVI). The two texts are evidently the working copy of an editor, who has added a date at the end of the second: Secundo die S⟨eptembris⟩ | 1550. (The bracketed letters are on the slope of the inner margin in my microfilm and illegible.) The hand is not that of Turnebus, to judge by the letter of his that survives (to Joachim Camerarius, in Greek with a Latin address; Paris, National Library, Greek Supplement, 1361). Meanwhile I assign the conjectures to Vasc.[2]

1552 The second Aldine of Aristotle, edited by Joannes Baptista Camotius. The sixth volume is entitled Theophrasti historiam de Plantis, et de Causis Plantarum, etc. . . .

1558 Theophrasti de causis Plantarum liber sextus, Graece & Latine. Parisiis, apud Guil. Morelium. M. D. LVIII.

1566 Iulii Caesaris Scaligeri . . . Commentarii, et Animadversiones, in sex libros de causis plantarum Theophrasti . . . Lugduni . . . M. D. LXVI. The book is posthumous; J. C. Scaliger died in 1558.

1574–1575 A manuscript Latin translation of *HP CP Scripta Minora Pl* (*inter alia*) by Jac. Dalecampius (1513–1588) or Dalechampius, as the copyist, his nephew, writes the name. Paris, National Library, Latin 11, 857. A note at the end of the *CP* (209ʳ, p. 433) gives the date: Inchoata haec uersio die primo Nouembris anno 1574· absoluta die -5· Septembris anno 1575·

1613 Theophrasti Eresii Graece et Latine opera omnia. Daniel Heinsius . . . Lugduni Batavorum . . . M. D. CXIII.

1644 Caspar Hoffmannus. On these unpublished papers, including a Latin translation of *HP* and *CP*, see J. G. Schneider, vol. I Praefatio, pp. XIII–XIV.

1818–1821 Theophrasti Eresii Quae Supersunt

Opera et Excerpta Librorum Quatuor Tomis Comprehensa. Ad Fidem Librorum Editorum et Scriptorum Emendavit Historiam et Libros VI de Causis Plantarum coniuncta Opera D. H. F. Linkii Excerpta Solus Explicare Conatus est Io. Gottlob Schneider, Saxo.

Tomus Primus Textum Graecum Continens (Lipsiae, 1818).

Tomus Secundus Versionem Latinam Librorum de Historia et de Causis Plantarum et Plerorumque Libellorum Physicorum Continens cum Curis Posterioribus Editoris Io. Gottlob Schneideri, Saxonis (Lipsiae, 1818).

Tomus Tertius Annotationes ad Historiam Plantarum Continens (Lipsiae, 1818).

Tomus Quartus Annotationes ad Libros de Causis Plantarum, Opuscula et Fragmenta Continens (Lipsiae, 1818).

Tomus Quintus Supplementa et Indicem Verborum Continens (Lipsiae, 1821). It was not until volumes I–IV had been published that Schneider obtained a collation of U.

1854 Theophrasti Eresii Opera Quae Supersunt Omnia. Ex Recognitione Friderici Wimmer.

Tomus Primus Historiam Plantarum Continens (Lipsiae, MDCCCLIV).

Tomus Secundus De Causis Plantarum Libros VI Continens (Lipsiae, MDCCCLIV).

Tomus Tertius Fragmenta Continens . . . (Lipsiae, MDCCCLXII).

1866 Theophrasti Eresii Opera, Quae Supersunt, Omnia. Graeca Recensuit, Latine Interpretatus est, Indices Rerum et Verborum Absolutissimos Adiecit Fridericus Wimmer Doct. Philos. Parisiis, Editore Firmin Didot . . . M DCCCLXVI.

Except for the addition of a Latin translation the edition is a virtual repetition of Wimmer's Teubner edition of 1854–62.

1910 Sprachliche Untersuchungen zu Theophrasts botanischen Schriften. Von Ludwig Hindenlang (Dissert. philol. Argentoratenses Selectae . . . Vol. XIV, Fasc. II [Strasburg, 1910]).

1927 Theophrastus: De Causis Plantarum Book One. Text, Critical Apparatus, Translation, and Commentary. Robert Ewing Dengler. University of Pennsylvania dissertation. Philadelphia 1927.

1941 Theophrastus On Plant Flavors and Odors. Studies on the Philosophical and Scientific Significance of *De Causis Plantarum* VI accompanied by Translation and notes. By George Raynor Thompson. Princeton dissertation (typescript); Princeton 1941.

1963 A Modern Translation of Theophrastus: De Causis Plantarum Book 2. With General Introduction and Commentary by Ursula Katherine Duncan (typescript).

1970 George Raynor Thompson: Theophrastus De Causis Plantarum I–VI Translated. Typescript.

It is a pleasure to thank Professor P. De Lacy, Professor A. Diller, Mrs. E. K. Ritter, and Mr. G. R. Thompson and B. Hillyard for help and information and the custodians of the manuscripts and rarer editions for their courtesies. Professor G. K. K. Link first suggested the translation, and without his partnership the work would never have been attempted or brought to the present state of completion. Much still remains to be done with a corrupt text and a difficult author.

The translations from the *HP* in the notes were made from a critical text of my own. For the convenience of the reader a few repetitions, where in different notes the same passage is translated in somewhat different language, have been allowed to stand.

CHICAGO
August 1975

Benedict Einarson

BOOK I

ΠΕΡΙ ΦΥΤΩΝ ΑΙΤΙΩΝ[1] Α′

τῶν φυτῶν αἱ γενέσεις ὅτι μέν εἰσι πλείους, καὶ πόσαι καὶ τίνες, ἐν ταῖς ἱστορίαις εἴρηται πρότερον· ἐπεὶ δ' οὐ πᾶσαι πᾶσιν, οἰκείως ἔχει[2] διελεῖν τίνες
5 ἑκάστοις καὶ διὰ ποίας αἰτίας, ἀρχαῖς χρωμένους ταῖς[3] κατὰ τὰς ἰδίας οὐσίας· εὐθὺ γὰρ χρὴ συμφωνεῖσθαι τοὺς λόγους τοῖς εἰρημένοις.

ἡ μὲν οὖν ἀπὸ τοῦ σπέρματος γένεσις κοινὴ πάντων ἐστὶν τῶν ἐχόντων σπέρμα, πάντα γὰρ
10 δύναται γεννᾶν. τοῦτο δὲ καὶ τῇ αἰσθήσει φανερὸν

1. φυσικων αιτιων Varro, *R.R.* i. 5. 1; φυτικὰ αἴτια Athenaeus ii. 45 (55 E), iii. 5 (74 AB), iii. 12 (77 C) [ἐν τῷ β′ τῶν αἰτιῶν ii. 12 (77 F)]; φυτικῶν αἰτιῶν α′ β′ γ′ δ′ ε′ [ϛ′] ζ′ η′ Diogenes Laertius, v. 46; Apollonius, *Mir.*, chap. xlvi (xlv) ἐν τῇ ε′ τῶν φυσικῶν; *cf.* M. Steinschneider, "Die arabischen Übersetzungen aus dem Griechischen" (*Zentralblatt für Bibliothekswesen*, Beiheft v [1889], pp. 129–130: ". . . von den Ursachen der Pflanzen, übersetzt von Ibrahim ibn Bakus, wovon sich nur ein Teil des I. Traktats findet."

[1] U has αἰτιῶν in the title here and at the end of book VI; αἰτίων at the end of book I and book II; αἰτιων at the end of books IV and V.
[2] ἔχει U[r] N HP: ἔχειν U[ar].
[3] ἀρχαῖς χρωμένους ταῖς Wimmer (*cf.* *CP* 6 11. 5): οὐχ αἷς χρωμένοις (-ους u) τὰς U.

---

[a] *HP* 2 1. 1–4 (cultivated trees); 3 1. 1–6 (wild trees); 2 2. 1 and 6 7. 1–6 (undershrubs and herbaceous plants); 7 2. 1–9 (herbaceous plants). *Cf.* especially *HP* 2 1. 1–2: "The generation of trees and of plants in general is either spontaneous or from seed or root or detached side-growth or from a branch or twig or the trunk itself or else from the wood chopped small . . . Of these ways the spontaneous is a pri-

BOOK I

THE MODES OF GENERATION

That plants have several modes of generation has been said earlier in the *History*,[a] where we have also enumerated and described them. Since not all occur in all plants, it is proper to distinguish the modes that occur in the different groups and give the reasons why, resting the explanations on the special character of the plant,[b] for the explanations must first of all accord with the account given there.[c]

(1) *From Seed*

Generation from seed is common to all plants that have seed, since all seeds are able to generate.

mary one, and those from seed and root would appear most natural, since they too are (as it were) ' spontaneous,' whereas the rest belong already to a choice made by human art. All plants come up in one of these ways, and most in several . . ." *Cf.* also Aristotle, *On the Generation of Animals*, iii. 11 (761 b 27–29): ". . . some (*sc.* plants) come from seed, some from pieces detached and planted, and a few from sending up side-growths, as onions." In the *CP* Theophrastus is almost exclusively concerned with seed-plants (Spermatophytes); in the *HP* a few references occur to plants which do not bear seeds: algae, fungi, lichens and ferns.

[b] Literally, " the distinctive essences " (*cf. CP* 2 19. 6), synonymous with " the distinctive natures " (*CP* 1 1. 3; *cf.* also 1 2. 4, 1 4. 2, 1 8. 2, 1 11. 6 [*bis*], 1 18. 4, 1 21. 3, 2 3. 4, 2 14. 5, 2 17. 10, 2 19. 6). The explanations are to rest on differences in the plants' own natures, not (except incidentally) on the environment or on human skill.

[c] That is, the details about what plants can be propagated from what part. All such parts will be seen to contain a centre of vital fluid.

ὅτι συμβαίνει, κατὰ δὲ τὸν λόγον ἴσως ἀναγκαῖον· ἡ γὰρ φύσις οὐδὲν μὲν ποιεῖ μάτην, ἥκιστα δ' ἐν τοῖς πρώτοις καὶ κυριωτάτοις, πρῶτον δὲ καὶ κυριώτατον τὸ σπέρμα· ὥστε τὸ σπέρμα μάτην
15 ἂν εἴη μὴ δυνάμενον γεννᾶν, εἴπερ τούτου χάριν αἰεὶ τὸ σπέρμα καὶ πρὸς τοῦτο [1] πέφυκεν.

1. 2 ὅπερ ἐκ τῶν ἄλλων πάντων λαμβάνειν ἐστὶν ὁμολογούμενον· ἀλλὰ διὰ τὸ μὴ χρῆσθαι τοὺς γεωργοὺς ἐπ' ἐνίων ὅτι θᾶττον ἀπ' αὐτομάτων παραγίνεται, καὶ διὰ τὸ μὴ εἶναι ῥᾳδίως ἐπί
5 τινων λαμβάνειν μήτε τῶν δένδρων μήτε τῶν ποιωδῶν τὰ σπέρματα, διὰ ταῦτ' οὐκ οἴονταί τινες πάνυ [2] ἀπὸ σπέρματος ἐνδέχεσθαι. καίτοι, καθάπερ ἐν ταῖς ἱστορίαις εἴρηται, καὶ φανερώτατον ἐπὶ τῶν ἰτεῶν.[3]

[1] τοῦτο u: τούτον U.
[2] πάνυ U: πάντα Gaza (*omnia*).
[3] ἰτεῶν ego: συκῶν U.

---

[a] Or perhaps "the nature (of the plant in question)," the usual sense in Theophrastus.

[b] The aim is to produce a plant like that from which the seed came.

[c] In fact some cultivated Spermatophytes have seedless fruit; others have sterile seeds.

[d] *Cf. HP* 2 2. 4 (cited in note *h* on *CP* 1 2. 2); *HP* 6 7. 1; *HP* 7 4. 10–11; *CP* 1 8. 1.

[e] This use of "spontaneous" for generation that is not from seed comes from Aristotle: *cf. History of Animals* v. 1 (539 a 15–21): "What applies to plants applies also to animals; thus some plants come from seed of other plants, whereas others arise spontaneously (as it were), when a starting-point of this character (*sc.* seed-like) has been formed, and of these some get their food from the earth, whereas others arise in other plants, as we said in our study of plants;" *On the Generation of Animals* i. 1 (715 b 25–30): "Some (*sc.* plants) are produced

That they do so is not only evident to sense, but in theory too it is perhaps a necessary conclusion: nature [a] not only does nothing in vain, but does so least of all in what immediately serves her aims [b] and is decisive for their achievement; now the seed has this immediacy and decisiveness; hence the seed, if unable to generate, would be in vain, since it is always aimed at generation and produced by nature to achieve it.[c]

1. 2 That all seeds are able to generate we can set down as a point of general agreement by all but a few persons. But because with some plants farmers do not use the seed (since the plant matures more rapidly [d] from spontaneous [e] growths, and again because in some it is difficult to secure the seed, seeds of trees [f] as well as of herbaceous plants),[g] some growers [h] for these reasons are not quite convinced that it is possible for the plants to come from seed. And yet, in the case of the willow (as was said in the *History*) [i] production from seed is in fact quite evident.

from seed and some when their nature behaves spontaneously (as it were), for they arise when either the earth or else certain parts in plants undergo decomposition (for some plants are not formed separately, away from the rest, but are produced in trees, as the mistletoe." (For Theophrastus' correction of this account of the mistletoe see *CP* 2 17. 5.)

[f] For the cypress *cf. CP* 1 5. 4; for the willow see note *i* below.

[g] As of thyme: *CP* 1 5. 3 with note *e*.

[h] Among them Homer and Aristotle (see the following note).

[i] *HP* 3 1. 2–3. This corrects Aristotle, *On the Generation of Animals*, i. 18 (726 a 6–7): "Some (*sc.* plants) bear no seed at all, as willow and black poplar." The authority was Homer: *cf. Odyssey*, x. 510, where the grove of Persephone has "tall black poplars and willows that lose their fruit."

10 ἔτι δὲ κατ' ἄλλον τρόπον ἡ βλάστησις αὐτῶν, εἴ πῃ[1] τι τῶν αὐτομάτων ἐκ συρροῆς καὶ σήψεως, μᾶλλον δὲ ἀλλοιώσεως γινομένης φυσικῆς.

ὅτι μὲν οὖν κοινὴ πᾶσιν ἡ διὰ τοῦ σπέρματος γένεσις φανερόν· εἰ δ' ἀμφοτέρως ἔνια, καὶ 15 αὐτόματα καὶ ἐκ σπέρματος, οὐδὲν ἄτοπον, ὥσπερ καὶ ζῷά τινα καὶ ἐξ ἄλλων[2] καὶ ἐκ τῆς γῆς.

1. 3 διὸ καὶ αἱ γενέσεις κατὰ τὰς ἰδίας ἑκάστων φύσεις· ὅσα μὲν γὰρ ξηρὰ καὶ μονοφυῆ καὶ ἀπαράβλαστα, ταῦτ' οὐδεμίαν (ὡς εἰπεῖν) δέχεται φυτείαν οὔτε ἀπὸ παρασπάδος οὔτε ἀπὸ ἀκρεμόνος· τὴν μὲν

[1] πῃ ego (Schneider reads δή or deletes): μή U.
[2] ἐξ ἄλλων ego: ἐξάλων U (ἐξ ἀλλήλων u HP; ἐξαλλήλων N).

---

[a] That is, in some part of the parent plant.

[b] The native fluid of the plant and the food collect in a permanent place, the food coming to a halt instead of flowing onward.

[c] The traditional term, used by Aristotle in the passage cited in note *e* (p. 5). He replaced it with " concoction " in another passage: *On the Generation of Animals*, iii. 11 (762 a 9–14) [of testacea]: "All that arise spontaneously . . . are observed to be produced to the accompaniment of decomposition . . . But nothing is produced by decomposition, but instead by undergoing concoction . . ." *Cf. CP* 2 9. 14.

[d] Modes of generation even more remote from one another than generation from seed and generation from a part in plants. *Cf.* Aristotle, *On the Generation of Animals*, iii. 11 (761 b 23–762 a 2): " The nature of testacea is formed in some kinds spontaneously, and in a few from the creatures' emitting a certain power from themselves, but in these last they often also arise from a spontaneous formation . . . Trumpet-shells and purple-fish and the creatures that produce the so-called 'honeycombs' emit phlegm-like fluids . . . But we must consider none of these fluids semen . . . This is why once

(2) *From " Spontaneous " Growth*

Seed-bearing plants come up in still another way when some spontaneous growth arises somewhere [a] from the collecting of a pool [b] and from decomposition [c] (or rather when a natural alteration occurs).

Now it is evident that generation through the seed is common to all; and if some are generated in both ways, spontaneously as well as by seed, there is no absurdity: so some animals similarly come from two sources, both from other animals and from the earth.[d]

Trees:
Reasons for the Two Types of Propagation, From Seed Alone and From Seed and Parts

1.3 Hence [e] the types of propagation depend on the distinctive natures of the several kinds of tree. So all trees that are (1) dry and that are (2) single-stemmed and without side-shoots allow of no propagation (practically speaking) either from a detached sucker [f] or from a branch: (2) from a sucker

one of the creatures has been formed, a great number of such creatures are produced. For all of them also happen to be produced spontaneously, but it is reasonable that they should also be formed to a greater extent from pre-existing specimens."

[e] Because of the pool of fluid required for propagation from the parts.

[f] " Detached suckers " (*paraspádes*) were taken to be literally " shoots (*blástai*) taken from the side; " *cf.* παρασπώμενον *HP* 2 1.3. These are slips taken from shoots growing from the root or from the base of the trunk. Perhaps the true literal meaning was active: "a shoot that draws the food aside;" *cf.* " sucker " in English, " Räuber " in German and *móschos* (" layer;" literally " calf ") in Greek.

5 ἀπὸ παρασπάδος τοῦ[1] μὴ ἔχειν τὰ μονοφυῆ, τὴν δὲ ἀπὸ ἀκρεμόνος καὶ κλωνὸς διὰ τὴν ξηρότητα, ἀεὶ γὰρ τὸ μέλλον βλαστάνειν ἔχειν τε δεῖ[2] τὴν ἔμβιον ὑγρότητα καὶ ταύτην δύνασθαι τηρεῖν. τὰ δὲ φύσει ξηρά, διὰ τὸ ὀλίγην ἔχειν, ὅταν ἀπὸ 10 τοῦ δένδρου ἀφαιρεθῇ, ταχὺ διαπνεῖται καὶ ἀτμίζεται συνεξάγον[3] ἅμα καὶ τὸ σύμφυτον θερμόν.

1. 4 καὶ ταῦτα μὲν διὰ τὰς εἰρημένας αἰτίας·

τὰ δὲ [ἀπαράβλαστα][4] συμμετρίᾳ[5] τινὶ τοῦ θερμοῦ καὶ ὑγροῦ [καὶ μονοφυῆ][6] δέχεται καὶ τὰς ἄλλας, δύναται ⟨γὰρ⟩[7] τηρεῖν τὴν ὑγρότητα 5 καὶ θερμότητα τὴν ξύμφυτον ὥστε βλαστάνειν μὴ μόνον ἀπὸ παρασπάδος, ἀλλ' ἔνια καὶ ἀπὸ τῶν ἄκρων, οἷον συκῆ καὶ ἄμπελος καὶ τῶν ἀκάρπων

[1] τοῦ U: τῷ Vasc.[2] (*quoniam* Gaza).

[2] ἔχειν τε δεῖ u: ἔχει τε δεῖν U.

[3] συνεξάγον U (supply τὸ ὑγρόν from ὑγρότητα): συνεξάγονθ' u.

[4] [ἀπαράβλαστα] Gaza.

[5] συμμετρία U: ἀσυμμετρίᾳ Wimmer.

[6] [καὶ μονοφυῆ] Gaza, Scaliger. (ἀπαράβλαστα καὶ μονοφυῆ was once a note explaining ταῦτα in line 1.)

[7] δύναται ⟨γὰρ⟩ De Lacy (*valeantque* Gaza; δυνάμενα or ⟨καὶ⟩ δύναται Vasc.[2]): δύναται U.

---

[a] *Cf. HP* 1 2. 4 (of the homoeomerous parts of plants): "First come the fluid and heat; for every plant has a certain fluidity and heat belonging to its nature, just as every animal does, and when these are giving out old age and decay occurs, and when they have given out entirely, death and withering."

because it belongs to the character of being single-stemmed to have none, and (1) from a branch or twig because these are too dry. For if a piece is to grow it must always possess and also be able to retain the fluid that gives it independent life.[a] But the parts naturally dry, since they possess but a small amount of the fluid, soon lose it to the wind and sun on removal from the tree, and the fluid carries off the native heat as well.[b]

1.4 These trees, then, behave in this way for the reasons given.[c]

The rest, owing to a certain adequacy in the amount of their heat and fluid, admit the other forms of propagation as well, since they are able to retain their native fluid and heat so as to grow not only from detached suckers, but some grow from cuttings as well that are taken from their extremities, as the fig [d] and vine [e] (and again among fruitless [f] and water-

[b] *Cf. CP* 5 12. 4, where it is the heat that takes the fluid along with it.

[c] *CP* 1 1. 3.

[d] *Cf. HP* 2 1. 2: "The fig grows in all these ways (*sc.* from root, detached sucker, twig, branch) except from the trunk and pieces of chopped wood."

[e] *Cf. HP* 2 1. 3 (it is held that most trees can be propagated from the twigs too, if these are smooth and young and well-grown) "for there are few at all that sprout and are propagated more from the upper parts, as the vine from the twigs (for the vine does not propagate from its 'prow' either, but from its twigs) . . ."

[f] That is, not bearing mature fruit: *cf.* also *CP* 5 12. 9. For the question whether these trees bear or do not bear *cf. HP* 3 1. 2–3 and 3 3. 4. It is to be noted that the willow and the black and white poplar are dioecious, so that the males bear no fruit.

δὲ καὶ φιλύδρων ὥσπερ ἰτέα καὶ ἀκτῆ καὶ λεύκη καὶ αἴγειρος.

10 καθόλου μὲν οὖν καὶ τύπῳ τοῦτον διωρίσθω τὸν τρόπον.

2. 1 οὐ μὴν ἀλλὰ μάλιστ' ἢ[1] μόνος τῶν ξηρῶν καὶ μονοφυῶν καὶ ἀπαραβλάστων ὁ φοῖνιξ δέχεται καὶ ἑτέρας γενέσεις παρὰ τὴν σπερματικήν· τάς τε γὰρ ῥάβδους φασὶ μοσχεύειν[2] περὶ Βαβυλῶνα 5 τὰς ἁπαλωτάτας καὶ ὅταν ἐμβιώσωσι μεταφυτεύουσιν· καὶ ἐν τοῖς περὶ τὴν Ἑλλάδα τόποις ἐὰν ἀποκόψας τις ⟨τὸ⟩ ἄνω[3] φυτεύσῃ ῥιζοῦσθαι,[4] καὶ βλαστάνειν οὐκ ἀπὸ τοῦ ἐγκεφάλου μόνον

3–12. Pliny, *N.H.* 13. 36: et ab radice avulsae ⟨*sc.* palmae⟩ vitalis est satus et ramorum tenerrimis. in Assyria ipsa quoque arbor strata in solo umido radicatur, sed in frutices, non in arborem. ergo plantaria instituunt anniculasque transferunt et iterum bimas.

3–6. *Cf.* Pliny, *N.H.* 17. 58: nam folia palmarum apud Babylonios seri atque ita arborem provenire Trogum credidisse demiror.

[1] μάλιστ' ἢ Schneider: μαλιστα U.
[2] μοσχεύειν U here, but at *HP* 2 2. 2 μωλεύειν.
[3] ⟨τὸ⟩ ἄνω ego (*cf. HP* 2 6. 2; τἄνω Schneider): ἄνω U.
[4] ῥιζοῦσθαι Schneider: ῥιζουσιν U.

---

[a] *Cf. HP* 3 1. 1 (of wild trees): "Their modes of propagation are simple: all grow either from seed or a root. This does not mean that it would not be possible for them to grow in other ways, but perhaps is due to the circumstance that nobody tries it out and plants them in other ways, and they would

loving trees willow, for example, elder and white and black poplar).[a]

Speaking generally and roughly, then, this is the line to be drawn.

An Exception: the Date-Palm

2.1 Nevertheless among trees that are dry and single-stemmed[b] and without side-growths the date-palm alone or chiefly allows of propagation in other ways than by seed. So it is reported that in Babylonia the tenderest of the branches[c] are layered,[d] and once they have developed independent life are trans-

grow up if they should get a suitable location and the proper care, just as nowadays the ones that grow in groves and are water-loving do, I mean for example plane, willow, white poplar, black poplar and elm: for all these and the like sprout when planted and do so most quickly and well when planted from detached suckers, even to the point that when the pieces are already tall and as high as a tree they survive the transference. Most of these trees are also planted by cuttings, as white and black poplar."

[b] *Cf. HP* 1 5. 1: "And again some trees are single-trunked (μονοστελέχη), some many-trunked, and this in a way is the same as having side-growths or having none . . ."

[c] Theophrastus' informant must have meant offsets by "branches" (ῥάβδοι). But he perhaps also used it of the leaves: *cf. HP* 2 6. 4: "When the date-palm is young it is not touched, except that the leaves are tied up so that the trees may grow straight and the branches (ῥάβδοι) may not get out of line. After this . . . the branches are cut off all around . . . about a span of the branches being left."

[d] *Cf. HP* 2 2. 2: "But some trees grow from seed alone, as silver-fir, pine, Aleppo pine and in general all conifers; and further the date-palm, except (it appears) that some say that in Babylonia it is also propagated from the branches by layering."

ἀλλὰ καὶ κάτωθεν αὔξειν τὰς ῥίζας· ἔτι δ' ἐὰν[1]
10 πλάγιος, ἐν ᾧ[2] καὶ ἔνικμος ἡ γῆ, τυγχάνῃ,[3] πολλαχόθεν ῥιζοφυεῖν καὶ ἐκβλαστάνειν, οὐ μέντοι μέγεθος τοῦτο λαμβάνειν.

2. 2 ἐπὶ δὲ πεύκης καὶ ἐλάτης καὶ εἴ τι παραπλήσιον τούτοις οὐδὲν συμβαίνει τοιοῦτον, οὐδ' ἐπὶ κυπαρίσσου, πλὴν εἴ τί που καὶ παραβλαστάνει, καθάπερ φασὶν ἐν Κρήτῃ περὶ τοὺς καλουμένους δρυΐτας·
5 ἐνταῦθα γὰρ οὐδὲν ἄλογον ὥσπερ[4] καὶ τὰς ἄλλας παραφυάδας, ὑπορρίζους[5] οὔσας καὶ ἐνύγρους[6]

[1] δ' ἐὰν ego: δε ἀν U.
[2] ἐν ᾧι u (ἐνῶι U): τεθῇ Schneider (after Gaza).
[3] τυγχάνῃ HP: -ει U N.
[4] ὥσπερ Schneider: ὥστε U.
[5] ὑπορρίζους HP: ὕπεροζους U; ὑπορίζους (ὑ- u) N.
[6] ἐνύγρους Vasc.[2] (*cf. CP* 1 3. 1): ἐνύδρους U.

---

[a] Presumably so called (in Greek τὸ ἄνω, simply "the upper," with which we may perhaps supply παραβλάστημα "side-growth") because it is opposed to a sucker. This "upper" is an offset containing a head. Possibly it was also called a "head" (*enképhalos*); so Palladius calls it a *cephalo* (v 5. 2): Nunc planta palmarum, quam cephalonem vocamus, locis apricis et calidis est ponenda.

[b] *Cf. HP* 2 6. 2: "The tree (*sc.* date-palm) is planted from itself (*sc.* and not from its fruit) when the upper growth, the one containing the head, is removed. They remove about two cubits, and after splitting this put the moist segment under ground."

[c] This is possible because each axil of a leaf either produces a bud primordium or a bud, and because the internodes are also able to produce adventitious roots.

[d] That is, conifers: *cf. HP* 2 2. 2: ". . . a few trees grow only from seed, as silver-fir, pine, Aleppo pine and in general all conifers;" *cf.* also *HP* 3 1. 2: ". . . except for those (*sc.* wild trees) that only grow from seed, as silver-fir, pine and Aleppo pine."

planted; and that in Greece if the upper growth [a] is cut off and planted it takes root [b] and not only sprouts at its head but also prolongs its roots below; furthermore, that when the tree lies on its side in a place where the ground is moist, it roots and sprouts at many points,[c] but that this growth does not attain to any size.

2. 2 But with pine, silver-fir and the like [d] nothing of the sort occurs, nor yet with cypress, except for countries where it also sends up side-growths, as is reported of the so-called "oak-Cypresses" [e] in Crete.[f] For there [g] it is not at all unreasonable that just as the other side-growths [h] sprout, so these detached suckers, since they have roots and possess

[e] This cypress was doubtless not pyramidal in habit but had spreading branches like the oak: *cf.* Pliny, *N. H.* 16. 141.

[f] *Cf. HP* 2 2. 2: "The cypress in other countries grows only from seed, but in Crete it also grows from the trunk, as with the trunk of the mountain cypress at Tarrha (for here is found the 'clipped' cypress, and this sprouts from the cut part no matter from where on the trunk the cut is made, whether at the base, the middle or the upper part; and in some districts it also sprouts from the roots, but rarely)." If the tree grows back when so cut, it can presumably propagate from suckers: both processes of growth are due to a conflux that is warmed.

[g] Crete is the "appropriate" country of the cypress: *HP* 4 1. 3.

[h] These are the new shoots sent out from the cut in the "clipped" cypress. But the word includes as well the suckers by which the tree can be propagated. The word rendered "side-growths" and "suckers" in this sentence also occurs at *HP* 2 2. 4: "Of all trees with several modes of propagation the quickest form of propagation, and one that produces rapid increase, is that from a detached sucker and still more than that from a side-growth (παραφυάδος), if the side-growth is from a root." Presumably the "side-growth" is here thought of as at some distance from the trunk.

καὶ ταύτας βλαστάνειν· ἐπεὶ εἴ γε μὴ ἦσαν ὑπόρριζοι,[1] χαλεπὸν ἢ ἀδύνατον. τοῦτο μὲν οὖν ὡς πρὸς ὑπόθεσιν λεγόμενον.

2. 3 ὁ δὲ φοῖνιξ ἀπὸ μὲν τῶν ῥάβδων βλαστάνει χώρας εὐβοσίᾳ καὶ εὐφυΐᾳ[2] πρὸς τὸ θᾶττον βλαστάνειν· τὸ δ' ἀπὸ τῶν ἁπαλωτάτων εὔλογον, εἰ ὑγρόταταί τε αὗται[3] καὶ τμητικώταται·[4] 5 ἀπὸ δὲ τοῦ ἐγκεφάλου διὰ τῶν ὑποκάτω μᾶλλον ἔτι τούτων εὔλογον, ἐντεῦθεν γὰρ καὶ ἡ τῶν ῥάβδων φύσις, καὶ ὅλως οἷον ἀρχή τις αὕτη ζωτική· διὸ καὶ ἐξαιρουμένου καὶ πονήσαντος θνῄσκει· ἐπεὶ καὶ ἡ ἐκ τῶν πλαγίων ἔκφυσις ἔχοντός 10 ἐστιν τὸν ἐγκέφαλον· ἐξαιρουμένου γὰρ αὐαίνεσθαι κατὰ λόγον, εἴπερ καὶ πεφυκότος καὶ ἐρριζωμένου 2. 4 τοῦτο συμβαίνει. ἡ δὲ ἔκφυσις δῆλον ὅτι πανταχόθεν· πανταχοῦ γὰρ διαμένει μέχρι τινὸς ἡ ὑγρότης καὶ ἡ θερμότης διὰ τὸ μὴ εὐξήραντον εἶναι τῇ πυκνότητι

[1] ὑπόριζοι u: ὑπόροιζοι U.
[2] εὐβοσία καὶ εὐφυΐα u: εὐβοσιαι καὶ εὐφυΐαι U.
[3] ὑγρόταταί τε αὗται ego (ὑγρόταται αὗται Wimmer): ὑγρότα|ταται αὐ(αῦ u)ται U.
[4] τμητικώταται U N (-τα HP) : γεννητικώταται Moldenhawer.

---

[a] *Cf. CP* 1 2. 1 *init.*

[b] For the fertility of Babylonia *cf. HP* 3 3. 5: "The date-palm is marvellous in Babylonia, but in Greece it does not even ripen its fruit . . .; *HP* 8 7. 4: ". . . the yield of cereals (*sc.* in Babylonia) when the farming has been negligent is fifty-fold, when it has been careful a hundred-fold."

[c] It is evidently the "appropriate" country of the date-palm: *cf. HP* 2 2. 8; 3 3. 5; *CP* 2 3. 7.

fluid, should sprout as well, since if they had no roots this would be difficult or impossible. This we say on the supposition that the report is true.

2.3 The date-palm grows from its branches[a] because the country is rich in food[b] and naturally tends[c] to promote more rapid sprouting. That the growth is from the tenderest of these is reasonable, since these have most fluid and are best able to pierce the ground.[d] It is even more reasonable that it grows from the head by way of the parts lower down,[e] for it is from the head that the branches grow too, and the head is in general a sort of starting-point of life; this is why the tree dies when the head is removed or damaged.[f] Indeed the tree sends out growth when lying on its side only when it retains the head. For it is reasonable that it should wither in this position when the head is removed, since it withers on such removal even when still standing and rooted. 2.4 The growth as the tree lies on its side is evidently from all parts, since the fluid and heat remain in all parts for a while because the tree does not readily dry out, owing

[d] *Cf. HP* 8 7. 4 (continued from note *b*): "The farming consists in letting the water remain on the plants as long as possible, so that it may produce a great deal of silt, for the soil, which is fat and close in texture, must be made open-textured." So salt was sprinkled around the date-palm to loosen the soil (*CP* 3 17. 3).

[e] *Cf. CP* 1 2. 1. These are offsets taken from lower in the trunk than the main head, but also containing heads.

[f] *Cf. CP* 5 16. 1. The dwarf palm is distinguished from the date-palm by living on removal of the head: *HP* 2 6. 11; 4 14. 8.

τῶν πόρων (διὸ καὶ δυσκάπνωτον).[1]

5 ὁ μὲν οὖν φοῖνιξ διὰ ταῦτα πλεοναχῶς γίνεται· τῶν δὲ ἄλλων τῶν ὁμοφυῶν[2] οὐδέν, διὰ τὸ μὴ τοιαύτην ἔχειν τὴν φύσιν.

3. 1 ὅσα δὲ πλείους γεννᾶται τρόπους ἔχει καὶ ἐν αὑτοῖς[3] διαφοράν· τὰ[4] γὰρ ἀπὸ παρασπάδος καὶ ῥίζης καὶ κλωνὸς οὐ πάντως δύναιντ᾽ ἂν καὶ ἀπὸ ξύλου καὶ ἀπὸ ἀκρεμόνος, καθάπερ ἄμπελος καὶ 5 συκῆ, τὰ δὲ πάλιν ἀπὸ τῶν βλαστῶν ὥσπερ καὶ ἡ ἐλάα.

τὸ δὲ αἴτιον ἐν ἀμφοῖν ταὐτὸν καὶ παραπλήσιον· ἀμπέλου μὲν γὰρ καὶ συκῆς καὶ τῶν ἄλλων τῶν τοιούτων ξηρὰ καὶ ξυλώδη τὰ μέσα, καὶ ἐνίων

[1] δυσκάπνωτον U^ar: δυσιάπνοτον U^r N H^ac(δύσ- H^t)P. U^r (that is, u) intended δυσδιάπνευστον but had or used no ink.
[2] ὁμοφυῶν U (*cf.* ὁμοφυὲς U at *HP* 4 2. 7): μονοφυῶν Vasc.[2]
[3] αὑτοῖς Scaliger: αὐτοῖς U.
[4] τὰ a: τὸ U N HP.

---

[a] *Cf. HP* 5 9. 4: " The wood of trees of a moist character makes in general evil smoke; this is why green wood does so too . . . The date-palm does so as a result of its peculiar nature, and it is this tree that some suppose makes the most evil smoke of all . . .;" Theophrastus, *On Fire*, chap. xii. 72: " Of woods those that are green, of crooked grain and fibrous (like that of the date-palm) produce evil smoke; for the wood should have no fibres and be easily divisible by the flame; . . ."
[b] Enumerated in *HP* 2 1. 1 (cited in note *a* on *CP* 1 1. 1).
[c] *Cf. HP* 2 1. 2: ". . . the olive grows in all ways except from a twig;" *HP* 2 1. 4: " The olive sprouts in the greatest number of ways, for it sprouts from the trunk, from the stock when chopped into pieces, from the root, from the wood and

to the dense crowding of its passages. (This moreover is why it produces an evil smoke.) [a]

These then are the reasons why the date-palm is generated in several ways. But no other single-stemmed tree is so generated, none having such a nature.

Distinctions in Trees Propagated from Parts as well

3.1 Trees propagated in several ways [b] also differ among themselves: those propagated from a detached sucker and a root and a twig are not necessarily capable of also being propagated from the wood [c] as well and from a branch (examples are the vine [d] and fig [e]), nor can the ones so capable be propagated in their turn from the shoots (for instance the olive).

In both groups the cause is the same and similar: [f] in the vine, fig and the other trees of this description the middle parts [g] are dry and woody, and in some

from a branch or stake . . . Of the rest the myrtle sprouts in most ways, for it too grows from the pieces of wood from the stock."

[d] *Cf. HP* 2 1. 3: "For there are in general few trees that sprout and are generated from the upper parts rather than the lower, as the vine from its twigs; for it grows not from the 'prow' (*sc.* its outermost and longest shoot) but from the stock."

[e] *Cf. HP* 2 1. 2: "The fig grows in all the other ways but not from the stock and pieces of wood."

[f] A favourite expression: *cf. CP* 2 2. 4; 2 5. 3; 5 9. 13; 6 14. 1 (*cf.* 6 11. 9).

[g] That is, the parts between the root and the twigs: the wood and branches.

10 τραχύ,[1] τὸ δὲ τραχὺ δυσβλαστές· τὰ δὲ ἄκρα καὶ ἁπαλὰ καὶ ἔνυγρα, καθάπερ τὸ κλῆμα καὶ ἡ κράδη,[2] καὶ ἡ[3] τῆς ἰτέας δὲ καὶ τῶν ἄλλων ὁμοίως.

3 2 ἐλαίας δὲ τὰ μὲν τῶν ἀκρεμόνων ἔνικμα διὰ τὴν λιπαρότητα καὶ πυκνότητα καὶ ἀμφοτέρως[4] δυσξήραντα,[5] τὰ δ᾽ ἄκρα διὰ λεπτότητα ξηρὰ καὶ οὐ δυνάμενα διαμένειν. ὡσαύτως καὶ ἀπίου καὶ 5 ἀμυγδαλῆς καὶ μηλέας καὶ δάφνης καὶ ἄλλων· ἀβλάστητα γὰρ τὰ τούτων διὰ τὴν αὐτὴν αἰτίαν, πλὴν καὶ εἴ τι σπάνιον, οἷον ἀμυγδαλῆς· δάφνην δὲ ἀδύνατον, ἐπεὶ οὐδὲ ἀπὸ παρασπάδος θέλει ῥᾳδίως. αἰτία δ᾽ (ὥσπερ ἐλέγομεν) ἡ ξηρότης, 10 καὶ τούτου γ᾽ ἔτι ἡ μανότης·[6] ἀσθενέστερον γὰρ καὶ πρὸς τὸ διατηρῆσαι ἐπὶ πλεῖον τὸ μανόν.[7]

3. 3 ὅσα δὲ καὶ ἀπὸ ξύλων ἔμβια καὶ δύναται βλαστάνειν (οἷον ἐλαία μύρρινος κότινος), καὶ αὐτὰ τῇ πυκνότητι (καθάπερ ἐλέχθη) τηρεῖ τὴν

[1] τραχύ U^c (-ὺ U^ac): τραχέα Gaza (*asperae*), Vasc.[2]
[2] κράδη Gaza (*crada*): καρδία U.
[3] ἡ u (*cf.* ἄλλη *CP* 6 14. 6; τὰ Schneider; ἐπὶ Wimmer): ἢ U.
[4] ἀμφοτέρως u: ὁ ἀμφοτέρως U.
[5] δυσξήραντα Gaza (*exsiccatu . . . difficiles*), <οὐκ> εὐξήραντα Itali: εὐξήραντα U.
[6] μανότης Gaza (*raritas*), Dalecampius: πυκνότης U.
[7] μανόν Gaza, Vasc.[2]: μᾶλλον U.

---

[a] Elder and white and black poplar (*CP* 1 1. 4); *HP* 3 1. 1 adds plane and elm.

[b] *Cf. HP* 2 1. 2: ". . . for an olive twig will not grow when set in the ground . . . And yet some assert that it has been known to happen that when an olive stake was set in the ground it came to live along with the ivy and became a tree . . ."

the part is rough (and nothing rough sprouts readily), whereas their extremities are both tender and full of fluid (as the twigs of the vine and fig, and so too of willow and the rest).[a]

3. 2 In the olive on the other hand branch cuttings are moist because of the oiliness and close texture of the tree and for both reasons do not dry out, whereas twig cuttings are so thin that they are dry and unable to survive.[b] So too with pear, almond, apple, bay and others:[c] for the same reason their twig cuttings will not sprout except rarely and as an anomaly. In some indeed growth even from branch cuttings is rare, as in the almond; and in the bay such growth is impossible, since the tree will not grow readily even from a detached sucker.[d] The cause (as we were saying)[e] is dryness, and in this tree open texture as well,[f] for what is open in texture is also too weak to preserve fluid for any time.

3. 3 But trees that can also establish independent life and sprout from pieces of their wood (as olive, myrtle and wild olive), these too preserve the vital starting-

[c] This is probably implied in *HP* 2 1. 2: " The fig . . . does not grow from the stock and pieces of wood; and the apple and pear grow rarely even from the branches. Still it is held to be possible for most or all (so to say) to grow from the branches too . . ."

[d] *Cf. HP* 2 1. 3: " They say that even the bay can be propagated from a detached sucker, if you plant the sucker after removing its branches. But the detached sucker should preferably have some part of the root or stock attached."

[e] *CP* 1 3. 1–2.

[f] *Cf. HP* 5 3. 3: " Of the wood of wild trees used for roof-timbers that of silver-fir especially is open in texture, and of other wood that of elder, fig, apple and bay."

ζωτικὴν ἀρχήν, ἔχοντά τινα ὑγρότητα τοιαύτην.
5 οὐ γὰρ ἱκανὸν ἐὰν ᾖ πυκνόν γε,[1] εἰ[2] ξηρὸν καὶ ἀπαράβλαστον (ἐπεὶ [δε][3] καὶ ἡ παράβλαστησις τῶν τοιούτων διὰ τὸ πρεμνώδη[4] καὶ ξυνεστραμμένην πως εἶναι τὴν ῥίζαν, ὥσπερ καὶ τῆς δάφνης· ἴσχει[5] γὰρ μᾶλλον (ᾗ[6] καὶ δυσώλεθρα τὰ τοιαῦτα,
10 τὸ γὰρ καταλειπόμενον ἀεὶ βλαστητικόν· τῶν δὲ μακρορρίζων διὰ τὴν τῶν ῥιζῶν ἰσχύν, ὥσπερ ἀπίου καὶ κοκκυμηλέας καὶ ἑτέρων).

3. 4 τάχα δὲ καθόλου περὶ πάντων ὧδε λεκτέον· πρῶτον μὲν ὅτι μετέωρα καὶ οὐ βαθύρριζα, ἢ εἰ καί τινας εἰς βάθος καθιᾶσιν, ἀλλ' ἐνίας καὶ ἐπιπολῆς. ἔπειθ' ὅτι πάντων ὅταν συρροῆς γενο-
5 μένης συνθερμανθῇ τοῦτο καὶ πεφθὲν ὑπὸ τοῦ ἡλίου καθάπερ κυῆσαν ἐκτέκῃ· καὶ γὰρ ἐκ τῶν ἀκρεμόνων οὕτω καὶ ἐκ τῶν ἄλλων οἱ βλαστοί. τούτων δὲ ὑποκειμένων κατὰ λόγον ἤδη τὸ παραβλαστάνειν. ἡ γὰρ ῥίζα μετέωρος οὖσα καὶ
10 συρροὴν λαμβάνουσα ταύτην ἐκθερμαίνουσα καὶ πέττουσα μεθίησι τὸν βλαστόν.

[1] γε HP: τε U N.
[2] εἰ Gaza: ἡ U.
[3] ἐπεὶ [δε] Vasc.[2] (*nam* Gaza; ἐπειδὴ Scaliger; ἔτι δὲ Wimmer): ἔπει δε U.
[4] πρεμνώδη HP: τερεμνώδη U N.
[5] ἴσχει ego: ἰσχύει U.
[6] ᾗ Gaza (*hac eadem de causa*), Vasc.[2]: ἡ U.

---

[a] *CP* 1 3. 2; *cf. CP* 1 2. 4 (of the date-palm).
[b] *Cf. HP* 1 6. 4: "Again some roots are straight and uniform, others crooked and crossing one another. For this occurs not merely on account of the location . . .; it may also belong to the natural character of the tree, as in bay and olive . . .;"

point by their close texture (as we said),[a] since they have a fluid of this vital sort in the pieces. For it is not enough for a tree to be close-textured if it is dry and develops no side-shoots. (In fact the sending up of side-shoots in dry trees is due to the stock-like and as it were concentrated character of the root, as in bay,[b] for the root has a greater tendency to get such centres of fluid, and this makes the tree hard to kill, since whatever part is left behind is capable of sprouting. Long-rooted trees, on the other hand, resist destruction because of the strength of the roots, as pear, plum [c] and others.)

Trees:
A General Formulation of the Conditions for Growing Side-Shoots

3. 4 One may perhaps generalize as follows about all trees that have side-shoots: (1) first, the roots are shallow and not deep, or at all events even if some go deep, others are near the surface; (2) second, the side-shoots appear in all when a conflux of fluid accumulates in a certain spot and this on being warmed and concocted by the sun becomes as it were pregnant and brings forth offspring; in fact shoots are produced from the branches and other parts in the same way. Once these points are laid down the sending out of side-growths becomes reasonable: the root is shallow and acquires a conflux, and when it thoroughly warms and concocts this conflux, sends up the shoot.

HP 4 13. 3: " Some trees quickly age and decompose, but send up new growths at the side of those lost from the same parts, as bay, apple, pomegranate and most water-loving trees."

[c] *Cf. HP* 3 6. 5: ". . . the plum is hard to kill."

3. 5 τὰ μὲν οὖν πλεῖστα παρ' αὐτὸ τὸ στέλεχος ἐκβλαστάνει, μετεωρόταται γὰρ αὗται· καὶ ἥ γε ἐλαία καὶ ἐκ τῶν πρέμνων. ἄπιος δὲ καὶ ῥόα καὶ ὅσα μὴ μόνον σύνεγγυς ἀλλὰ διὰ πολλοῦ, μακρόρ-
5 ριζα ὄντα, ᾗ [ὧν][1] ἂν ἡ ῥίζα μετέωρος ᾖ, ταύτῃ τὸν βλαστὸν ἀφίησιν· ἐνταῦθα γὰρ ἡ συρροὴ καὶ πέψις θερμαινομένης. διὸ καὶ ἄτακτος ὁ τόπος· ἄτακτος ⟨γὰρ⟩[2] ὁ μετεωρισμὸς καὶ ἡ συρροή. τὰ δ' ἄκαρπα καὶ εὔζωα, καθάπερ αἴγειρος καὶ
10 λεύκη, καὶ συμπληροῦν δύναται βλαστάνοντα·[3] τὰ δ' ἄλλα οὐχ ὁμοίως, ἀλλ' ἐπὶ τῶν ὑλημάτων καὶ ποιωδῶν ἐνίων τοῦτο συμβαίνει. πρὸς ἃ καὶ δεῖ[4] μεταβῆναι τὸν λόγον· ὑπὲρ γὰρ τῶν δένδρων ἱκανῶς εἴρηται.

4. 1 τούτων δὲ τὰς μὲν γενέσεις ἐκ τῶν αὐτῶν θεωρητέον, κοινοτάτην πᾶσιν τὴν ἀπὸ σπέρματος τιθέντας. οὐ μὴν ἀλλὰ καὶ πλείους εἰσὶν καὶ τούτων· ᾗ[5] δὲ ἕκαστα τῶν προειρημένων ἐφάπτε-
5 ται, ταύτῃ[6] διαιρετέον.

οἷον καὶ ὑπὲρ ἧς νῦν λέγομεν, τῆς ἀπὸ τῶν

[1] ᾗ [ὧν] Gaza (*quacunque*): ἡ ὧν U.
[2] ⟨γὰρ⟩ Gaza (*enim*), Vasc.[2]
[3] βλαστάνοντα Wimmer (τὸ [*i.e.* τῷ] βλαστάνειν Scaliger): βλαστάνειν U.
[4] δεῖ u HP: δὴ U N.
[5] ᾗ u: ἡ U; εἰ N HP.
[6] ταύτῃ u: ταῦτα (?) U.

---

[a] *Cf.* note *f* on *CP* 1 1. 4.
[b] *Cf.* *HP* 7 13. 4 (of bulbous plants): "All these grow in

Now most trees produce these suckers next to the trunk, the roots being here most shallow; and the olive produces them from the base of the trunk as well. But the pear, pomegranate and all trees that produce suckers not only close to the trunk but at a distance from it, have long roots, and send up the shoot wherever the long root comes near the surface, for it is here that the conflux is formed with the resulting concoction as it is warmed. This is why there is nothing fixed about the place of the sucker, for there is nothing fixed about the approach of the root to the surface and the site of the conflux. Trees that are fruitless [a] and full of life, like the black and white poplar, can even fill out the space around them with shoots. The rest cannot do this to the same extent; but it does happen with certain woody and herbaceous plants.[b] To these the discussion must now turn, since trees have been sufficiently dealt with. 3. 5

The Modes of Propagation in Woody and Herbaceous Plants

We must study the modes of generation of these [c] in the light of the same considerations, laying it down that the mode most common to all is generation from seed. Still here too several modes occur, and we must distinguish the groups as they touch on the groups that have been discussed. 4. 1

So with the mode under discussion,[d] generation

masses, as onion and garlic, for they produce side-growths from the root . . ."

[c] The modes of generation of undershrubs and herbaceous plants are discussed in *HP* 6 6. 6; 6 6. 8–6 7. 4; 7 2. 1–9.

[d] *CP* 1 3. 4–5.

ῥιζῶν. ἔνια γὰρ καὶ αὐτόματα βλαστάνει καὶ φυτεύουσιν ἀπὸ τῶν ῥιζῶν, ὥσπερ τὰ κεφαλόρριζα καὶ ὅλως ὧν παχεῖα καὶ σαρκώδης ἐστὶν ἡ ῥίζα.
10 δεῖ δὲ καὶ τὸ[1] μὲν ὑδατῶδες μηδεμίαν ἔχειν, ὥσπερ τῆς γογγυλίδος καὶ ῥαφανῖδος, εὐξήραντοι γὰρ αὗται καὶ ἀσθεν⟨εῖς⟩ εἰς[2] διαμονήν· ἀλλ' ἤτοι χιτῶνας ἔχειν πλείους καὶ ἅμα γλισχρότητά τινα, καθάπερ αἱ τοῦ βολβοῦ καὶ ⟨τῆς⟩[3] σκίλλης,
15 ἢ[4] ὅλως εὔχυλόν τινα καὶ εὔσαρκον εἶναι, καθάπερ αἱ τοῦ ἀμαράκου[5] τοῦ χλωροῦ καὶ τοῦ λειρίου καὶ τῶν ὁμοειδῶν. αἱ γὰρ τοιαῦται φυτείαν μόνον
4. 2 δέχονται καὶ μέταρσιν. καὶ τούτων αἱ μὲν καὶ πλείω χρόνον διαμένουσιν, αἱ δὲ ἐλάττω, κατὰ τὰς ἰδίας ἑκάστων φύσεις·[6] ἄλλαι δ' αὖ πάλιν εἰσί τινες αἳ μένουσαι μὲν ἐν ταῖς ἑαυτῶν χώραις ἀφιᾶσιν βλαστόν,
5 οἷον αἱ τῶν ἐπετειοκαύλων, μεταιρόμεναι δ' οὐ δύνανται διὰ τὴν ξηρότητα· τὴν γὰρ αὐτὴν αἰτίαν

[1] τὸ u: τῷ U.
[2] ἀσθεν⟨εῖς ss. u⟩εῖς U.
[3] ⟨τῆς⟩ Schneider.
[4] ἢ ego (*denique* Gaza): καὶ U.
[5] ἀμαράκου U: ἀράκου Gaza; κρόκου Vasc.[2]
[6] ἑκάστων φύσεις ego (φ. ἑκάστων Schneider): φύσεις U[t]; ἑκάστω U[lm] with an index over the ς of ἰδίας (that is, ἑκάστω φύσεις U[c]).

---

[a] That is, the root or bulb survives the winter in the ground and sends up a new plant in the next season.

[b] *Cf. HP* 7 2. 1: " From the root are planted garlic, onion, purse-tassel, cuckoo-pint and in short such bulbous plants as resemble them." This refers to the practice of sowing seeds late in the season, then taking up the small bulbs and storing them through winter, planting them in spring and thus obtaining an earlier crop, with consequent advantages in marketing. The bulb is regarded as a root, and Theophrastus

from the root. Some not only send up shoots from the roots spontaneously,[a] but are also propagated from these by growers, as bulbous plants and in general all with a thick and fleshy root.[b] But no such root should also have a watery fluid, as turnip and radish,[c] since these roots dry out easily and are too weak to survive. The root instead must either have several coats together with a certain viscosity (as in purse-tassel and squill), or else be quite succulent and with plenty of flesh (as in fresh sweet marjoram,[d] narcissus [e] and plants of the same kind). For only such roots as these can be planted and removed from their place. In these furthermore the survival of some is longer, of others shorter, the duration depending on the distinctive nature of each kind. Again, other roots send up a shoot when left in place (as roots of plants with an annual stem),[f] but the root is too dry to do so when taken up (since we must sup-

4. 2

can thus call such planting " generation from the root," even though in onion the " lower roots " were removed (*CP* 1 4. 5).

[c] Turnip and radish are always spoken of not as " planted " but as " sown " (*HP* 7 1. 2; 7 1. 7).

[d] *Cf. HP* 6 7. 4: "Sweet marjoram grows in both ways, from a detached sucker and from seed." " Fresh " indicates that the root was left in the ground.

[e] *Cf. HP* 6 6. 9 (of the fruit of narcissus): " This drops and produces a spontaneous sprouting; still it is also gathered and set in the ground. And the root is planted. It has a root that is fleshy, round and large."

[f] *Cf. HP* 7 2. 1–2 and especially 7 2. 1: " By root are planted garlic, onion, purse-tassel, cuckoo-pint and in general such bulbous plants. Such propagation is also possible in cases where the roots live for more than a year, although the shoots are annual; " 7 2. 2: " Of those propagated from the root, the root is long-lived, although the plant itself may be annual . . ."

ὑποληπτέον καὶ ἐπὶ τούτων ἥνπερ ἐπί τινων.[1]

ἔστι δέ τινων καὶ ἀπὸ παρασπάδος καὶ ἀπὸ τῶν ἄκρων φυτεία[2] καὶ γένεσις· ἀπὸ παρασπάδος[3]
10 μὲν καὶ ῥαφάνου καὶ πηγάνου, τῶν δὲ στεφανωτικῶν[4] οἷον ἀβροτόνου καὶ σισυμβρίου καὶ ἑρπύλλου· καὶ ἀπὸ τῶν ἄλλων[5] ἐνίων τῶν αὐτῶν, πηγάνου τε καὶ ἀβροτόνου καὶ τῶν στεφανωτικῶν·
4. 3 καὶ γὰρ ἔχει ταῦτά γε καὶ καθίησιν εὐθὺς ῥίζας ἐκ τῶν βλαστῶν, ὥσπερ ὁ κιττός· οὗτος γὰρ δὴ μάλισθ' ὅλως ἔμβιος καὶ εἰσδυόμενος εἰς αὐτὰ τὰ δένδρα καὶ ἐν τῇ γῇ κρυπτόμενος.
5 τῶν δὲ λαχανωδῶν τὸ ὤκιμον· καὶ γὰρ τὰς[6] ἀποφυτείας[7] ἐκ τῶν ἄνω δέχεται, καίπερ ξυλῶδες ὄν, ἀλλ' ὅτι[8] δυσξήραντόν ἐστιν, ταύτῃ φύεται, διὸ καὶ πολὺν χρόνον διαμένει καὶ κολουόμενον

[1] ἐπί τινων U: ἐπὶ τῶν ἄλλων Gaza (*in caeteris*), Vasc.[2]; ἐπ' ἐκείνων Wimmer.

[2] φυτεία Moldenhawer: φύεται U[1m] (in an omission in U[t]).

[3] παρασπάδος—παρασπάδος U[1m]: παρασπάδος U[t].

[4] στεφανωτικῶν a: στεφανητικῶν U N HP.

[5] ἄλλων U: ἄκρων Vasc.[2]; ἄκρων δὲ Schneider.

[6] τὰς Itali: τοὺς U.

[7] ἀποφυτείας U N: ἀπὸ φυτείας u HP.

[8] ὅτι Vasc.[2] (*qua* Gaza): ἔτι U or U[c] (ἔτι U[ac?] N HP).

---

[a] He does not specify the roots too dry to be treated in this way, since they in fact include most of the ones he has not mentioned.

[b] *Cf. HP* 7 2. 1: "Cabbage grows from a detached sucker, for one must include some of the root."

[c] For rue planted in a fig-tree *cf. CP* 5 6. 10.

[d] *Cf. HP* 6 7. 3: "Southernwood grows better from seed than from a root or a detached sucker . . ."

[e] *Cf. HP* 2 1. 3: "For . . . there are few plants that grow

pose the same explanation to apply to these roots as applies to certain others).[a]

Some can be planted and propagated both from a detached sucker and from their extremities. From a detached sucker grow cabbage [b] and rue,[c] and among the coronaries southernwood [d] for example, bergamot mint and tufted thyme; and some of the same—rue and southernwood with some coronaries—also grow from the other parts.[e] Indeed these last (at any rate), like ivy, have roots that come from their shoots and send them down at once, ivy being the plant which in general is best at living when cut off, both when it penetrates the trees themselves [f] and when it is stuck in the ground and covered with earth. 4. 3

Of vegetables basil does this best, for it will even grow from cuttings taken from the upper parts,[g] in spite of being woody. It does so because it does not readily dry out; this is why it not only survives for

and are propagated more readily from the upper parts, as the vine from its twigs . . . and an occasional tree or undershrub of this description, as is held to be the case with rue, stock, bergamot mint, tufted thyme and calamint;" *HP* 7 2. 1: "From the shoots are planted rue, marjoram and basil . . ."

[f] *Cf. HP* 3 18. 10: "Ivy . . . constantly sends out roots from its shoots in the interval between the leaves, and with these roots it penetrates trees and walls . . . Hence, by removing and drawing to itself the fluid it causes the tree to wither; and if you sever the ivy below it is able to survive and live."

(This is due to incorrect observation. Ivy does not send its roots below the surface of the tree. It kills by shading the leaves. When the stem is severed it can survive for a while in a humid climate but eventually dies.)

[g] *Cf. HP* 7 2. 1: "From the shoots are grown rue, marjoram and basil; for basil too is propagated by cuttings when it has reached the height of a span or more, about half the shoot being cut off."

πάλιν βλαστάνει.[1] ξυλῶδες δὲ καὶ τὸ ἀβρότονον,
10 ἀλλ᾽ ἔχει τινὰ τῇ πυκνότητι καὶ δριμύτητι φυλακήν,
ὥσπερ ὁ κιττός· καὶ γὰρ οὗτος φύεται καταπηγνύμενος.

αὗται μὲν οὖν κοιναὶ πλειόνων·

4.4 αἱ δὲ τοιαῦται καὶ σπάνιαι καὶ ἐλάττους,
ὥσπερ τῆς κρινωνιᾶς καὶ ῥοδωνιᾶς[2] καὶ ὁ
καυλὸς[3] ὁ σχισθεὶς φύεται καὶ βλαστάνει· ταῦτα
δὲ ὅμοια καὶ παραπλήσια καὶ τὰ τῆς ἐλαίας καὶ
5 εἴ τι ἄλλο βλαστητικὸν ἀπὸ τοῦ ξύλου, διὸ καὶ
ὑπὸ τὴν αἰτίαν πίπτειν δύνανται καὶ[4] διατηρεῖν
τὴν ὑγρότητα καὶ θερμότητα τὴν γόνιμον. καὶ
τό τε κατακόπτειν καὶ κατασχίζειν εὔλογον· ἐκ
γὰρ τοῦ ἐλάττονος καὶ ἀνεῳγμένου θάττων[5] ἡ
10 ἀρχὴ καὶ εὐκολωτέρα, τὸ δὲ μέγα καὶ συμπεπτωκὸς οὐχ ὁμοίως παθητικὸν οὐδὲ βλαστητικόν.

4.5 διὸ καὶ τὰ σκόρδα διαιροῦσιν εἰς τὰς γέλγεις καὶ

[1] βλαστάνει U^r N HP: -ειν U^ar.

[2] κρινωνιᾶς καὶ ῥοδωνιᾶς Vasc.[2]: κρινωνίας καὶ ῥωδωνιας U.

[3] καυλὸς Wimmer (*cf. CP* 1 4. 6; *HP* 2 2. 1): κάλαμος U.

[4] πίπτειν δύνανται. καὶ U: πίπτει καὶ δύναται or πίπτει ὅτι καὶ δύνανται Schneider; πίπτει· δύναται γὰρ Wimmer.

[5] θάττων u HP: θᾶττον U N.

---

[a] *Cf. HP* 7 2. 4: "When the stems are broken off in practically all (*sc.* vegetables), the stem sprouts again . . . and most obviously . . . in basil, lettuce and cabbage;" *CP* 2 15. 6.

[b] *Cf. HP* 2 2. 1: "The rose and the lily are also generated when the stems are cut up, and so also dog's-tooth grass;" *HP* 6 6. 6: "The rose also grows from the seed . . .; nevertheless, since it then matures slowly, they cut up the stem and propagate it in this way."

[c] These shoots may develop from buds at the nodes or from adventitious buds which may develop from the callus which forms at the cut surface of the stem.

a long time, but also sprouts again when cut back.[a] Southernwood too is woody, but (like ivy) is protected by its close texture and pungency (for ivy too will grow from a cutting stuck into the ground).

These then are common to a large number of plants.

4.4 We pass to forms of propagation that are both rare in occurrence and found in fewer plants.

In lily and rose [b] even the split stem grows and sends out shoots.[c] This is very similar to what happens with the olive [d] and other trees [e] that can grow from a cut piece of wood, and this is why these plants can come under the cause that was given [f] and preserve their generative fluid and heat. It is moreover reasonable that the wood should be cut up and the stem split: a start can be made more rapidly and easily from a smaller and open piece, whereas a large and closed piece is not so readily affected and thus does not sprout so easily. 4.5 (This is why garlic when planted is separated into its cloves [g] and the lower

[d] *Cf. CP* 1 3. 3.

[e] Myrtle (*cf. CP* 1 3. 3; *HP* 2 1. 4) and wild olive (*CP* 1 3. 3).

[f] *CP* 1 3. 3: "they . . . preserve the vital starting-point by their close texture."

[g] *Cf. HP* 7 4. 11: "Garlic is planted . . . divided into cloves."

The garlic of the market and kitchen is a cluster of bulbs. Each of these bulbs arises as a "side-growth" in the axils of the leaves of the parent bulb, which sends up a flowering stem. Each bulb when planted develops roots from the base and leaves from the sides of its very short stem. Buds develop in the axils of the leaves and grow into short stems with fleshy leaves, that is, into bulbs each of which is a clove. The apical end of the stem of the mother bulb grows into a floral stem. The scales which enclose each bulb and the cluster of bulbs are the bases of leaves of the parent bulb. The floral stems were used to braid the "garlic" into the chains of garlic which once festooned the vegetable markets.

τῶν κρομμύων ἀφαιροῦσι τάς τε ῥίζας τὰς κάτω καὶ τὰ κελύφη, ταῦτα γὰρ παρέχει πᾶν τὸ ⟨ἀλλότριον⟩· ἀλλότριον δὲ[1] τῷ ζῶντι τὸ μὴ ζῶν
5 (ὥσπερ καὶ τῶν δένδρων τὰ ἀφαυαινόμενα).[2] τῆς μὲν οὖν ἐλαίας καὶ τῶν μυρρίνων οὐ δεῖ[3] περιαιρεῖν τὸν φλοιόν, ἀποστέγει γὰρ καὶ τηρεῖ τὴν ζωήν· τῶν δὲ κρομμύων καὶ τῶν σκόρδων δεῖ, διὰ τὸ μὴ τοὺς ζῶντας μηδὲ τοὺς κυρίους ἀφαιρεῖν
10 (ἐπεὶ κἀκείνων εἴ τις ἀφαιροίη τούτους, ⟨οὐ⟩ βλαστάνειν).[4]

4. 6 ἰδιωτάτη δὲ βλάστησις ἡ ἐκ τῶν δακρύων, οἷον τοῦ θ' ἱπποσελίνου καὶ τοῦ κρίνου καὶ ἐνίων ἑτέρων. οὐκ ἄλογος δέ, ἀλλ' ὁμολογουμένη τῇ ἐκ τῶν καυλῶν· οὐδὲν γὰρ ἕτερον ἀλλ' ἢ[5]
5 ξυνηθροισμένην εἶναι δεῖ[6] καὶ τὴν ἐκεῖθεν ἀρχὴν τὴν γόνιμον, οὐ γὰρ ἄνευ θερμότητος ἡ τοιαύτη καὶ ὑγρότητος.[7]

ὅτι δ' οὐ πάντων οἱ ὀποὶ καὶ τὰ δάκρυα γεννητικά, πρὸς τὰς ἐπάνω καὶ προτέρας αἰτίας

[1] ἀλλότριον· ἀλότριον δὲ a: ἀλλότριον δὲ U N; ἀλλότριον HP.
[2] ἀφαυαινόμενα u HP: ἀφαυηνόμενα U N.
[3] οὐ δεῖ a: οὐδὲ U N HP.
[4] οὐ βλαστάνειν Heinsius (*germen nullum poterit emitti* Gaza; οὐ βλαστάνει Vasc.[2]): βλαστάνειν U.
[5] ἢ HP: εἰ U N.
[6] εἶναι δεῖ ego: δεῖ εἶναι U.
[7] ὑγρότητος Gaza: ὑγρότης U.

---

[a] That is, the true roots, since the Greeks took the bulb to be a root: *cf. HP* 1 6. 8–9. The roots that are removed are the dried remnants of the roots of the parent bulb.

roots [a] and outer scales of onions are removed, since these furnish everything that interferes with propagation, what is not alive (as the withered parts of trees) interfering with what is. Now in the olive and myrtle one must not peel off the coating when one cuts the pieces of wood,[b] since it seals off the piece and preserves the life; whereas we must do this with onion and garlic, because here we are not removing coats that are alive or that determine growth. Indeed if here too we should remove this sort of coating, we are told that the pieces will not sprout.)

4.6 The most distinctive mode of generation is that from exudations,[c] as in the alexander, lily and a few others.[d] It is however not unreasonable, but accords with generation from split stems: all that is needed is that the generative starting-point [e] from this source as well should have been accumulated, since this kind of generation too is not without heat and fluid.

The explanation why sap and exudations [f] are not generative in all must be referred to the reasons

[b] *Cf. HP* 2 1. 4: "But both in this tree (*sc.* myrtle) and in the olive one must divide the pieces of wood into sizes not smaller than a span in length and not remove the bark."

[c] That is, from bulbils or bulblets. A bulblet is a small bulb formed above ground on some plants, as in the axils of the leaves of the common bulbiferous lily, and often in the flower clusters of leek and onion.

[d] *Cf. HP* 2 2. 1, 6 6. 8, 9 1. 4. The "others," mentioned also in *HP* 9 1. 4, are never specified.

[e] That is, a bud primordium must be formed. *Cf.* the discussion of "conflux" at *CP* 1 3. 5.

[f] *Cf. CP* 6 11. 16.

10 ἀνακτέον,[1] ὅτι[2] [δ'][3] οὐδ' οἱ καυλοί, οὐδ' αἱ ῥίζαι.

κατὰ λόγον δὲ καὶ τοῦτό ἐστιν, ὅπερ ἐλέχθη πρότερον, ὅτι πλείους αἱ γενέσεις καὶ τῶν ἄλλων, ᾗ τῶν ἄλλων[4] ῥᾷον[5] γεννῆσαι τὰ ἀτελέστερα καὶ 15 ἀπ' ἐλάττονος ἀρχῆς.

αἱ μὲν οὖν ἐκ τῶν μορίων ὁποῖαι, καὶ διὰ τίνας αἰτίας, ἐκ τούτων θεωρείσθω· καὶ γὰρ εἴ τι παραλέλειπται, προσθεῖναι καὶ συνιδεῖν οὐ χαλεπόν.

5. 1 αἱ δ' αὐτόματοι γίνονται μέν (ὡς ἁπλῶς εἰπεῖν) τῶν ἐλαττόνων καὶ μάλιστα τῶν ἐπετείων καὶ ποιωδῶν· οὐ μὴν ἀλλὰ καὶ τῶν μειζόνων ἔστιν ὅτε συμβαίνουσιν, ὅταν ἢ ἐπομβρίαι κατάσχωσιν 5 ἢ ἄλλη τις ἰδιότης [τίς][6] γένηται περὶ τὸν ἀέρα καὶ τὴν γῆν· οὕτω γὰρ καὶ τὸ σίλφιον ἀνατεῖλαί φασιν ἐν Λιβύῃ, πιττώδους τινὸς ὕδατος γενομένου καὶ παχέος, καὶ τὴν ὕλην δὲ τὴν νῦν οὖσαν ἐξ ἑτέρας τινὸς τοιαύτης αἰτίας· οὐ γὰρ ἦν πρότερον.

[1] ἀνακτέον u: ἀνεκτέον U N; ἀνενεκτέον HP (but *cf.* ἀνοιστέον *CP* 4 11. 8).
[2] ὅτι U: ἔτι Keil.
[3] [δ'] Vasc.[2] (*idest quod* Gaza).
[4] ᾗ τῶν ἄλλων ego (Gaza and Schneider omit): ἡ τῶν ἄλλων U.
[5] ῥᾷον Gaza (*facilius*): ῥάδειον U.
[6] [τίς] N HP.

---

[a] *CP* 1 4. 4 (why the cut stems are not generative).
[b] *CP* 1 4. 1–2 (why the roots of lesser plants are not generative). For the need for adequate heat and fluid *cf.* *CP* 1 1. 3–4; 1 2. 3–4; 1 3. 1–5; 1 4. 2–3.
[c] *CP* 1 4. 1.
[d] The parts are seed, sucker (from root or stem), root, twig, branch, stem, pieces of wood, pieces of stem, and exudations.

mentioned above [a] and earlier,[b] which explain why the stems too, and the roots too, are not generative.

The fact mentioned earlier [c] is also reasonable, that the lesser plants too have several modes of generation, in so far as it is easier to generate less perfect plants than the rest, and a smaller starting-point is needed.

The characters, then, and the causes of the modes of generation from the parts [d] are to be studied from this discussion. Indeed if anything has been omitted, it is not difficult to supply it [e] and perceive the explanation.

(3) *Spontaneous Generation*

5. 1 Cases of spontaneous generation occur in the smaller plants (broadly speaking), especially in annuals and herbaceous plants. They nevertheless sometimes also occur in larger plants, either after spells of rain or when some other special condition has arisen in the air and the ground. For it is thus that silphium is said to have come up in Libya, when there had been a fall of rain described as "pitch-like" and thick, and the forest now existing there is said to have come from another such cause, not having existed before.[f]

[e] *Cf.* Aristotle, *On Sophistical Refutations*, chap. xxxiv (183 b 25–28); *Nicomachean Ethics*, i. 7 (1098 a 24–25); *CP* 6 15. 1.

[f] *Cf. HP* 3 1. 6: ". . . in some places they say that after a rain a special kind of forest has come up, as at Cyrene after a rain described as 'pitch-like' and thick; for in this way the forest near the town sprang up. And they say further that silphium, which did not exist before, made its appearance from such a cause."

5. 2 αἱ δ' ἐπομβρίαι καὶ σήψεις τινὰς ποιοῦσιν καὶ ἀλλοιώσεις, ἐπὶ πολὺ διικνουμένου τοῦ ὑγροῦ, καὶ τρέφειν καὶ ἐπαύξειν δύνανται τὰ συνιστάμενα, θερμαίνοντος τοῦ ἡλίου καὶ καταξηραίνοντος, 5 ὥσπερ καὶ τὴν τῶν ζῴων γένεσιν οἱ πολλοὶ ποιοῦσιν.

εἰ δὲ δὴ καὶ ὁ ἀὴρ σπέρματα δίδωσιν συγκαταφέρων, ὥσπερ φησὶν Ἀναξαγόρας, καὶ πολλῷ μᾶλλον· ἄλλας γὰρ ἂν ποιοῖεν [1] ἀρχὰς καὶ τροφάς. 10 ἔτι [2] δ' οἱ ποταμοὶ καὶ αἱ συρροαὶ καὶ ἐκρήγματα τῶν ὑδάτων πολλαχόθεν ἐπάγουσαι [3] σπέρματα [4]

7–8. Anaxagoras, Fragment A 117 Diels-Kranz, *Die Fragmente der Vorsokratiker*, vol. ii [8], p. 31.

7–11. Varro, *R.R.*, i. 40. 1: latet si sunt semina in aere, ut ait physicos Anaxagoras, et si aqua quae influit in agrum inferre solet, ut scribit Theophrastus.

[1] ποιοῖεν HP: ποιεῖ ἐν U[ar]; ποιεῖεν U[r]; ποιεῖ N.
[2] ἔτι HP: εἴ (εἰ U) τι u N.
[3] ἐπάγουσαι U: -σι Gaza, Vasc.[2]
[4] σπέρματα HP (σπέρματων U[ac]): σπερμάτων U[c] N.

---

[a] *Cf. HP* 3 1. 5 (of spontaneous generation, as presented by the natural philosophers): "But this kind of generation is somehow beyond the reach of our sense. There are other kinds that are admitted and evident to sense, as when a river overflows or makes a new bed . . . And again when there is a long spell of rainy weather, for here too plants sprout. It appears that the invasion of the rivers imports seeds and fruits . . ., and rainy weather does the same, for it deposits many kinds of seeds, and together with this it produces a certain decomposition of the earth and the water; indeed the mere mixture of water (*sc.* when there is no rain) with the Egyptian earth is held to produce a certain vegetation."

[b] *Cf.* note *c* on *CP* 1 1. 2.

[c] *Cf. HP* 3 1. 4 (of generation from seed, root, sucker,

Rainy spells [a] not only bring about certain cases of decomposition and alteration,[b] the water penetrating far and wide, but they can also feed what is formed and make it grow larger, while the sun warms and dries it, this [c] being also how most authorities account for the generation of animals as well. 5. 2

False Spontaneous Generation: (a) *From Imported Seeds*

And if the air too provides seeds which it carries down with the rain, as Anaxagoras [d] says, the rainy spells will be all the more prolific, since they would then produce an additional set of starting-points possessing supplies of food.[e] Rivers again and collections of water and streams bursting forth from the ground would do so too, importing from many sources

extremity): " We must suppose that these forms of generation belong to wild trees and also the spontaneous ones, of which the natural philosophers speak; as Anaxagoras says that the air has seeds of all and that these are carried down with the rain water and generate plants; Diogenes says plants are produced when the rain water decomposes and acquires a certain mixture with the earth, and Clidemus that plants are formed of the same components as animals . . . And certain others as well speak of the generation (*sc.* of plants)." *Cf.* also Anaximander, Fragments A 11 and A 30 (Diels-Kranz, *Die Fragmente der Vorsokratiker*, vol. i[10], p. 84. 15–16, p. 88. 31) and Lucretius, v. 797–798.

[d] *Cf. HP* 3 1. 4, cited in note *c*.

[e] As in plants the seed, so in animals the egg and larva contain not only the starting-point, but food as well. *Cf.* Aristotle, *On the Generation of Animals*, iii. 11 (762 b 18–21): " Now the formation of plants that are generated spontaneously is from uniform substance: for they come from a certain portion of their source-substance, and one portion of it becomes the starting-point, the other the initial food for the plant that grows out."

καὶ δένδρων καὶ ὑλημάτων (διὸ καὶ ⟨αἱ⟩ [1] μεταστάσεις τῶν ποταμῶν πολλοὺς τόπους ποιοῦσιν ὑλώδεις τοὺς πρότερον ἀνύλους).[2] ἀλλ' αὗται μὲν οὐκ αὐτόματοι δόξαιεν ἄν, ἀλλ' ὥσπερ σπειρόμεναί τινες ἢ φυτευόμεναι.

5. 3 τὰς [3] δὲ τῶν ἀκάρπων [4] οἰηθείη τις ἂν μᾶλλον 5 αὐτομάτους [5] εἶναι, μήτε φυτευομένων μήτε ἀπὸ σπέρματος γινομένων, ὅπερ ἀναγκαῖον ⟨εἰ⟩ [6] μηδέτερον τούτων.

ἀλλὰ μή ποτ' οὐκ ἦν τοῦτ' ἀληθὲς ἐπί γε τῶν μειζόνων, ἀλλὰ μᾶλλον λανθάνουσιν αἱ πᾶσαι τῶν 10 σπερμάτων φύσεις, ὅπερ καὶ ἐν ταῖς ἱστορίαις ἐλέχθη περί τε τῆς ἰτέας καὶ τῆς πτελέας. ἐπεὶ καὶ τῶν ἐλαττόνων πολλαὶ διαλανθάνουσιν [7] τῶν ποιωδῶν, ὥσπερ καὶ περὶ τοῦ θύμου καὶ ἑτέρων

[1] ⟨αἱ⟩ u HP.
[2] ἀνύλους Ur N HP: ἀναύλους Uar.
[3] τὰς ego: τὸ U.
[4] ἀκάρπων Gaza (*sterilia*): καρπῶν U.
[5] αὐτομάτους U N: αὐτόματον HP.
[6] ⟨εἰ⟩ Schneider.
[7] διαλανθάνουσι u HP: διαλαμβάνουσιν U (-ανοῦσι N).

[a] *Cf. HP* 3 1. 5, cited in note *a* (p. 34).

[b] By seeds (and " seed " can include any part from which a plant is propagated) imported by air and surface water.

[c] *Cf. HP* 3 1. 5: " It appears that the invasion of rivers imports seeds and fruits, and they say that irrigation ditches import the seeds of herbaceous plants; and rainy spells do the same, for they carry down many seeds . . ."

[d] *HP* 3 1. 2: "All (*sc.* wild trees) that have a seed and fruit, even if they grow from a root, grow also from these. So they say that even the trees considered to be fruitless generate (*sc.* from seed and fruit), as the elm and willow." Contrast Aristotle, *On the Generation of Animals*, i. 18 (726 a 6–7), cited in note *i* on *CP* 1 1. 2.

[e] *Cf. HP* 3 1. 3: " The occurrence (*sc.* with elm and willow,

seeds both of trees and of woody plants (which is why rivers that shift their course make many regions wooded that were unwooded before).[a] These last forms of generation,[b] however, would not appear to be spontaneous, but a kind of propagation by sowing seeds (as it were) or setting pieces in the ground.[c]

5. 3

False Spontaneous Generation: (b) *From Unnoticed Seeds*

One might fancy that the generation of the fruitless trees is rather a spontaneous one, since these trees are neither set in the ground nor produced from seed, and it is a necessary consequence that they are produced spontaneously if they are not produced in either of these ways.

But perhaps it is not true, at least of the larger plants, that they bear no seed, the truth being that we fail to observe all the cases of growth from seed, as we said in the History[d] of the willow and elm. Indeed among the smaller plants too we do not observe many cases of this among herbaceous plants, as we said[e] of thyme and others, whose seeds are not

which appear to have no fruit, but yet generate themselves from seed) appears to be similar to what is found in certain undershrubs and herbaceous plants: they have no visible seed, some having a kind of down, some a flower (like thyme), and sprout from these;" *HP* 6 2. 3: "But no such seed (*sc.* evident to the eye) can be found in thyme. Instead it is mixed up somehow in the flower, since the flower is sown and the plant comes up from it;" *HP* 6 7. 1–2: "All the rest (*sc.* the undershrubs not grown especially for their flowers) flower and bear seed, but not all are held to do so because the fruit in some is not visible; indeed even the flower in some is hard to see . . . And yet some insist that they have no fruit . . . Nevertheless what we said first is truer, . . . and the nature of the wild congeners testifies to this . . ."

[ὧν] εἴπομεν, ⟨ὧν⟩ [1] κατὰ μὲν τὴν ὄψιν οὐ φανερά,
15 κατὰ δὲ τὴν δύναμιν φανερά (σπειρομένων γὰρ τῶν
5. 4 ἀνθῶν γεννᾶται). καὶ δυσόρατα καὶ μικρὰ καὶ τῶν
δένδρων ἔνια σπέρματα τυγχάνει, καθάπερ καὶ τῆς
κυπαρίττου· ταύτης γὰρ οὐχ ὅλος ὁ καρπὸς ⟨ὁ⟩ [2]
σφαιροειδής ἐστιν, ἀλλὰ τὸ ἐγγινόμενον ἐν τούτῳ
5 λεπτὸν καὶ ὥσπερ πιτυρῶδες [3] καὶ ἀμενηνόν, ἅπερ
ἐκπέταται διαχασκόντων τῶν σφαιρίων· [4] διὸ [5] καὶ
ἐμπείρου τινός ἐστιν συλλέξαι, τήν θ' ὥραν [6] παρατηρεῖν [7] αὐτό ⟨τε⟩ [8] τὸ σπέρμα γνωρίζειν δυνάμενον. [9]
10 ἐπὶ πολλῶν μὲν οὖν καὶ τοῦτο συμβαίνει καὶ
μάλιστα ὅσα συνεχῶς ἐν ταῖς ὕλαις ταῖς ἀγρίαις
καὶ τοῖς ὄρεσίν ἐστιν· οὐ γὰρ ῥᾴδιον αὐτομάτως
συνισταμένων διαμένειν τὸ συνεχές, ἀλλὰ δυοῖν
θάτερον, ἢ ἀπὸ ῥίζης ἢ ἀπὸ σπέρματος βλαστάνειν.

1–9. Varro, *R.R.* i. 40. 1: illud quod apparet ad agricolas, id videndum diligenter. quaedam enim ad genendum propensa [Schöll; propterea] usque adeo parva ut sint obscura, ut cupressi. non enim galbuli qui nascuntur, id est tanquam pilae parvae corticiae, id semen, sed in iis intus.

3–5. Pliny, *N.H.* 17. 71–72: . . . sic cupressos semine satas et ipsas. minimis id granis constat, vix ut perspici quaedam possint . . .

[1] [ὧν] εἴπομεν· ⟨ὧν⟩ Vasc.²

[2] ⟨ὁ⟩ Scaliger.

[3] πιτυ⟨ρ⟩ῶδες Vasc.²

[4] σφαιρίων Vasc.² (*pilula* Gaza): σφαιρῶν U.

[5] διὸ HP: διοῦ U; δι' οὗ u; διὸ ἢ N. (A predecessor took the breathing and accent on διὸ for a superscribed υ.)

[6] τήν θ' ὥραν Vasc.²: τηῖ θεωρίᾳ U.

[7] παρατηρεῖν Vasc.²: παρατηρῶν U.

[8] αὐτό ⟨τε⟩ Wimmer (⟨καὶ⟩ αὐτὸ Vasc.²): αὐτὸ U.

[9] δυνάμενον U N HP: δυναμένου u.

evident to the eye, but evident in their effect, since the plant is produced by sowing the flowers. Further 5. 4 in trees too some seeds are hard to see and small in size, as in the cypress. For here the seed is not the entire ball-shaped fruit, but the thin and unsubstantial bran-like flake produced within it. It is these that flutter away when the balls split open. This is why an experienced person is needed to gather it, by his ability to observe the proper season and recognize the true seed.

Here then is one point, propagation from unnoticed seed, and it applies to many trees, especially those that succeed each other without a break in wild forests and on mountains, since the succession could not easily be maintained if the trees were formed spontaneously.[a] Instead there are two alternatives: to come from a root or from seed.[b]

[a] The so-called fruitless trees are all wild. *Cf. HP* 3 1. 2: ". . . since it is asserted that also the trees reputed fruitless generate seed, such as elm and willow. In proof is cited not only the fact that many of them grow separate from the roots of the parent, no matter where they are found, but certain localized occurrences are also taken into account, as at Pheneus in Arcadia, after the water broke out that had flooded the plain when the underground channels were blocked. Where willow had been growing near the lake, a willow grew up (they say) the next year in the part that was drained; and where elms had been growing, elms grew up, just as pines and silver-firs came up where pines and silver-firs had been growing before the lake was drained. This implies that the willows and elms were doing the same thing as the pines and silver-firs." (*Cf. HP* 3 1. 2 [of wild trees]: ". . . . excepting those that grow only from seed, as silver-fir, pine . . .")

[b] *Cf. HP* 3 1. 1 (of wild trees): "Now their modes of generation are of an uncomplicated sort: all come from a seed or a root."

5. 5 ὀλίγα δὲ ἄκαρπα τῶν ὁμογενῶν (οὐχ ὁμοιογενῶν δ')[1] οἱ ὑλοτόμοι φασὶν εἶναι. ταῦτα[2] ἤτοι λανθάνειν εἰκὸς ἢ διὰ τὸ καταναλίσκειν εἰς τὰ ἄλλα τὴν τροφὴν ἄκαρπα γίνεσθαι, καθάπερ τὰς
5 ἀμπέλους τὰς τραγώσας[3] καὶ ὅσοις ἄλλοις τοῦτο συμβαίνει· γιγνόμενον ⟨δ'⟩[4] ἐπὶ τῶν καρπίμων ἢ καρποφόρων[5] τί κωλύει τοῦτο συμβαίνειν ἐπὶ τῶν ὅλων,[6] ὥσπερ πηρουμένων[7] πρὸς καρπογονίαν;
10 ἀλλὰ τοῦτο μὲν ὡς ἐπιδοξαζόμενον εἰρήσθω· δεῖ δὲ ἀκριβέστερον ὑπὲρ αὐτοῦ σκέψασθαι καὶ ἀνιστορῆσαι τὰς αὐτομάτους γενέσεις. ὡς δὲ ἁπλῶς εἰπεῖν ἀναγκαῖον γίνεσθαι διαθερμαινομένης τῆς γῆς καὶ ἀλλοιουμένης τῆς ἀθροισθείσης
15 μίξεως ὑπὸ τοῦ ἡλίου, καθάπερ ὁρῶμεν καὶ τὰς τῶν ζῴων.

[1] οὐχ' ὁμοιογενῶν δ Uar (Ur erases δ): N HP omit.
[2] ταῦτα ⟨δ'⟩ Vasc.[2]
[3] τραγώσας u: τρυ- U N HP.
[4] ⟨δ'⟩ Vasc.[2] (*nam* Gaza).
[5] ἢ καρποφόρων U: Dalecampius deletes; Gaza has *fructiferis*, deleting either καρπίμων ἢ or ἢ καρποφόρων; ἢ (καὶ Wimmer) ἀνθοφόρων Moldenhawer.
[6] ὅλων U: ἄλλων Gaza (*aliis . . . omnibus*), Itali.
[7] πηρουμένων Vasc.[2]: τηρουμένων U.

---

[a] *Cf. HP* 3 3. 6–8.
[b] *Cf.* Aristotle, *History of Animals*, v. 14 (546 a 1–3): " He-goats when fat are less fertile, and from them vines are said to ' get goatish ' when they fail to bear;" *On the Generation of*

On the other hand woodcutters report that among trees of the self-same (and not just of a similar) kind a few individuals are fruitless.[a] Here it is likely that either the seed passes unnoticed or else that the tree becomes fruitless because it expends all its food on the other parts, as with vines that "get goatish"[b] and other trees[c] where this occurs. And when failure to bear is found in individuals of kinds that can or do bear fruit, what is to keep it from happening in whole kinds, which are maimed as it were in their capacity to engender fruit? 5. 5

This however is to be taken as a mere opinion thrown in. We must examine the question more exactly and gather information about the cases of spontaneous generation. Broadly speaking it must occur when the earth is thoroughly warmed and the accumulated mixture[d] is qualitatively altered by the sun, which is what we observe when animals are spontaneously formed.[e]

Animals, i. 18 (725 b 34–726 a 3): "Similar (*sc.* to failure of animals and plants to produce semen or seed because they expend their provision on bodily growth) is what happens to the vines that 'get goatish,' which get out of hand because of their feeding; thus he-goats when fat do less copulating, which is why they are previously made thin, and the vines are said to be goatish from what happens to them."

[c] *Cf. CP* 1 17. 10 (almond, fig, vine) and *HP* 2 7. 6–7 (vine, fig, almond, pear, sorb).

[d] Of earth and fresh water.

[e] *Cf.* Aristotle, *On the Generation of Animals*, ii. 6 (743 a 35–36) [of animals]: "To those produced spontaneously the movement and heat imparted by the season is the cause;" iii. 11 (762 a 9–12) [cited in note *c* on *CP* 1 1. 2].

6. 1 λοιπὸν δ' εἰπεῖν ὑπὲρ τῶν ἐν ἄλλοις γενέσεων οἷον τῶν κατὰ τὰς ἐμφυτείας καὶ τοὺς ἐνοφθαλμισμούς. ἁπλοῦς δέ τις λόγος καὶ σχεδὸν εἰρημένος πρότερον· ὥσπερ γὰρ[1] γῇ χρῆται τὰ
5 ἐμφυτευόμενα. καὶ φυτεία δέ τις καὶ ὁ ἐνοφθαλμισμός, οὐ μόνον παράταξις, ἀλλ' ἐνταῦθα δῆλον ὅτι καὶ τὸ βλαστάνον καὶ τὸ γεννῶν ἡ ὑγρότης[2] ἐστὶν ἡ γόνιμος ἥνπερ ⟨ὁ⟩[3] ὀφθαλμὸς ἔχων ἁρμόττεται θατέρῳ, καὶ τὴν τροφὴν ἔχων, ἀποδί-
10 δωσι τὴν οἰκείαν βλάστησιν.

6. 2 εὐαξῆ δὲ πάντα τὰ τοιαῦτα διὰ τὸ κατειργάσθαι τὰς τροφάς, καὶ ταῖς διὰ τῶν ἐνοφθαλμισμῶν ἔτι μᾶλλον· καθαρωτάτη γὰρ αὕτη καὶ ὥσπερ ἐν τοῖς συνεχέσιν[4] ἤδη τῶν καρπῶν· εὐπρόσφυτον δ'
5 ἀεὶ τῷ ὁμοίῳ[5] τὸ ὅμοιον, ὁ δὲ ὀφθαλμὸς ὥσπερ ὁμογενές.

εὐλόγως δὲ καὶ ἡ ἀντίληψις μάλιστα τῶν ὁμοφλοίων,[6] ἐλαχίστη γὰρ ἡ ἐξαλλαγὴ τῶν ὁμογενῶν, καὶ ὥσπερ μετάθεσις γίνεται μόνον·

§ 2. 7–13. Pliny, *N.H.* 17. 104: facillime coalescunt quibus eadem corticis natura quaeque pariter florentia eiusdem horae cognationem succorumque societatem habent.

[1] *γὰρ* U^css: U^ac omits.
[2] *ὑγρότης* U^c: *γρότης* U^ac.
[3] ⟨ὁ⟩ Schneider.
[4] *ἐν τοῖς συνεχέσιν* ego (*ἡ ἐν τοῖς τετελειωμένοις* Schneider): *ἐν τοῖς στελεχεσιν* U.
[5] *τῶι ὁμοίωι* u: *τῶν ὁμοίων* U.
[6] *ὁμοφλοιων* U: *ὁμοιοφλοίων* Schneider.

---

[a] *HP* 2 1. 4 refers to the present chapter: ". . . for twig-grafting and bud-grafting are as it were mixtures or generations

Propagation in Another Tree: Grafting

It remains to discuss the cases where propagation occurs in other trees, namely in twig and bud-grafts.[a] What we have to say is simple and has (so to speak) been said already,[b] since the twig uses the stock as a cutting uses the earth.[c] So bud-grafting too is a kind of planting, and not a mere juxtaposition; here however it is evident that what produces both the sprout and the fruit is the generative fluid: the bud possesses this when it is fitted into the stock, and getting its food from the latter produces its own type of sprout. 6. 1

All grafts grow rapidly because their food has already been worked up; and this applies still more to the bud-grafts, for their food is the purest and just as it is already in the fruits [d] that are continuous with the stock. Like always coalesces readily with like,[e] and the bud is as it were of the same variety.[f] 6. 2

It is also reasonable that grafts should best take hold when scion and stock have the same bark, for the change is smallest between trees of the same kind, and what occurs is as it were a mere shift in position.

occurring in a different way, and of these we must speak later."

Aristotle treats grafting with growth from seed and from a cutting as forms of generation: *cf. On Youth and Age, Life and Death and Respiration*, chap. iii (468 b 17–28).

[b] In the discussion of propagation from cuttings at *CP* 1 1. 3–1 3. 5.

[c] *Cf. CP* 2 14. 4.

[d] "Fruit" (*karpós*) can be used of a fruiting shoot: *cf. CP* 1 12. 10.

[e] *Cf. CP* 5 5. 2.

[f] No great differentiation (as of flavour) has as yet occurred.

10 ἅμα γὰρ συμβαίνει καὶ τοὺς ὀποὺς ὀργᾶν καὶ τὰ ὅλα δένδρα πρὸς τὴν βλάστησιν, ὥσθ' ὅταν ὅμοιόν τε ᾖ καὶ ὁμοιοπαθὲς τοῖς καρποῖς,[1] ἐξ ἀμφοτέρων εὔλογον τὸ τάχος τῆς αὐξήσεως· ἐν δὲ τοῖς ἄλλοις ὅσῳ ἂν[2] ἧττον ᾖ[3] καὶ τοῖς γένεσι 15 καὶ τοῖς ὀποῖς[4] καὶ τοῖς καιροῖς παραλλαγή.[5]

6. 3 εὔλογοι δὲ καὶ αἱ ὧραι, μᾶλλον δὲ ἴσως ἀναγκαῖαι, καθ' ἃς καὶ ὅλως ἐπιβλαστήσεις γίνονται· μετόπωρόν τε καὶ ἔαρ καὶ Κυνὸς ἐπιτολή· δεῖ γὰρ ὀργῶν[6] φέρειν.[7] παραπλήσιοι δὲ καὶ οἱ λόγοι καὶ 5 περὶ ἑκάστης οἵπερ καὶ περὶ τῆς φυτείας· οἱ μὲν γὰρ τὴν ἐαρινὴν ἐπαινοῦσιν, ἐπ' ἰσημερίαις ἔτι κυόντων, ἅμα γὰρ τῇ ἐγκυήσει[8] βλαστήσει, καὶ ὁ φλοιὸς ἐπιφύεται καὶ περιλαμβάνει· οἱ δὲ περὶ τὴν ἐπ' Ἀρκτούρῳ, παραχρῆμα μὲν γὰρ οἷον 10 ῥιζοῦται καὶ οἷον ἐπισημαίνει,[9] προσφυὴς δὲ γενομένη, πρὸς τὸ ἔαρ ἀθρόον ἀποδίδωσι τὴν βλάστην ἀπὸ ἰσχυροτέρας ἀρχῆς.

1–3. Varro, *R.R.* i. 40. 3: Tempus enim idoneum (*for planting from* vivae radices), quod scribit Theophrastus, vere et autumno et caniculae exortu.

[1] καρποῖς U: ὀποῖς Moldenhawer.
[2] ὅσῳ ἂν Schneider (*quo* [*minus*] Gaza): ὅτ' ἂν U.
[3] ᾖ Vasc.[2], Moldenhawer: ὃ U.
[4] ὀποῖς U[c] (ὁ- U[ac]): τόποις Wimmer.
[5] παραλλαγὴ HP: -ῇ U N.
[6] ὀργῶν U: ὀργῶντα Gaza (*turgida* [sc. *germina*]), Scaliger.
[7] φέρειν U: μεταφέρειν Wimmer.
[8] ἐγκυήσει u (ἐγκύσει ⟨καὶ τῇ⟩ Gaza, Itali): ἐγκύσει U.
[9] ἐπισημαίνει U: *maiorem*[*que*] *in modum coire posse* Gaza; ἐπισυμμύει Heinsius.

[a] At *HP* 2 5. 1 Theophrastus says of planting "the seasons when one should plant have been mentioned earlier;" the

For the impulse not only of the saps of the two but of the whole trees toward sprouting is then simultaneous, so that here, when graft and scion are like and have fruit with like responses, both circumstances make the rapidity of growth reasonable. In the rest the growth is more rapid as the difference in the kind of tree, the character of the sap and the seasons of development diminishes.

6. 3 The seasons of grafting are also reasonable, or rather perhaps are necessarily the ones that they are, when all further sprouting in general takes place: autumn, spring and the rising of the dog-star; for we must take a graft that feels the urge to sprout. The arguments in favour of each season are much like the arguments in favour of each as a time for planting.[a] Some persons recommend spring, the trees being still pregnant at the time of the vernal equinox, since the graft in that case will sprout at the time of the pregnancy,[b] and meanwhile the bark grows over the graft and encloses it. Others recommend the season at the rising of Arcturus, for the graft at once " takes root " (as it were) [c] and (as it were) " seals over;" and once it has coalesced with the stock it puts forth its sprouts all at once at the coming of spring, having as it does a more powerful basis to start from.

reference is perhaps to *HP* 2 4. 2, where he speaks of sowing vetch in spring and in autumn, but it is perhaps as likely that the passage has been lost. For spring and autumn as seasons for planting *cf. CP* 3 2. 6–3 3. 2 (for autumn *cf.* also *CP* 1 12. 2); for the dog days *cf. CP* 3 3. 3. For these as the seasons of further sprouting *cf. CP* 1 13. 3–7.

[b] And not wait until the subsequent year.

[c] *Cf.* the argument for autumn planting in *CP* 1 12. 2.

6. 4 εὐλόγως δὲ καὶ τὸ τὰς μασχάλας ἐνοφθαλμίζειν τὰς λειοτάτας[1] καὶ νεωτάτας· ἀντιλαμβάνεται γὰρ ἐντεῦθεν μάλιστα διά τήν τε λειότητα[2] καὶ ἡλικίαν, εὔζωα γὰρ καὶ εὐβλαστῆ [ἐστι][3] τὰ νέα.
5 μάλιστα δὲ εὐφυῆ πρὸς ἐνοφθαλμισμὸν (ὥς γ' ἑνὶ[4] λαβεῖν) ὅσων[5] ἡ ὑγρότης ἔχει τι γλίσχρον, ἔτι δὲ[6] μαλακόφλοια καὶ ὁμόφλοια[7] καὶ ὁμοιοπαθῆ (διὸ καὶ εἰς τὰ παραπλήσια φύσει καὶ ἡλικίᾳ κάλλιστος ὁ ἐνοφθαλμισμός)· ἥ τε γὰρ γλισχρότης 10 καὶ[8] ἀντιληπτική, ὅ τε φλοιός, μαλακὸς ὢν καὶ ⟨ὅμοιος⟩,[9] ὁμοίως εὐμενὴς καὶ[10] οὐ ποιεῖ μεγάλην τὴν μεταβολήν.

6. 5 ἔστι δὲ τοῖς μὲν ἄλλοις βραχὺς ὁ καιρὸς διὰ τὸ ταχεῖαν εἶναι τὴν βλάστησιν, τῇ ἐλαίᾳ δὲ πλείω χρόνον,[11] διὰ τὸ πλείω χρόνον ποιεῖν τοὺς ὀφθαλμούς· ἔτι δὲ[12] ἁπαλὰ ⟨τὰ⟩ ἠρινά,[13] καὶ εὔροα[14] 5 διὰ τέλους, καὶ τὸν τόπον αὐτῆς ὑγρὸν εἶναι πᾶν τὸ θέρος. καὶ ἀπὸ τούτων μάλιστα πάντων βλαστά-

§5. 1–6. Pliny, *N.H.* 17. 113: verno inserentes tempus urget, incitantibus se gemmis praeterquam in olea, cuius diutissime oculi parturiunt, minimum suci habet sub cortice, qui nimius insitis nocet.

[1] λειοτάτας N P (τελειοτάτας H): λειοτάτατας U.
[2] τήν τε λειότητα Gaza, Vasc.[2]: τὴν τελειότητα U.
[3] [ἐστι] ego.
[4] ἑνὶ Vasc.[2]: ἔνι U.
[5] ὅσων HP: ὅσον U N.
[6] δε U: δὲ τὰ Schneider; δ' ἃ Wimmer.
[7] ὁμόφλοια U: ὁμοιόφλοια Gaza, Schneider.
[8] [καὶ] Gaza, Schneider.
[9] ⟨ὅμοιος⟩ ego.
[10] [καὶ] Gaza, Schneider.
[11] πλείω χρόνον U: πλείων χρόνος Gaza (*plus temporis destinatum est*), Vasc.[2]

The advice to graft buds on the smoothest and youngest axils is also reasonable. For here the buds best take hold because of the smoothness and youth of the axils, since what is young is full of life and sprouts well. 6. 4

The stocks best fitted for bud-grafting, to put it in a word, are those with a certain stickiness in their fluid; further, those with bark that is soft and of the same kind and that have similar responses (which is why the best bud-grafting is on stocks close to the bud in nature and age). For the stickiness also establishes a hold;[a] and when the bark is soft and similar it favours the bud equally with the bud's own bark[b] and makes the change no great one.

In the rest the time for grafting is short because of their rapid sprouting, but lasts longer for the olive, which keeps producing buds longer.[c] Further we are told that the new wood produced in spring stays tender and has a flow of fluid throughout the period, and the site of the graft remains moist all summer; 6. 5

[a] As well as prevents drying: *cf. CP* 1 4. 1.

[b] Since the bud is always young its bark is always soft.

[c] Columella (v. 11. 2, *de arb.* 26. 2) lets the time for grafting the olive last from the vernal equinox to the ides of April (April 13). The *Geoponica* at iii. 4. 3 give the month as April; at ix. 16. 3 they give the time from May 24 (?) to June 1. Palladius (v. 2. 3) says that the Greeks set the time as from March 25 to July 5.

The point of all this is that the shoots of the olive remain meristematic longer than those of other plants, and growth is not so rapid as in other plants but persists longer.

---

[12] ἔτι δὲ U: δεῖ δὲ Schneider (*Ad haec . . . petendi* Gaza); ἔτι δεῖ Wimmer; ὅτι δὲ (adding δεῖ after εἶναι below) Keil.

[13] ⟨τὰ⟩ ἠρινά ego (*petendi surculi sunt* Gaza; δεῖ ἔρνη Vasc.[2]; τὰ ἔρνη Schneider): ερινὰ U.

[14] εὔροα ego (*humidi* Gaza): ἀθρόα U.

νειν·[1] διὰ γὰρ τούτων οἴονταί τινες καὶ τέτταρας καὶ πέντε μῆνας δεύεσθαι τὴν ἐμφυτείαν.

6. 6 τὸ δ' ὕδωρ τῷ μὲν ἐνοφθαλμισμῷ πολέμιον (ἐκσήπει γὰρ καὶ ἀπόλλυσιν παραρρέον διὰ τὴν ἀσθένειαν), διὸ καὶ ἀσφαλέστατος ἐπὶ Κυνὶ δοκεῖ[2] (καίτοι νῦν[3] γέ τινες οὕτως περιδοῦσιν 5 τοῖς φλοιοῖς ὥστε μὴ παραρρεῖν)· τῇ δ' ἐμφυτείᾳ χρήσιμον, ἂν μὴ ᾖ ὑγρὰ[4, 5] τῇ φύσει (διὸ καὶ οἱ μὲν αὐτῶν[6] πηλὸν ἐπικολλαίνουσιν, οἱ δὲ χύτραν προσβάλλουσιν ὕδατος ὥστε κατὰ μικρὸν ἐπιρρεῖν· ἀναξηραίνεσθαι[7] γὰρ ἂν ταχύ, μὴ ἔχον ὑγρότητα, 10 διὰ τὸ μέγεθος τῆς ἑλκώσεως).

6. 7 ὀρθῶς δὲ καὶ διατηρεῖν ἀρραγῆ τὸν ὀφθαλμὸν καὶ τὸν φλοιόν, καὶ τὸ ἔνθεμα οὕτως ἀποξύειν ὥστε μὴ γυμνοῦν τὴν μήτραν· ῥαγέντος γὰρ ἢ γυμνωθείσης ἀναξηραίνεται καὶ διαφθείρεται. διὰ 5 τοῦτο γὰρ καὶ περιδοῦσιν φιλύρας ἔνδοθεν φλοιοῖς[8] καὶ ἐπὶ τούτοις περιαλείφουσι πηλῷ τετριχωμένῳ, ὅπως ἔμμονος ἡ ὑγρότης ᾖ καὶ μήθ' ἥλιος μήθ'

§ 6. 1–6. *Cf. Geoponica*, x. 75. 19: . . . ὄμβρος τῷ μὲν ἐγκεντρισμῷ χρήσιμος, τῷ δὲ ἐμφυλλισμῷ ἀσύμφορος.

1–10. Varro, *R.R.* i. 41. 1: aqua recenti insito inimica; tenellum enim cito facit putre. 2. itaque caniculae signo commodissime existimatur ea inseri. quae autem natura minus sunt mollia, vas aliquod supra alligant, unde stillet lente aqua, ne prius exarescat surculus quam colescat.

1–10. Pliny, *N.H.* 17. 117: aptissima insitis siccitas; huius enim remedium: adpositis fictilibus vasis modicus umor per cinerem destillat. inoculatio rores amat lenes.

§ 7. 1–8. Varro, *R.R.* i. 41. 2: cuius surculi corticem integrum servandum et eum sic exacuendum ut non denudes medullam. ne extrinsecus imbres noceant aut nimius calor, argilla oblinendum ac libro obligandum.

[1] βλαστάνειν ego: βλαστάνει U.

and that with these advantages the graft grows better than that of any other tree; since some suppose that all this keeps the graft steeped in fluid for as long as four or even five months.[a]

Rain is harmful to a bud-graft, seeping in and decomposing it and killing it because of its weakness, and this is why it is considered safest to graft buds in the dog days, although nowadays some growers tie bark around the site to prevent rain from seeping in. For a twig-graft on the other hand rain is helpful if the graft is not naturally moist. This is why some growers plaster it with mud and others set a pot of water over it and let the water drip, in the belief that the wound is large enough for the scion to dry out quickly unless it gets fluid. 6. 6

We are rightly told (1) to keep the bud and bark from getting torn and (2) to trim the insert in such a way that no core wood is exposed at the site; for when the bark is torn[b] or the core exposed the scion dries out and perishes. This is why cultivators also first bandage the site with layers of lime bark and then plaster mud over it mixed with hair: to make the fluid remain and keep sun, rain and cold from doing 6. 7

[a] That is, for the three months of summer and one or two months of spring, depending on when the graft is made.

[b] Especially if a break occurs in the continuity of the cambium of the stock.

---

[2] *ἀσφαλέστατος ἐπὶ Κυνὶ δοκεῖ* ego (Hindenlang suggests transposition): *ἐπὶ κυνὶ δοκεῖ ἀσφαλέστατος εἶναι* U.

[3] *καίτοι νῦν* Wimmer: *καίτοινυν* U; *καὶ τοίνυν* u.

[4] For *ᾖ ὑγρὸν* of the editions Scaliger reads *ὑγρὸν ᾖ*.

[5] *ὕγρὰ* u: *ὕγρᾳ* U; *ὑγρῶ* N; *ὑγρὸν* HP.

[6] *αὐτῶν* U: *αὐτῷ* (or *αὐτῇ*) Schneider.

[7] *ἀναξηραίνεσθαι* ego: *-νεται* U N; *-νοιτο* HP.

[8] *φιλύρας ἔξωθεν* (*ἔνδοθεν* ego) *φλοιοῖς* Vasc.[2] (*foliis et corticibus insuper* Gaza): *φιλυρᾳ κενωθεν φλοῖοι* U.

6. 8 ὕδωρ μήτε ψῦχος παραλυπῇ. καὶ ὅταν σχίσαντες ἐντιθῶσι τὸ ἔνθεμα σφηνοειδὲς ποιήσαντες, σφύρᾳ συνελαύνουσιν,[1] ὅπως ὅτι μάλιστα προσαχθῇ.

χρὴ δὲ καὶ τῆς ὑγρότητος τῆς αὐτῶν συμμε-5 τρίαν [2] τινὰ ὑπάρχειν. διὸ καὶ τὴν μὲν ἄμπελον προαποτέμνουσιν ἡμέραις τρισὶ [3] πρότερον, ὅπως προαπορρυῇ τὸ δάκρυον καὶ μὴ σήπηται [4] μηδ' εὐρωτιᾷ· ῥόα δὲ καὶ συκῇ καὶ ὅσα τούτων ἐστὶ ξηρότερα παραχρῆμα.

6. 9 δεῖ δὲ καὶ πρὸς τὰς χώρας προσλαμβάνειν τὰς οἰκείας ὥρας καὶ πρὸς τὰς τῶν δένδρων φύσεις· ἐπεὶ τὰ [5] μὲν ἔνυδρα, τὰ δὲ ξηρά. καὶ [6] λεπτογείῳ [7] καὶ [8] ἄμεινον τὸ ἔαρ, οἰκεῖον γὰρ οὕτως, διὰ τὸ 5 ὀλίγον ἔχειν τὸ ὑγρόν· ἐν δὲ τῇ εὐγείῳ καὶ πηλώδει τὸ μετόπωρον, τοῦ γὰρ ἦρος πολλὴ λίαν ἡ ὑγρότης πρὸς τὸ διατηρεῖν ἕως ἔτι διαμένει [9] τὸ δάκρυον, ὁρίζονται δέ τινες τοῦτο [10] τριάκονθ' ἡμέραις.

§ 8. 5–9. Varro, *R.R.* i. 41. 3: itaque vitem triduo antequam inserant desecant, ut qui in ea nimius est umor defluat antequam inseratur; aut in quam inserunt, in ea paulo infra quam insitum est, incidunt, qua umor adventicius effluere possit. contra in fico et malo punica, et siqua horum natura aridiora, continuo.

§ 9. 1–9. Varro, *R.R.* i. 40. 3: tempus enim idoneum quod scribit Theophrastus, vere et autumno et caniculae exortu, neque omnibus locis ac generibus idem. in sicco et macro loco et argilloso vernum tempus idoneum, quo minus habet umoris; in terra bona ac pingui autumno, quod vere multus umor, quam sationem quidam metiuntur fere diebus XXX.

any harm. So too after slitting the stock and giving the scion a wedge-like shape [a] they drive it in with a mallet to make the fit as tight as possible. 6. 8

There must also be no excess of their own fluid in the scions. This is why in the case of the vine scions are cut two days before grafting, to allow the exudation that collects at the cut first to run off and save the scion from decomposition and mould. On the other hand scions of the pomegranate and fig and of trees drier than these are grafted at once.

One must choose the proper seasons for grafting with both the country and the nature of the trees in view, since some combinations are too wet, others too dry. For thin soil spring is in fact [b] the better season; for what makes this combination appropriate is that thin soil contains but little fluid. For rich and muddy soil on the other hand the better season is autumn, since in spring there is far too much wetness to preserve the graft so long as bleeding still persists. Some set this autumnal season at thirty days. 6. 9

[a] In cleft grafting.
[b] And not the dog days, as was believed (*CP* 1 6. 6).

---

[1] *σφύρᾳ συνελαύνουσιν* ego (*σφύρᾳ ἐλαύνουσιν* Scaliger; *ἐν . . σφύρᾳ ἐλαύνουσιν* Schneider): *ἐν σφύρᾳ ἐλαύνουσιν* U.
[2] *συμμετρίαν* u HP: *-ας* U N.
[3] *τρισὶ* Itali (*triduo* Varro): *τισι* U.
[4] *σήπηται* HP: *σ̄ήπεται* U N.
[5] *ἔπεὶ τὰ* u: *ἔπειτα* U.
[6] *ξηρὰ. καὶ* N (*ξηρὰ καὶ* u HP): *ξηρᾶ καὶ* U.
[7] *λεπτογειωι* U: *λεπτόγεια* u.
[8] *καὶ* U: Schneider deletes; *μὲν* Wimmer.
[9] *διαμένει* u HP: *διαμένηι* U; *διαμένειν* N.
[10] *τοῦτο* U^{r} N HP: *τοῦτω* U^{ar}.

6. 10 εὔλογον δὲ καὶ τὸ καλλικαρπότερα ταῦτα γίνεσθαι, καὶ μάλιστα ἐὰν τὰ ἥμερα εἰς τὰ ἄγρια τιθῆται τῶν ὁμοφλοίων·[1] εὐτροφία γὰρ συμβαίνει πλείων διὰ τὴν ἰσχὺν τῶν ὑποκειμένων. διὸ καὶ 5 κελεύουσιν κοτίνους φυτεύσαντας[2] ἐνοφθαλμίζειν ἢ ἐμφυτεύειν ὕστερον, ἀντιλαμβάνονται γὰρ μᾶλλον ἰσχυροτέρου[3] καὶ τροφὴν ἐπισπώμενον[4] πλείω καλλίκαρπον τὸ δένδρον ποιεῖ· ἐπεὶ εἴ γέ τις ἀνάπαλιν τὸ ἄγριον εἰς τὸ ἥμερον ἐμβάλλοι, 10 διαφορὰν μέν τινα ποιήσει, τὸ δὲ καλλικαρπεῖν οὐχ ἕξει.

καὶ τὰ μὲν περὶ τὰς φυτείας καὶ τὰς ἐμφυτείας ἱκανῶς εἰρήσθω.

7. 1 τὰ δὲ σπέρματα πάντων ἔχει τινὰ τροφὴν ἐν αὑτοῖς, ἣ[5] συναποτίκτεται τῇ ἀρχῇ καθάπερ ἐν τοῖς ᾠοῖς· ᾗ[6] καὶ οὐ κακῶς Ἐμπεδοκλῆς εἴρηκεν φάσκων "ᾠοτοκεῖν μακρὰ δένδρεα,"[7] παρα- 5 πλησία[8] γὰρ τῶν σπερμάτων ἡ φύσις τοῖς ᾠοῖς.

§ 1. 3–4. Empedocles Frag. B 79 Diels-Kranz, *Die Fragmente der Vorsokratiker*, Vol. i[10], p. 340, from Aristotle, *De Generatione Animalium*, i. 23 (731 a 1–9).

[1] ὁμοφλοίων ego (φυτῶν Vasc.[2]): φλοιῶν U.
[2] φυτεύσαντας Schneider: φυτεύσαντες U.
[3] ἰσχυροτέρου Wimmer: ἴσχυρότερον U.
[4] ἐπισπώμενον U N: ἐπισπώμενα HP.
[5] ἣ u HP: ἧ U N.
[6] ᾗ HP: η U; ἧ u N.
[7] δένδρεα Wimmer: δένδρα U.
[8] παραπλησία u HP: -ήσια U N.

[a] *Cf. CP* 4 3. 6 and Aristotle, *On the Generation of Animals*, i. 23 (730 b 33–731 a 9): "Now in all animals capable of locomotion the female is separate from the male, and there is one

6.10 It is also reasonable that trees so grafted should bear finer fruit, especially when the scion is from a cultivated tree and the stock from a wild tree of the same bark, since the scion is better fed because the stock is strong (this is why it is recommended to plant wild olives first and later graft them with cultivated buds or twigs). For the grafts hold better to the stronger tree, and since this tree attracts more food they make it a finer producer. Indeed if one should reverse the procedure and graft wild scions on a cultivated stock, there would be a certain improvement in the wild crop but no fine fruit.

Let this suffice for the discussion of planting in the sense of grafting.

The Provision of the Seed for Survival: Food and Protection

7.1 The seeds of all contain within themselves a certain amount of food,[a] which is brought forth together with the starting-point, as in eggs. Thus Empedocles has not put it badly when he says

> the tall trees lay their eggs,

since the nature of seeds is close to that of eggs. He should however have spoken not just of trees, but

female animal and another male, though the same in kind . . .; but in plants these two capacities are combined in the same individual, and the female does not exist apart from the male. Hence they generate out of themselves, and discharge not semen but a fetation, the so-called seeds. Empedocles puts this well in the verse

> So tall trees lay their eggs; and first the olive.

For the egg is a fetation, and the animal comes from a portion of it, the rest being food; and the plant comes from a part of the seed, the rest becoming food for the shoot and the first root."

πλὴν ἔδει περὶ πάντων εἰπεῖν καὶ μὴ μόνον [1] τῶν δένδρων· ἅπαν γὰρ ἔχει τινὰ τροφὴν ἐν αὑτῷ,[2] διὸ καὶ δύνανται διαμένειν εἰς χρόνον, οὐχ ὥσπερ τὰ τῶν ζῴων εὐθὺ φθείρεται χωριζόμενα πλὴν τὰ
10 τῶν ᾠοτόκων· ταῦτα γάρ, ὥσπερ εἴρηται, τροφὴν ἔχοντα, καὶ φυλακὴν ἅμα τῆς ἀρχῆς, διαμένει.

7. 2 χρονιώτερα δὲ ἑτέρων ἕτερα, καὶ μάλιστα ⟨τὰ⟩ [3] πυκνὰ καὶ ξηρὰ καὶ ξυλώδη (καθάπερ τὰ τοῦ φοίνικος)· οὐκ ἔχει γὰρ οὔτε ἔξωθεν οὐδεμίαν παρείσδυσιν οὔτε ἐν αὑτοῖς [4] ὑγρότητα τὴν διαφθειρομένην·
5 ὅθεν καὶ οὔτε θηριοῦται (καθάπερ τὰ σιτηρὰ τῶν σπερμάτων) οὔτε ἀναξηραίνεται (καθάπερ τὰ τῶν λαχάνων), ἀλλ' ἐν αὑτῷ περιστέγον [5] σῴζει τὴν ἀρχήν.

ὅτι δ' ἐν ἅπασιν [σπερμάτων] [6] κἀκεῖθεν δῆλον·
10 ἃ γὰρ δοκεῖ ξηρὰ καὶ ὥσπερ κελυφανώδη πάμπαν, οἷα τὰ τῶν λαχάνων, ταῦτα κινεῖται κατὰ τὰς οἰκείας ὥρας,[7] ἐὰν καὶ ὁτιοῦν ἰκμάδος λάβῃ (καὶ

[1] μόνον U^r N HP: -ων U^ar. [2] αὑτῶι u: αὐ- U.
[3] ⟨τὰ⟩ u HP; ⟨καὶ⟩ N.
[4] αὑτοῖς N HP: αυ- U; αὐ- u.
[5] αὑτῷ (N HP; αυτῶι U; αὑτῶι u) περιστέγον U N HP: αὑτοῖς περιστέγοντα Vasc.[2]
[6] [σπερμάτων] ego (σπέρμασι τροφή Vasc.[2] after Gaza): σπερμάτων U N; σπέρματα HP.
[7] ὥρας Gaza, Itali: χωρας U.

---

[a] Semen. [b] Eggs.
[c] In the first sentence of the paragraph.
[d] *Cf. CP* 5 18. 4 and *HP* 1 11. 3: ". . . in some plants the seeds are immediately enclosed in a stone or something stone-like, and are (as it were) dry . . .; most evidently so are the seeds of the date-palm. For this seed does not even have a hollow inside but is all of it straight (*sc.* without the curves

of all plants, since every seed contains in itself a certain amount of food. This is why they are able to survive for some time, and do not, like the seed [a] of animals, perish directly on separation from the parent (except for the seeds [b] of oviparous animals, for these survive, since they contain food, as we said,[c] and at the same time a protection for the starting-point).

7. 2

Some seeds however survive longer than others, especially when close-textured, dry and woody (like those of the date-palm); [d] for they allow no entrance from without nor contain within themselves a fluid liable to corruption. Hence they neither get wormy (like the seeds of cereals) nor dry out (like those of vegetables), but the seed preserves the starting-point by sealing it off within itself.

But the presence of food in all can also be seen from this: seeds that appear quite dry and as it were husk-like, like those of vegetables, start to grow at their proper seasons if they get even the slightest amount of moisture,[e] and are on this account kept in upper

making a hollow). Nevertheless it must have some fluid and heat, as we said (*sc.* at *HP* 1 11. 1: ' The seed contains in itself natural fluid and heat . . .')."

[e] Theophrastus speaks of the behaviour of seeds in storage because here it is more easily noticed. *Cf. CP* 4 3. 3: " In general the driest seeds as a class are those of coronary plants and vegetables, which is why they are the quickest to attract moisture, and for this reason they are hung up away from the ground and the rooms are not sprinkled or any water brought into them at all." *Cf.* also *HP* 7 10. 1 (of herbaceous plants): " There being differences between the various plants in the seasons of sprouting, flowering and maturing of fruit, none of them comes up before its proper season, either of those grown from a root or of those grown from seed. Instead each awaits its proper season, and is not affected in the least even by the rains . . ."

διὰ τοῦτο ἐν ὑπερῴοις τιθέασιν καὶ κρεμαννύουσιν[1] ἐν ἀρρίχοις[2] καὶ οὔτε ῥαίνουσιν οὔθ'
15 ὕδωρ ὅλως εἰσφέρουσιν[3] εἰς τὰ οἰκήματα).

7. 3 τῶν δὲ δὴ λοιπῶν[4] καὶ τῇ αἰσθήσει φανερὰ τὰ προσόντα (καὶ τά γε δὴ τῶν σιτωδῶν[5] ὡσπερεὶ γαλακτοῦται διαβλαστανόντων). σχεδὸν δὲ ὁμολογούμενον τοῦτο καὶ ἐπὶ τῶν ζῴων ἐστίν· οὐδὲ γὰρ
5 ἐν τούτοις ἅπαν τὸ κατὰ πρόεσιν[6] σπέρμα καθαρὸν καὶ εἰλικρινὲς ὑποληπτέον.

ὅσα δὲ ξυλώδη καὶ πυρηνώδη[7] περίκειταί τισι, φυλακῆς χάριν οἰητέον, ὥσπερ καὶ τὰ δερματικὰ καὶ ὑμενώδη· πάντα γὰρ ταῦτα πρὸς
10 τὴν σωτηρίαν ἐστίν (ὑγρὰ γὰρ ἡ ἀρχὴ δυνάμει πάντων).

7. 4 ἔχουσι δέ τινων τὰς ζωτικὰς ἀρχὰς (ὥσπερ ἐλέχθη) καὶ ῥίζαι καὶ ἀκρεμόνες καὶ ξύλα καὶ καυλοὶ χωριζόμενοι τῶν φυτῶν, ὥστε κινεῖ καὶ[8] μέχρι τοῦ[9] βλαστάνειν, οἷον αἵ τε σκίλλαι καὶ

[1] κρεμαννύουσιν H² (-ουσιν in an erasure) P^css (κρεμανύουσιν P^ac): κεκρεμάννυσιν U; κεκραμάννυσιν N.
[2] αρριχοις U: ἀναρρίχοις u (*cf.* Et. Gud. *s. v.* Ἀρρίχων, where ἄρριχοι is derived from ἀναίριχοι); ἀξαρρίχοις N; ἀζαρίχοις HP.
[3] εἰσφέρουσιν Gaza (*cf. CP* 4 3. 3), Vasc.²: εκφέρουσιν U.
[4] λοιπῶν u: -ὸν U.
[5] σιτωδῶν ego: σαρκωδων U.
[6] κατὰ πρόεσιν u: κατὰ προαίρεσιν U; καταπρόεισι N; κάτω προϊὸν HP.
[7] πυρηνώδη Basle ed. of 1541: πυρρινώδη U; πυρρηνώδη u, πυρρονώδη N(ω from ο?) HP.
[8] κινεῖ καὶ (κινεῖ και U) N: κινεῖν καὶ HP; κινεῖσθαι καὶ Gaza, Schneider; κινεῖσθαι Wimmer.
[9] τοῦ U: του Schneider.

stories and hung in baskets, and the rooms are neither sprinkled nor is any water at all brought into them.

7. 3 In the rest the presence of an addition is also plain to the eye,[a] the seeds of cereals even becoming (as it were) milky when the plants come out. This presence of food is also (one might say) admitted in the case of animals,[b] for here too we must not suppose that all the ejaculatory seed is pure and unmixed.

As for the woody and kernel-like enclosures of certain seeds, we must take them to be present for protection, as in the case of leathery and membranous seeds too;[c] for all these are for the preservation of the starting-point, since in all seeds it is potentially fluid.

Food in Other Generative Parts

7. 4 In some plants (as we said)[d] even the roots and branches and wood and stalks possess on removal the

[a] It had been inferred (*cf. CP* 1 7. 1–2) from the parallel with eggs, and from germination and growth in storage.

[b] *Cf.* Aristotle, *On the Generation of Animals*, ii. 4 (740 b 5–8): "Or is not the answer this? That it is not true that all the food comes from without; instead some is present initially, and just as in the seeds of plants something of the sort is present which first appears as something milky, so in the matter of animals what is left over from the formation of the animal is food."

[c] For leathery and woody envelopes and kernels *cf. HP* 1 11. 3; for a leathery membrane *cf. HP* 1 11. 5.

[d] *CP* 1 3. 1–1 4. 6.

5 ὅσα σκιλλώδη καὶ τὰ τῆς ἐλαίας ξύλα καὶ οἱ τῶν κρίνων καυλοὶ καὶ οἱ τῆς βλήχρου κλῶνες· ἀνθεῖ γὰρ καὶ αὕτη [1] περὶ τροπάς, ὃ δὴ καὶ μάλιστα θαυμαστόν, ἐπεὶ τά γ' ἄλλα φαίνεταί τινα ἔχειν ὑγρότητα κούφην καὶ γλίσχραν, τὰ δὲ λιπαρά,[2]
10 καὶ ἔτι περιέχεσθαι τὰ μὲν χιτῶσιν πλέοσιν, τὰ δὲ πυκνότητι τῇ αὐτῶν,[3] ὥστε μὴ εὐξήραντα εἶναι· τοῦ ⟨δ'⟩ [4] ἀέρος μεταβάλλοντος καὶ τῆς οἰκείας ὥρας,[5] συμπαθῇ τε γίνεται καὶ βλαστάνει.

7. 5 ἡ δὲ βλῆχρος [6] ξηρὰ φαίνεται παντελῶς, ἀλλὰ δῆλον ὡς ἔχει τινὰ τοιαύτην ἀρχήν, ᾗ [7] κινεῖται τῇ τοῦ ἀέρος ἅμα μεταβολῇ καὶ ἀλλοιώσει. θαυμαστὸν

[1] αὕτη Gaza (*ea*), Schneider: αὐτὴ U HP; αὐτῇ N.
[2] λιπαρὰ U: λιπαράν Vasc.[2]
[3] αὐτῶν U N: αὐ- HP.
[4] ⟨δ'⟩ Schneider (*tum* Gaza).
[5] ὥρας Vasc.[2], Scaliger: χώρας U.
[6] βλῆχρος HP: βληχρᾶ U (-ὰ u N).
[7] ᾗ u HP: ἡ U (ῄ N).

---

[a] *Cf.* [Aristotle], *Problems*, xx. 26 (926 a 1–10): "Why is it that some (*sc.* plants or parts of plants) sprout when they are not in the ground but have been cut off or are in storage, as lily stalks, garlic and onion? Is the answer this? That they all contain food within themselves, although no plant has its food in a definite place, and each grows not by possessing food, but only when the food is concocted and distributed. Now they had the food before they started growing, but they grow when the season arrives when concoction and distribution occurs from the concocting effect of the season, as do also the eggs of crocodiles. But the growth is not continued, since no further food flows in;" *ibid.*, xx. 28 (926 a 16–20): "Why do garlic and onion alone of plants sprout in storage? Is the answer this? They are full of fluid and food. So it is this extra supply of food that makes them sprout. This is evident,

vital starting-points,[a] so that it[b] sets them growing to the point of sprouting, as the roots of squill and squill-like plants, pieces of olive wood,[c] lily stalks and twigs of pennyroyal, for pennyroyal flowers too, at the winter solstice.[d] This last is the most astonishing case, since the rest appear to have a certain light and sticky (or in some cases, oily) fluid, and some moreover appear to be enveloped in several coats, others in their own close texture, and so are kept from readily drying out,[e] and so when the air changes and their proper season arrives, they are caught up in the change and sprout.

7.5 The pennyroyal on the other hand appears completely dry. But it evidently possesses some such starting-point, and this is set in motion by the change of weather and the alteration attendant on the

since squill and grape-hyacinth do the same. And each grows when its season arrives."

[b] The food present in them.

[c] *Cf. HP* 2 1. 2: "And yet some assert that it has been known to happen that even when a stake of olive was set in the ground it came to live along with the ivy that it supported and grew into a tree . . .;" *CP* 1 12. 9.

[d] *Cf.* [Aristotle], *Problems*, xx. 21 (925 a 19–24): "Why is it that pennyroyal and lilies and onions when hung up flower at the solstice [summer solstice E. S. Forster; Sommersonnenwende H. Flashar]? Is the answer this? They contain unconcocted food, which is not concocted in winter because of the cold, whereas it is concocted at the solstice because of the season. But the growth quickly dies down because it has no basis and no influx of food." *Cf.* Cicero, *On Divination*, ii. 14. 33: ". . . dry pennyroyal is said to flower on the very day of the winter solstice;" Pliny, *N. H.* 2. 108, 18. 227, 19. 160.

[e] For stickiness and several coats in the root of purse-tassel and squill *cf. CP* 1 4. 1; for oiliness and close texture in the olive *cf. CP* 1 3. 2.

δὲ καὶ τὸ τῆς ὥρας·[1] οὐ γὰρ ἀνιεμένης, ἀλλὰ[2]
5 μᾶλλον ἐπιτεινούσης, εἰ μὴ ἄρα περὶ τὰς ἁλκυονίδας,[3] ἱκανὸς γὰρ ὁποσοσοῦν χρόνος εἰς τὰς τῶν ἀσθενῶν καὶ μικρῶν μεταβολάς.

αἱ μὲν οὖν γενέσεις πόσαι τε καὶ ποσαχῶς, καὶ τίνες ἑκάστοις οἰκεῖαι, φανερὸν ἐκ τῶν εἰρημένων.

8. 1 εὐβλαστῆ δὲ καὶ εὐαξῆ τὰ ἐκ τῶν φυτευμάτων μᾶλλον ἢ τῶν σπερμάτων εὐλόγως, ἄλλως τε καὶ ὑπόρριζ' ἂν ληφθῇ· προϋπάρχει γὰρ πολλὰ τῶν μορίων ἃ δεῖται μόνον τροφῆς· τῶν δ' ἐκ τῶν
5 σπερμάτων ἅπαντα ταῦτα ἀνάγκη γεννᾶσθαι[4] πρῶτον,[5] ἔπειθ' οὕτως αὐξηθῆναι. τὸν αὐτὸν δὲ τρόπον καὶ ὅσα ῥιζοφυῆ[6] τυγχάνει, καθάπερ τὰ κεφαλόρριζα· καὶ γὰρ ἐν τούτοις προ⟨εν⟩ίσταται[7] τῆς φύσεως ἐξ ὧν πλείων ἡ ὁρμὴ πρὸς τὴν
10 βλάστησιν ἢ[8] τῶν σπερμάτων· ῥιζωθῆναι γὰρ ἐκεῖνα δεῖ πρότερον.

[1] ὥρας Gaza (*temporis*), Itali: χώρας U.
[2] ἀλλὰ Uᶜ (-ά Uᵃᶜ): ἀλλ' εὖ N HP.
[3] ἁλκυονίδας Schneider: ἀλ- U.
[4] γεννᾶσθαι a: γενᾶσθαι U N; γενέσθαι HP.
[5] πρῶτον Uʳ N HP: πρώτων Uᵃʳ.
[6] ῥιζοφυῆ u: ῥιζοφύει U.
[7] προενίσταται ego (*cf. HP* 7 10. 4 ἐνιστῶνται; προϋφίστατα Schneider): προΐσταται U.
[8] ἢ u: ἡ U.

[a] *Cf.* Aristotle, *History of Animals*, v. 8 (542 b 4–16) " The halcyon lays its eggs at the time of the winter solstice This is why, when there is fair weather at the solstice, the nam ' halcyon days ' is given to the seven days preceding the solstic and the seven that follow . . . It is said that the halcyo

change. Astonishing too is the time when it flowers, for this happens not when the season is getting milder, but rather when it is getting more severe (unless the case is this: that it flowers in the halcyon days,[a] since even the briefest period is sufficient to produce changes in plants that are weak and small).

The number of modes of generation, then, and how they occur, and what modes are proper to what plants, is clear from the preceding discussion.[b]

Comparative Speed of Growth:
(1) *Dependent on Mode of Propagation*

8. 1 It is reasonable that plants propagated from slips should sprout and grow faster than those propagated from seed, especially if the slips are taken with some root attached;[c] for in slips many parts are already present, needing only to be fed, whereas in a plant produced from seed all these parts must first be generated and only then can grow. The same holds of plants propagated from the roots, as bulbous plants:[d] here too parts of the nature of the plant have already begun, and the impulse toward sprouting that comes from these is more extensive than that coming from the seeds, since the seed must first get roots.[e]

takes seven days to make her nest, and in the remaining seven lays her eggs and rears her young."

[b] *CP* 1 1. 1–1 7. 5.

[c] *Cf. HP* 2 1. 3; 2 5. 3; *CP* 1 2. 2; 3 5. 3.

[d] *Cf. CP* 1 4. 1. A sucker is also usually from the root, but is always distinct from it; with the bulbous plants a part of the " root " itself is planted and little or nothing else.

[e] *Cf. HP* 1 7. 1: " The roots of all plants are held to grow before the upper parts."

8. 2 ὅσα δὲ κατὰ τὰς ἰδίας φύσεις, ὡς ἂν γένος πρὸς γένος ὁ [1] συγκρίνων λάβοι τις, πότερα [2] κατὰ τὰς εὐθύτητας τῶν πόρων ληπτέον, ὥσπερ Δημόκριτος (εὔρους γὰρ ἡ φορὰ καὶ ἀνεμπόδιστος,[3]
5 ὥς φησιν), ἢ μᾶλλον ὅσα μανότερα καὶ ὑγρότερα; τὰ μὲν γὰρ πυκνὰ καὶ ξηρὰ δυσαύξητα· πᾶν γὰρ ἐν μικρῷ πολὺ [4] τὸ πυκνὸν καὶ ξηρόν [καὶ πικρόν],[5] ὥστε βραδεῖαν εἶναι τὴν ἐπίδοσιν· θάτερον [6] δὲ ὀλίγον ἐν πολλῷ [7] διὰ τὴν μανότητα,
10 καὶ ὅλως ἡ ὑγρότης εὔβλαστόν τι καὶ εὔτροφον.

8. 3 σημεῖον δὲ καὶ ⟨τὸ⟩ [8] κατὰ τοὺς τόπους [9] καὶ τὰς φυτείας συμβαῖνον· ἐν μὲν γὰρ τοῖς εὐδιεινοῖς καὶ ἀπνευμάτοις, ἔτι δ' ὅταν ἡ φυτεία πυκνή, εὐαξῆ μὲν τὰ δένδρα, μανὰ δὲ καὶ ὑγρὰ μᾶλλον·
5 ἐν δὲ τοῖς πνευματώδεσιν καὶ ψυχροῖς, καὶ τῇ φυτείᾳ μανῇ, ἀναυξέστερα μέν, πυκνότερα δὲ καὶ ξηρότερα. συνίστησιν γὰρ τὰ πνεύματα καὶ τὰ ψύχη, καὶ ὅλως ὁ προσπίπτων ἀήρ· ἅμα δὲ καὶ συστέλλεται καὶ οὐ λαμβάνει τὸν ἴσον ὄγκον.
10 ὡσαύτως δὲ ὅταν ἡ φυτεία μανή·[10] καὶ γὰρ ἐνταῦθα

§ 2. 3–4. Democritus Frag. A 162 Diels-Kranz, *Die Fragmente der Vorsokratiker*, vol. ii [8], p. 128.

[1] [ὁ] Schneider.
[2] πότερα Schneider: ποτέρα U.
[3] ἀνενπόδιστος U.
[4] ἐν μικρῷ πολὺ ego (ἔμπηρον Vasc.[2]): ἔμπυρον U.
[5] [καὶ πικρόν] ego.
[6] θάτερον N HP (-ττ- U): θατέρου Vasc.[2]
[7] ὀλίγον ἐν πολλῷ ego (ἐν ὀλίγῳ πολλὴν [-ὴ Schneider] Vasc.[2]): ὀλίγον μὲν πολλῶν U; ὀλίγον μὲν πολλῶ u.
[8] ⟨τὸ⟩ HP.
[9] τόπους U N HP: τρόπους u.

Comparative Speed of Growth:
(2) Dependent on the Nature of the Tree

8. 2 Passing to matters dependent on the distinctive natures of the trees, and comparing rapid and slow growers as one would compare two natural classes, is one to take the determining character to be the straightness of the passages, like Democritus,[a] who says that in this case the flow is plentiful and unimpeded? Or is one rather to take the character to be greater openness of texture and more fluidity? For plants of close texture grow slowly, since everything close-textured has much substance in a narrow compass and is dry, so that its increase takes a long time, whereas the other class has little substance in a wide compass in virtue of its open texture; and fluidity is in general good at producing sprouts and rearing them.

8. 3 This is shown by what happens in different locations and with different types of planting. In locations with fair weather and no wind, and again with close planting, the trees grow quickly but are looser in texture and more supple; whereas in windy and cold locations, and when the planting is spaced, they do not grow to this extent but are closer in texture and more rigid, since winds and cold spells and in general the contact of the air makes them compact, and with this goes a reduction in height and failure to attain to the same bulk. So too when planting is spaced: here too they are closer in texture and grow

[a] *Cf.* the views of Democritus presented at *CP* 2 11. 7.

10 μανὴ u: -ῆ U N HP.

πυκνότερα καὶ εἰς βάθος αὐξανόμενα[1] μᾶλλον, ἐν δὲ ταῖς πυκναῖς ἀνάπαλιν.

8. 4 φανερὸν δὲ καὶ ἐπὶ τῶν ἄλλων ζῴων τοῦτο συμβαῖνον καὶ μάλιστα ἐπὶ τῶν ἀνθρώπων, εὐαξέστερα[2] γὰρ τὰ θήλεα τῶν ἀρρένων, ὑγρότερα καὶ μανότερα τὴν φύσιν ὄντα· δῆλον δὲ καὶ ἐπ' 5 αὐτῶν τῶν δένδρων, ῥόα μὲν ⟨γὰρ⟩[3] καὶ συκῆ καὶ ἄμπελος εὐαξῆ, φοῖνιξ δὲ καὶ κυπάριττος καὶ δάφνη καὶ πεύκη καὶ ἐλαία δυσαυξῆ. καίτοι τό γ'[4] εὐθυπορεῖν ὑπάρχει τισὶ τούτων, ἀλλ' ἤτοι πυκνότης ἢ ξηρότης ἢ ἄμφω κωλύει (καὶ γὰρ ἡ ξηρότης 10 ἀναυξής). ἔτι δὲ πρὸς τούτοις ἔνια τῆς[5] ἀσθενείας· καὶ γὰρ τοιαῦτα δυσαυξῆ καὶ δύστροφα, δεῖ γὰρ μὴ κρατεῖσθαι μηδὲ κωλύεσθαι τὸν κλάδον[6] ὑπὸ τοῦ περιέχοντος, ἐπεὶ καὶ τὸ εὐθυπορεῖν προσδεῖται δυνάμεως καὶ τῆς κατεργαστικῆς 15 καὶ τῆς ἀπαθοῦς· ἄλλως[7] οὐδὲν ὄφελος.

εὐαξῆ μὲν οὖν καὶ δυσαυξῆ τοῖς τοιούτοις ἀφοριστέον.

3–7. Varro, *R.R.* i. 41. 4: . . . omnia enim minuta et arida ad crescendum tarda, ea quae laxiora, et fecundiora, ut femina quam mas et pro portione in virgultis item: itaque ficus, malus punica et vitis propter femineam mollitiam ad crescendum prona, contra palma et cupressus et olea in crescendo tarda: in hoc enim umidiora quam aridiora.

1 αὐξανόμενα HP: -ξεν- U; -ξαιν- u; -ξυν- N.
2 εὐαυξέστερα a: εὐαξέστατα U N H: εὐαυξέστατα P.
3 ⟨γὰρ⟩ Schneider (*enim* Gaza).
4 το γ' U^c (το ss.): γ' U^t.
5 ⟨ὑπὸ⟩ τῆς Scaliger.
6 τὸν κλάδον ego (τὴν τροφὴν Wimmer): τὴν κράδην U.
7 ἄλλως Scaliger (ἄλλως δὲ Vasc.²): ἀλλ' ὅμως U.

laterally more than in height, whereas the reverse occurs when they are planted close.[a]

This is also observed in animals, and especially in man, for the females grow faster than the males,[b] being in their nature more fluid and more loose in texture. It can also be seen by looking at the trees themselves: pomegranate, fig and vine are rapid growers, whereas date-palm, cypress, bay, pine and olive are slow. Yet some of the latter have straight passages.[c] But a close texture or dryness or both prevents rapid growth (dryness too being bad for growth). Furthermore some cases belong to weakness, since weak trees too are poor at growing and at rearing what is grown. For the twig [d] must not be overpowered or checked by the environment; indeed even the possession of straight passages, to be of any avail, requires to be supplemented by power, both the power to work up the food and the power to remain unaffected. 8. 4

So by such points as these we must draw the line between rapid growers and slow.

[a] *Cf. HP* 1 9. 1: ". . . the same trees which when growing close together become tall and slender, become stouter and shorter when growing far apart . . .;" *cf.* also *CP* 2 3. 1 and 2 9. 1–2.

[b] *Cf.* Aristotle, *On the Generation of Animals*, iv. 6 (775 a 12–14): ". . . but after the child is born, everything is completed earlier, such as puberty, prime and old age, for the females than for the males . . ."

[c] So the pine (*HP* 1 5. 1), cypress (*HP* 1 5. 1) and date-palm (*CP* 5 17. 3). The fig on the other hand and pomegranate are crooked (*HP* 1 5. 1).

[d] *Cf. HP* 1 1. 9: "I call the 'twig' (*kládos*) the shoot coming as a single whole from these branches, as especially the annual shoot."

9. 1 ἅπαντα δὲ χείρω τὰ ἐκ σπέρματος ὡς ἐπίπαν ἔν γε[1] τοῖς ἡμέροις, οἷον ῥόα συκῆ ἄμπελος ἀμυγδαλῆ· καὶ γὰρ ὅλα γένη μεταβάλλει καὶ ἀπαγριοῦται[2] πολλάκις ἔνια (καθάπερ ἐν ταῖς 5 ἱστορίαις εἴρηται).

τούτου δ' αἴτιον ἡ ἀσθένεια τῶν σπερμάτων· κρατεῖται γὰρ ὑπὸ τῆς ἐπιρροῆς, πλείονος οὔσης· καὶ ὥσπερ ἄκαρπα γίνεται διὰ πλῆθος τροφῆς οὐ δυνάμενα πέττειν, οὕτως καὶ χείρω γίνεται διὰ τὸ 10 μὴ κρατεῖν. καὶ διὰ τοῦτο Θάσιοι τὰς ἀμυγδαλᾶς ὅταν προσαυξηθῶσιν ἐνοφθαλμίζουσιν·[3] ἐκ γὰρ μαλακῶν σκληραὶ γίνονται μετὰ τὴν φυτείαν. εἴη δ' ἂν τοῦτο καὶ ἐπὶ τῶν ἄλλων ποιεῖν.

9. 2 ὅσα δὲ ἰσχυρὰ τῶν σπερματικῶν,[4] ταῦτα διαμένει μᾶλλον, ὥσπερ ὅ τε φοῖνιξ καὶ πεύκη ἡ κωνοφόρος καλουμένη καὶ πίτυς ἡ φθειροφόρος. ὡσαύτως δὲ καὶ τὰ[5] τῶν ἀγρίων· ἴσως ⟨δ'⟩[6] 5 οὐκ ἔχει ταῦτά γε μετάβασιν εἰς τὸ χεῖρον (ἡ γὰρ

[1] γε Scaliger: τε U.
[2] ἀπαγριοῦται ego (ἐξαγριοῦται Wimmer): ἐκπαππουται U.
[3] ἐνοφθαλμίζουσιν u Hr?P (ἐνὸ- N Har?): -ωσιν U.
[4] σπερματικῶν U: σπερμάτων Schneider.
[5] καὶ τὰ u: κατα U.
[6] ⟨δ'⟩ HP.

---

[a] *HP* 2 2. 4–6: "And those that . . . are planted from slips are all held to breed true. But those that propagate from the fruit (*sc.* the seed) among trees that can also grow in this way are practically all inferior, and some depart completely from their kind, as vine, apple, fig, pomegranate and pear; for from the fig seed no cultivated tree at all is produced, but either a wild-fig or fig gone wild . . ., and from the noble vine comes an ignoble one, and often one of a different kind, and sometimes no cultivated tree at all but a wild one, and

Growth: Degeneration from Seed

9. 1 All trees grown from seed are as a rule inferior, at least among the cultivated (as pomegranate, fig, vine and almond; some indeed often undergo a mutation of their entire kind and become wild, as was said in the History).[a]

The cause of this is the weakness of the seeds, for they are mastered by the influx of food, which is too plentiful for them under cultivation; and just as trees become non-bearing because the food is too abundant for them to be able to concoct, so too they deteriorate from inability to master it. This is why the Thasians graft buds on their almonds when these are full grown, since a soft almond tree when planted from seed turns into a hard one. This could also be done with the rest.

9. 2 Among trees grown from seed the strong preserve their character better, as the date-palm and the so-called cone-bearing and small-seeded pines. So too with the seeds of wild trees; but these perhaps allow of no transition to the worse, wildness being as far in that direction as a tree can go. Instead when wild

occasionally of such a sort that it cannot bring its fruit to concoction, and some cannot even form fruit but only get as far as flowering. 5. From the stones of the olive grows an olive run wild, and from the berries of the sweet pomegranate ignoble pomegranates, and from those of the stoneless kind hard ones, and often sour ones . . . The almond too becomes inferior both in flavour and in turning from soft to hard, which is why we are told to graft on it when it is grown, or else to transplant the layered slip repeatedly. 6. . . . These examples, then, are found in trees subjected to cultivation . . ."

ἀγριότης ἔσχατον), ἀλλ' ἡ εὐχυλία καὶ ἡ εὐσαρκία[1] τούτων πρὸς ἄλληλα γίνεται διὰ τὸν ἀέρα καὶ ἁπλῶς τοὺς τόπους. ἐπεὶ καὶ τὰ ἥμερα εἰς τοῦτο διαφέρει· δύνανται γάρ τινες χῶραι διατηρεῖν
10 τὰς φύσεις, αἱ μὲν μέχρι τινός, αἱ δ' ἁπλῶς, αἱ δὲ καὶ μεθιστάναι[2] πρὸς τὸ βέλτιον, καθάπερ εἴρηται περί τε τῶν ἐν Αἰγύπτῳ καὶ Κιλικίᾳ[3] ῥοῶν· αἱ μὲν γὰρ γλυκεῖαι καὶ οἰνώδεις, αἱ δ' ἀπύρηνοι καὶ καλλίκοκκοι γίνονται περὶ τὸν Πίναρον ποταμόν. ἁπλῶς δ' ὅταν οἰκείαν χώραν λάβωσιν οἱ καρποὶ μᾶλλον δύνανται τὰ γένη διατηρεῖν, ὅταν μάλιστ'[4] εὐθενῇ[5] καὶ καλλικαρπῇ τὰ δένδρα.

9. 3 ἐν δὲ τοῖς ἐπετείοις σπέρμασιν πανταχοῦ πρὸς

[1] εὐσαρκία Gaza, Itali: συσαρκία U.
[2] μεθιστάναι Schneider: μεθιστανται U.
[3] Κιλικίᾳ Gaza: κοιλίᾳ U; ἀκυλίᾳ u; ἀκυλία N HP.
[4] ὅταν μάλιστ' N HP: ὅτ' ἂν (ὅταν u) μάλισθ' U; μάλισθ' ὅταν Schneider.
[5] εὐθενῇ ego: εὐσθενῆι U.

[a] *HP* 2 2. 6–7: "Most of all these trees the date-palm is held to preserve its character as it were completely among the trees produced from seed, and the cone-bearing and small-seeded pines. These cases are from trees subjected to cultivation. Among the wild there are evidently more in proportion that preserve their character, because they are stronger; since the alternative would be in fact strange, if trees became inferior both among the rest and among those produced from seed alone (unless they are able to improve because of cultivation). 7. Difference of locality and weather make a difference here, for the country in some places is held to make a tree breed true, as at Philippi; but the reverse change is held to occur with few trees and in few places, the production of a cultivated tree from a wild seed or simply of a better tree from a worse. For we have heard of this last occurring only with

rees show relative succulence and fleshiness of fruit
mong themselves the improvement is due to the
veather and in a word to the locality. Indeed culti-
ated trees also differ here, for certain countries have
he power to preserve the tree's nature either for a
ertain time or in other cases indefinitely, and yet
thers can even introduce a mutation for the better,
s was said[a] of the pomegranates in Egypt and
ilicia: in Egypt they become sweet and get a wine-
ike taste, whereas the stoneless ones with the fine
erries grow by the Pinarus river. In a word, when 9. 3
ree-fruits find an appropriate country they are better
ble to maintain their kinds, this being when the
rees thrive best and bear the finest fruit.[b]

In grains[c] the change is in all countries in the

he pomegranate in Egypt and Cilicia; that in Egypt the sour
ree both when sown and planted becomes sweet after a fashion
r of wine-like taste; and at Soli in Cilicia by the river Pinarus,
here the battle with Darius was fought, all come to be stone-
ss."

[b] *Cf. HP* 2 2. 8 (continued): "It would also be reasonable
hat if one planted our date-palm in Babylonia it would become
bearer and get assimilated to the date-palms there. So it is
ith any other region with a fruit so well suited to it: the
egion by itself outdoes cultivation and care. A proof is that
lants of that country planted here fail to bear, and some
ven to sprout at all."

[c] Literally "annual seeds." *Cf. HP* 2 4. 1: ". . . one-
eeded wheat and rice-wheat change to wheat if bruised before
owing; and the change does not take place at once but in the
hird year. Here we may say we have something that re-
embles the change in cereals that depends on the country,
ereals too changing with every different country in about the
me time as one-seeded wheat. So too wild wheat and wild
arley when tended and cultivated change in the same time;"
8. 1: "Foreign seeds change in about three years to the
ative variety;" *CP* 2 13. 1–5; 4 1. 6.

5 τὴν χώραν ἡ μεταβολὴ γίνεται, πλὴν οὐκ εὐθὺς σπαρέντων, ὀλίγος γὰρ ὁ ἐν τῇ γῇ χρόνος, ἀλλὰ τρίτῳ δὴ ἔτει· τότε γὰρ ἀλλοιοῦται πρὸς τὴν ἐκτελείωσιν (ὥσπερ καὶ τὰ ζῷα, καὶ γὰρ ταῦτα τριγονήσαντα συνεξομοιοῦται)· οὐ μὴν ἀλλ' ἐπίδη-
10 λόν γέ τι ποιεῖ καὶ ὁ πρῶτος ἐνιαυτός.

καὶ τὰ μὲν ἐκ τῶν σπερμάτων χείρω διὰ ταύτας τὰς αἰτίας.

10. 1 ἡ δ' ἐπέτειος βλάστησις, αὕτη γὰρ οἷον δευτέρα γένεσίς ἐστιν, καὶ [1] οὐχ ἅμα γίνεται πᾶσιν, ἀλλὰ παραλλάττει ταῖς ὥραις, ὥστε σχεδὸν ἐν ταῖς ἐναντίαις ἐνίων εἶναι καὶ τὰ μὲν θέρους, τὰ δὲ
5 χειμῶνος βλαστάνειν. ὁμοίως δὲ καὶ ἡ καρπογονία· καὶ γὰρ αὕτη διέστηκε τοῖς χρόνοις. ὡς μὲν οὖν ἁπλῶς εἰπεῖν, τὰ πρὸς ἑκάστην ὥραν σύμμετρον

[1] [καὶ] HP.

---

[a] *Cf.* Plato, *Republic* vi 497 B 3–5: ". . . as a foreign seed sown in a new country tends to become overpowered, and losing its distinction, to go native." *Cf. CP* 2 13. 3–4, 3 24. 1, 4 11. 5; *HP* 2 4. 1 and Aristotle, *On the Generation of Animals*, ii. 4 (738 b 27–35): ". . . wherever male and female of different kinds produce offspring . . ., the first generation shares in appearance in both parents, as with the offspring of fox and dog and of partridge and chicken, but as time goes on and one generation comes from another, the descendants finally end up resembling the female in bodily conformation, as foreign seeds change to the local character."

It may be added that Athenian citizens were on certain occasions required to swear that their parents on both sides had been Athenians for three generations: *cf.* Pollux viii. 85 and Aristotle, *Constitution of Athens*, chap. lv. 3 (of the testing of the nine archons for fitness in the Council): " When they are tested the question is first put: ' Who is your father and of what deme, and who is your father's father, and who is your

direction of the regional character, except that it does not occur at the first sowing, since the time spent in the ground is too short, but only in the third year, for it is then that the plant undergoes the alteration that gives it its final character, as with animals too, which also become assimilated in three generations.[a] Still even the first year produces a noticeable difference.

So the trees grown from seed deteriorate for these reasons.

Annual Sprouting: The Early and Late Sprouters

10. 1 The annual sprouting, which is as it were a second generation, like the first does not occur at the same time in all trees but differs in its season,[b] so that for some it occurs (one might say) at opposite seasons,[c] in summer for one set, in winter for the other. Similarly with fruiting: this too occurs at different times.[d] Now to put it in a word, when the tempering

mother and who is your mother's father and from what deme? " Other passages could be cited to show that a character (usually for rascality) was held to be bred in the bone if it had endured for three generations (see Jebb on the *Oedipus Tyrannus* 1062).

[b] *Cf. HP* 3 4. 1–2 and especially 1: " Now in some wild trees the sprouting occurs at the same time as that of their cultivated counterparts, whereas in others it is somewhat later, and in still others considerably later; but in all it occurs in the spring season."

[c] *Cf. CP* 1 13. 3: " It is reasonable that both planting and additional sprouting occur at several seasons. And autumn, spring and the rising of the dog-star are held to be opposite in a fashion . . ."

[d] *Cf. HP* 3 4. 1 (after the words cited in note *b*): " But the difference in time of fruiting is greater . . ."

ἔχοντα τὴν κρᾶσιν,[1] ταῦτα καὶ τὰς βλαστήσεις ἐν ἑκάστῃ καὶ τὰς τελειώσεις τῶν καρπῶν ἀποδίδωσιν.

10. 2 οὐ μὴν ἀλλὰ δεῖ καὶ τοῖς καθ' ἕκαστα πειρᾶσθαι διαιρεῖν· ἔνια μὲν γὰρ εὐβλαστῆ καὶ ἡ βλάστησις δι' ἰσχὺν καὶ πολυτροφίαν, ὥσπερ ἀμυγδαλῆ ῥοιά, καὶ ὅλως τὰ ἄγρια μᾶλλον τῶν ἡμέρων· ἔνια δὲ 5 δι' ἀσθένειαν, ὥσπερ τὰ ποιώδη καὶ ἐπέτεια, καὶ γὰρ ἀνθεῖ πολλὰ τούτων κατὰ χειμῶνα, καθάπερ καὶ ἡ ἀνεμώνη·[2] φαίνεται δὲ οὐδ' ἡ μηλέα[3] προανθεῖν[4] δι' ἰσχύν, ἀλλὰ δι' ἄλλην αἰτίαν. ἑκάτερα δ' εὐλόγως πρωϊβλαστῆ[5] καὶ τὰ ἰσχυρὰ 10 καὶ τὰ ἀσθενῆ· τὰ μὲν γὰρ τῇ δυνάμει καὶ τῷ πλήθει προωθεῖ, τὰ δὲ ὑπὸ μικρῶν εὐκίνητα τῶν[6] κατὰ τὸν ἀέρα.

10. 3 μέγα δὲ καὶ ⟨ἡ⟩[7] ὑγρότης καὶ ἡ μαλακότης[8] εἰς τὸ προϊέναι, ὥσπερ καὶ ἐπὶ τῆς πρώτης

[1] κρᾶσιν HP (κράσιν U^r N): κράτησιν U^ar.
[2] ἀνεμώνη U^ar HP: ἀνεμόνη U^r N.
[3] μηλέα u: μιλία U.
[4] προανθεῖν U: πρωϊανθεῖν Wimmer.
[5] πρωϊβλαστῆ U: -εῖ u.
[6] τῶν u: τὸν U.
[7] ⟨ἡ⟩ HP.
[8] μαλακότης U: μανότης Vasc.[2]

---

[a] *Cf. HP* 1 9. 6: ". . . some trees sprout early . . . as the almond."
[b] *Cf. CP* 1 20. 5.

of its qualities is adjusted to a given season the plant will sprout or mature its fruit in that season.

10. 2 We must nevertheless also endeavour to distinguish the early and late sprouters by their special features. In some the early sprouting is due to strength and plentiful feeding, as with almond [a] and pomegranate; [b] and wild trees as a whole are earlier than cultivated. Some plants sprout early from weakness, as herbaceous plants and annuals.[c] So too the apple appears to flower early not from strength but for some other reason.[d] The early sprouting of each group, the strong and the weak, is reasonable; the strong push out early because of their power and the abundant food that they attract, whereas the weak are easily set going by slight changes in the air.

10. 3 Also important for early sprouting are fluidity and softness,[e] as was said [f] of the initial growth; for

[c] *Cf. HP* 6 8. 1, where the so-called " mountain " anemone, a wild flower, is among those which appear after stock and wild wallflower, which appear earliest. In *HP* 7 10. 2 anemone is called a winter plant.

[d] That is, from weakness. The " spring apple," a weak tree (*cf. CP* 2 11. 6), is no doubt meant. Some apple trees are early, some late (*cf. CP* 1 18. 3; 4 11. 2).

[e] " Soft " is much the same as " open in texture: " *cf.* Aristotle, *Physics*, iv. 9 (217 b 17–18): ". . . for the heavy and the hard are held to be close in texture, and their opposites, the light and the soft, to be open . . .; " viii. 7 (260 b 7–10): " Further the starting-point of all affections is closing and opening of texture; for heavy and light, soft and hard and hot and cold are held to be kinds of density and rarity; " [Aristotle], *Problems*, xi. 58 (905 b 21–22): ". . . the open in texture and the soft being either the same or proximate in their nature . . ."

[f] *CP* 1 8. 2.

φύσεως ἐλέχθη· τὸ γὰρ ξηρὸν καὶ πυκνὸν οὔτ' εὐδίοδον ὁμοίως, οὔθ' ὕλης ἔχει πλῆθος.

5 ὅσα δὲ ἐν ταῖς ἐναντίαις ὥραις ἐκβλαστάνει, τούτων αἰτιῶνταί τινες τὴν ψυχρότητα [1] καὶ θερμότητα· τὰ μὲν ⟨γὰρ⟩ [2] ψυχρὰ τοῦ θέρους, τὰ δὲ θερμὰ τοῦ χειμῶνος βλαστάνειν, ὥστε ἑκατέραν τὴν φύσιν σύμμετρον εἶναι πρὸς ἑκατέραν τῶν 10 ὡρῶν· οὕτω γὰρ οἴεται καὶ Κλείδημος. οὐ κακῶς μὲν οὖν ἴσως οὐδὲ τοῦτο λέγεται, δεῖ δὲ καὶ τὴν ὀλιγότητα καὶ ⟨τὴν⟩ [3] ἀσθένειαν καὶ εἴ τι ἄλλο συναίτιον προστιθέναι, καθάπερ καὶ ἐπὶ τῶν λαχανωδῶν ὁρῶμεν.

10. 4 ἡ [4] μὲν γὰρ ἀνδράχνη [5] καὶ ὁ σίκυος [6] καὶ ὅλως τὰ τοιαῦτα κάθυγρα καὶ ψυχρά, τὸ δ' ὤκιμον ξηρὸν καὶ ξυλῶδες· ἀλλ' ὅμως οὐ δύναται πρὸ τῶν θερμημεριῶν βλαστάνειν εἴς γε διαμονὴν 5 καὶ τελέωσιν. αἴτιον δὲ ἡ ἀσθένεια, καθάπερ καὶ τῆς σικύας καλουμένης. αὕτη γάρ, δένδρων [7] ὕψος λαμβάνουσα καὶ σχιζομένη τοῖς κλωσὶν

10. Kleidemos, Frag. 4 Diels-Kranz, *Die Fragmente der Vorsokratiker*, vol. ii⁸, p. 50.

[1] ψυχρότητα Itali: ὑγρότητα U.
[2] ⟨γὰρ⟩ Gaza (*enim*), Vasc.²
[3] ⟨τὴν⟩ Wimmer.
[4] ὁρῶμεν· ἡ a: ὁρωμενη U (-ένη u N HP).
[5] ἀνδράχνη (no acc. N) HP: ἀνδράχλη (no acc. U) u.
[6] σίκυος P (σύκιος H): σίκυμος U N.
[7] δένδρων Itali: -ον U N H¹P; -ου u H²a.

---

[a] The order is chiastic: the dry has little matter (that is, food), and the close in texture offers it no passage.

[b] Vegetables were sown at three different seasons (*HP* 7 1. 1–2), some at all three (*HP* 7 1. 2). In the case of the latter

what is dry and dense does not offer so easy a passage or contain so much matter.[a]

Some account for the trees that sprout at opposite seasons by the coldness and heat of the trees, the cold sprouting in summer, the hot in winter, each of the two natures being so adjusted to each of the seasons that there is no excess. For such is also Clidemus' view. This too is perhaps not ill said, but we must also add smallness, weakness and other contributory causes as we observe them in vegetables as well.[b]

10. 4 So purslane, cucumber and in general all plants of this sort are full of fluid and cold, whereas basil is dry and woody; yet none is able to come up before the warm days,[c] at least so as to survive and mature. The cause is their weakness, just as it is with the so-called *sikýa*.[d] For although this grows as high as trees and branches out like the tree-climbing vine, it

one cannot say that the heat or coldness of the plant is adjusted to that of the season, since the three seasons differ. *Cf.* *HP* 7 1. 5: "For putting it broadly one must look for the causes of this in a number of things: in the seeds themselves, in the locality, in the air and in the times of wintry and fair weather in which the various plants are sown."

[c] Basil, purslane and cucumber are all "summer vegetables," sown in the "summer sowing" (so called because the crops come up in summer) in the month Munychion (*HP* 7 1. 2), which begins roughly with the vernal equinox. Basil comes up on the third day, cucumber on the fifth to seventh, purslane later (*HP* 7 1. 3). If a fluid character alone determined growth purslane and cucumber should be earlier than basil.

[d] *Cf. CP* 2 11. 4. Sir A. Hort and Liddell-Scott-Jones identify it as the bottle-gourd, *Lagenaria vulgaris.* "So-called" is odd; perhaps it was thought to be named from *sikýa*, a cupping iron, although the cupping iron was actually named from the gourd.

ὥσπερ ἡ ἀναδενδράς, ὅμως ἐπίκηρόν τε καὶ οὐ δύναται πρὸ τῶν θερινῶν, ἀλλὰ δεῖται πορρωτέρω 10 ταῖς ὥραις [1] παρατείνειν πρὸς τὸ μετόπωρον.

10. 5 καὶ τὰ ἀκανθώδη καὶ ποιώδη καὶ ῥιζοκέφαλα, καθάπερ ὅ τε σκορπίος καλούμενος καὶ τὰ τίφυα καὶ ἡ ἄκανθα ⟨ἡ⟩ [2] βασιλικὴ καὶ τὸ λείριον· ἔνια δὲ καὶ μετ' Ἀρκτοῦρον, εἴτε οὖν ὕστερα χρὴ 5 καὶ ὀψιαίτερα ταῦτα λέγειν, εἴτε καὶ πρωϊαίτερα, γινομένων τῶν πρώτων ὑδάτων· οὐδὲν γὰρ διαφέρει πρός γε [3] τὸ νῦν.

ἀλλ' ἐκεῖνο φανερόν, ὡς οὔτε μεγέθει καὶ μικρότητι τὸ πρωϊβλαστὲς καὶ ὀψιβλαστὲς ἀφορι- 10 στέον, οὔτ' [4] ἴσως θερμότητι ⟨καὶ ψυχρότητι⟩ [5] καὶ ὑγρότητι καὶ ξηρότητι μόνον, ἀλλὰ δεῖ τινα συμμετρίαν ἕκαστον ἔχειν πρὸς τὴν ὥραν· γίνεται

10. 6 γὰρ αἰτιωτάτη.[6] φαίνεται γὰρ ἅπαντα τὴν οἰκείαν ἀναμένειν, [τελείωσιν. τὰ] ⟨τέως⟩ δὲ κἀβλαστῆ

[1] ταῖς ὥραις U (the plural avoids hiatus): τῆς ὥρας Schneider.
[2] ⟨ἡ⟩ ego.
[3] γε HP: τε U N.
[4] οὔτ' HP: οὔθ' U N.
[5] ⟨καὶ ψυχρότητι⟩ Schneider.
[6] γίνεται γὰρ αἰτιωτάτη ego (*ut tenerrima effici possint* Gaza; Wimmer deletes): γίνεται γὰρ ἁπαλωτάτη U.

---

[a] Which are evidently much weaker and smaller, yet come up earlier.

[b] Here the four great classes of plants are reduced to three, shrubs being grouped with trees. The three are trees, undershrubs (including spinous plants) and herbaceous plants (including vegetables and bulbous plants). Scorpion and

is nevertheless delicate and unable to come up before the summer vegetables,[a] requiring the season to be further advanced towards autumn.

10. 5 So too with spinous and herbaceous and bulbous plants,[b] as the so-called scorpion,[c] autumnal squill, royal thorn [d] and narcissus.[e] Some even sprout after the rising of Arcturus, whether we should say that they sprout later than the rest and call them " late " or else call them " early,"[f] since they come out with the first rains, for it makes no difference at the moment.

Another point, however, is clear: we must neither distinguish early and late sprouters by the greater or smaller size nor yet perhaps by greater heat or coldness or fluidity or dryness alone. Instead every plant must possess a certain adjustment to the season, since the season turns out to be more responsible than anything else. 10. 6 For all are seen to await their own appropriate season, meanwhile not sprouting at all

royal thorn are spinous; narcissus and autumnal squill are bulbous.

[c] A wholly spinous plant (*HP* 6 1. 3; 6 4. 1); it flowers after the autumnal equinox (*HP* 6 4. 2).

[d] " Fish-thistle " (*Cnicus acarna*) according to Liddell-Scott-Jones.

[e] *Cf. HP* 6 8. 3: " In autumn flower the other kind of lily (*sc.* narcissus) . . ."

[f] *Cf. HP* 6 6. 10: " The crocus . . . flowers and sprouts very late or flowers very early, depending on how you take the season, for it flowers at the rising of the Pleiades . . .;" *HP* 7 10. 3 (of certain herbaceous plants): " Because of the unbroken succession and overlapping by one another it does not appear easy in certain cases even to determine which sprouts first and which are late sprouting, unless one should lay down a beginning of the year at their generation." (The last three words translate a conjecture, πρὸς τῇ γενέσει.)

τελέως τε[1] καὶ ἀκίνητα ὄντα ὁμοίως δένδρα καὶ ὑλήματα καὶ ποιώδη· φανερώτατον γὰρ τοῦτ' ἐν
5 τοῖς ἀγρίοις ἐν οἷς δὴ καὶ ἡ γένεσις αὐτοφυὴς καὶ ἄνευ παρασκευῆς. πάντα δ' (ὡς εἰπεῖν) ἢ τά γε πλεῖστα τῶν μετοπωρινῶν ἐν τοῖς ὑλήμασι καὶ τοῖς φρυγανικοῖς καὶ ποιώδεσίν ἐστιν· ἐπεὶ δένδρον γε οὐδέν (πλὴν εἰ μή τι κατὰ τὴν ἐπι-
10 βλάστησιν, ἐπιβλαστάνει γὰρ ἔνια μετ' Ἀρκτοῦρον).

10. 7 ὡς δ' ἐπίπαν καὶ καθόλου λαβεῖν, τὰ ἀείφυλλα τῶν δένδρων καὶ ὀψιβλαστότερα καὶ ὀψικαρπότερα, διά τε πυκνότητα καὶ ξηρότητα, κατὰ μικρὸν γὰρ ἡ ἐπιρροή, καὶ διὰ τὸ συνεχὲς ἀεὶ τῆς εἰς τὰ φύλλα
5 διδομένης τροφῆς· οὐ γίνεται γὰρ ἀθροισμὸς ὥστε προορμᾶν, ἀλλ' ἀπὸ τῆς ὥρας κινεῖται τῆς οἰκείας τὸ καθῆκον. σπάνια γὰρ ἄν τις λάβοι τὰ πρωϊβλαστῆ καὶ πρωΐκαρπα[2] τῶν ἀειφύλλων, οἷον τόν τε κιττὸν καὶ τὴν ῥάμνον καὶ εἴ τι τοιοῦτον· ἡ δὲ
10 ἰδιότης εἴτε διὰ θερμότητα τούτων, εἴτε καὶ

[1] [τελείωσιν. τὰ] through τε ego (*perfectionem . . . : nec plus germen nullum effundunt sed* Gaza; ⟨ὥραν πρὸς τὴν⟩ τελείωσιν, ἀβλαστῆ τε τέως [δὲ] Schneider): τελείωσιν. τὰ δὲ καὶ βλαστῆ. τελέως δὲ U. (For τελέως δὲ Itali read τέως δὲ, Vasc.[2] proposes τελέως or τέως.)

[2] πρωΐκαρπα HP: πρωϊόκαρπα U[ar]; προϊόκαρπα U[r] N.

---

[a] *Cf. HP* 7 10. 1: "There being a distinct season in every plant for its sprouting, its flowering and the maturing of its fruit, no plant comes out before its own season, either among those generated from a root or among those generated from

and not being set in motion,[a] trees and woody and herbaceous plants alike; for the fact is plainest in wild plants, where generation is the plant's own doing and not promoted by man. But practically all, in any case most, of those that sprout in autumn belong to woody plants, undershrubs and herbaceous plants, since no tree sprouts then [b] (except in the course of its additional sprouting, some sprouting in this way after the rising of Arcturus).[c]

10. 7 As a rule and speaking generally, evergreen trees both sprout and fruit later than the rest [d] because of their (1) close texture and dryness, for these make the influx of food gradual, and of (2) the uninterrupted distribution of food to the leaves,[e] for no accumulation of food is formed that might lead to a spurt of growth, but each part is in turn set in motion by its own appropriate season. For one could point to but few evergreen plants that sprout and fruit early, as ivy, buckthorn and the like.[f] Whether the peculi-

seed, but each awaits its own season and is not even affected by the rains . . ."

[b] *Cf. HP* 7 10. 2 (of trees): " For these all sprout either at the same time or very close to one another, or else one might say at a single season . . ."

[c] *Cf. HP* 3 5. 4 (of trees): " The sproutings that take place after the spring sprouting and occur at the rising of the dog-star and at that of Arcturus are (one may say) common to all trees."

[d] *Cf. CP* 1 17. 6: ". . . for among evergreens all, practically speaking, fruit late;" 1 22. 4: ". . . for the evergreens, practically speaking, fruit latest."

[e] *Cf. CP* 1 11. 6; 2 17. 2. The Greek for " evergreen " is " everleaf " (ἀείφυλλον).

[f] Ivy and buckthorn are shrubs: *cf. HP* 1 9. 4: " Of shrubby plants the following are evergreen: ivy, bramble, buckthorn, reed and juniper . . ."

μανότητα καὶ ὑγρότητα, εἴτε καὶ διὰ ταύτας πάσας συμβαίνει τὰς αἰτίας, διαιρετέον αὐτὸ τοῦτο [1] πρῶτον ἴσως εἰπόντας ποῖα θερμὰ καὶ ποῖα ψυχρὰ καὶ τοῖς ποίοις ληπτέον.
15 ἀλλὰ περὶ μὲν τούτων ἐν τοῖς ὕστερον πειρατέον εἰπεῖν· διὰ τί δὲ τὰ μὲν πρωϊβλαστῆ, τὰ δὲ ὀψιβλαστῆ, ταύτας ἄν τις ἀποδοίη τὰς αἰτίας.

11. 1 ὅσα δὲ κατὰ πᾶσαν ὥραν βλαστάνει τε καὶ ἀνθεῖ καὶ καρποτοκεῖ, καθάπερ ἡ Περσικὴ μηλέα [2] καὶ εἴ τι ἄλλο τοιοῦτον, ἀπορήσειεν ἄν τις οὐ μόνον ὅτι [3] πρὸς πάσας ἁρμόττει τὰς ὥρας ἡ
5 κρᾶσις, ἀλλ' ὅτι οὐδ' αὐτὰ αὑτοῖς ἅμα βλαστάνει [4]

[1] τούτο U^cm: U^t omits.
[2] μηλέα U^c: μηλία U^ac.
[3] ὅτι Wimmer (*quoniam* Gaza): ὅ | U (ὃ u N); ὅτι οὐ HP.
[4] ἅμα βλαστάνει ego (ὁμοβλαστεῖ Liddell-Scott-Jones; ὁμοιοβλαστεῖ Vasc.[2]): ὁμοβλαστάνει U^c (ν from ρ).

[a] As Menestor maintained: *cf. CP* 1 21. 7.

[b] Aristotle lets evergreens retain their leaves because of the presence of oily (that is, warm) fluid: *cf. On the Generation of Animals*, v. 3 (783 b 8–22): "Men become noticeably bald most of all animals. This affection is something general: so among plants some are evergreen, some deciduous, and hibernating birds shed their feathers. Baldness too, among men to whom it occurs, is an affection of this sort; for whereas not only the leaves among all the plants, but also the feathers and hair in the animals that have them, are shed gradually, on the other hand when the affection occurs all at once it gets the names mentioned, the names used being 'balding' and 'leaf-shedding.' The cause of the affection is lack of warm fluid, and of fluids it is the oily that has most of this character, and hence of plants the oily are more often evergreen. But the causes of these matters are to be discussed elsewhere [*i.e.*, perhaps in the lost work '*On Plants*']; for there are other contributory causes of this sort of affection in them."

arity of these last is due to their heat [a] or else to open texture and fluidity [b] or to all of these causes together must perhaps be decided [c] only when we have discussed what plants are hot and what are cold and by what points this is to be ascertained.

This we must endeavour to do later.[d] Meanwhile these are the causes that one would assign for the distinction between early and late sprouters.

Sprouting (and Flowering and Fruiting) at all Seasons: Problems

11. 1 About trees that sprout, flower and fruit at all seasons, like the citron [e] and any other such tree there may be,[f] one might object that not only is their balance of qualities adjusted to all the seasons, but also that each kind of tree does not even sprout at the same time with itself or have fruit at the same stage of

[c] It turns out that evergreens are (1) hot and (2) close-textured and dry: *cf. CP* 1 22. 5.

[d] *CP* 1 21. 4–1 22. 7.

[e] *Citrus Medica. Cf. HP* 4 4. 3: "It (*sc.* citron) bears its fruit at every season: thus at any time some fruit has already been gathered, some is in flower and some is being concocted to ripeness."

[f] Theophrastus has the garden of Alcinous in mind (*Odyssey*, vii. 114–121):

On one side grow tall trees and flourish
Pears, pomegranates and apples with gleaming fruit,
Sweet figs and flourishing olives.
Their fruit is never lost or fails
Winter or summer, lasting the year; but in succession
The west wind, blowing, makes some to grow and ripens others.
Pear grows old upon pear, apple upon apple,
And again cluster of grapes upon cluster, fig upon fig.

τὰ γένη καὶ ὁμοιοκαρπεῖ·[1] τοῦτο γὰρ ἐν τοῖς κεχωρισμένοις τῶν ὁμογενῶν ἀξιοῦμεν καὶ ὁρῶμεν συμβαῖνον.

τὸ μὲν οὖν πάθος ὅμοιον φαίνεται τοῖς ἐπικυϊσκο-
10 μένοις ζῴοις, πλὴν αἰτία[2] τοῖς μὲν φανερά, τοῖς δὲ
11. 2 λόγου δεομένη. τὰ μὲν γάρ, οὐχ ἅμα λαμβάνοντα[3] τὰς παρ' ἑτέρων ἀρχάς, οὐδ' ἴσως εἰς τὸν αὐτὸν τόπον, οὐδ' ἅμα κυΐσκεται (εἰ μὴ καθάπερ ἅμα καταλάβοι)· τῶν δὲ διὰ τί ἡ βλάστησις οὐχ ἅμα οὐ
5 ῥᾴδιον εἰπεῖν. ἀνάγκη γὰρ διὰ τὸ τὰς ῥίζας μηδ' ἅμα ἢ[4] μηδ' ὁμοίως ἔχειν, ἢ τοὺς ἀκρεμόνας μὴ δέχεσθαι μηδ' ἐκπέττειν[5] ἐξ ὧν ἡ πρώτη βλάστησις· ταῦτα δὲ διὰ τί καὶ τίνος ἕνεκα γίνοιτ' ἂν οὐκ εὔλογον εἰπεῖν.

11. 3 ἡ μὲν γὰρ ἄμπελος ἡ μαινομένη[6] τάχ' ἂν[7] δόξειεν οὐκ ἀλόγως δέχεσθαι· τῷ[8] γὰρ μὴ ἐκπέττειν τὸν καρπὸν ὑπολείμματα[9] πολλὰ ποιεῖται ὑγρότητος γονίμου, ταῦτα[10] δ' ὅταν ἀὴρ ἐπιλάβῃ
5 μαλακὸς ἐκτίκτει, καθάπερ καὶ ἄλλοις τῶν

[1] ὁμοιοκαρπεῖ U: ὁμοκαρπεῖ Schneider.

[2] πλὴν αἰτία Gaza (*sed causa*), Vasc.²: πλάγια U.

[3] λαμβάνοντα HP: λαμβάνουσα U N; λαμβάνουσι Schneider (M³?).

[4] μηδ' ἅμα ἢ ego (μὴδ' ἅμα Itali): μηδε ἀμῆ U (μηδαμὴ u).

[5] ἐκπέττειν HPᶜ(τ² ss.; -ν in an erasure): ἐπετειν U (-εῖν N).

[6] ἡ μαινομένη Gaza (*quam insanam vocamus*), Itali: σημαινομένη U.

[7] τάχ' ἂν Wimmer: τάχα U.

[8] τῷ Vasc.² (*quod* Gaza): τὸ U.

[9] ὑπολείμματα N HP: ὑπόλειμμα τὰ U.

[10] ταῦτα (*sic*) U.

[a] Discussed in Aristotle, *On the Generation of Animals*, iv. 5 (773 a 32–774 b 4).

development, whereas we expect agreement here in separate individuals of the same kind and observe it to occur.

Now what happens here appears similar to superfetation [a] in animals, except that in animals there is an evident cause, whereas in plants the cause needs explaining. So if animals do not receive the starting-points from other animals [b] at the same time,[c] nor yet into the same place, they are also not simultaneously pregnant with the different broods (unless conception is practically simultaneous). In plants on the other hand it is not easy to say why the sprouting fails to be simultaneous. For failure must occur because (1) the roots do not have simultaneously, or at the same stage, or (2) because the branches do not receive, or do not concoct, simultaneously the wherewithal for the initial sprouting. But why all this should occur and what end it could serve are questions to which it is not easy to find a reasonable answer. 11. 2

Now the "mad" vine [d] might not unreasonably be held to receive the food at different times, since its failure to concoct its fruit fully makes it possess many left-overs of unused generative fluid, and these bring forth progeny when a spell of mild weather 11. 3

[b] That is, the males.

[c] *Cf.* Aristotle, *On the Generation of Animals*, iv. 5 (773 b 7–9) [of multiparous animals]: ". . . all that are large, like man, mature the embryo by superfetation if the one copulation occurs very close in time to the other . . .;" iv. 5 (773 b 13–16): "But when copulation occurs after the embryo has grown to some size, superfetation sometimes takes place . . .;" iv. 5 (774 a 17–20): "For some animals in which superfetation occurs are able to complete the embryos even when there is a long interval between the copulations . . ."

[d] *Cf. CP* 1 18. 4.

δένδρων αἱ πρῶται [1] βλαστήσεις ἐπιγίνονται. ἡ δὲ μηλέα [2] καὶ ὅσα ἄλλα τοιαῦτα τελεογονεῖ καὶ ἐκπέττει, διὸ τὸ μερίζεσθαι ⟨καὶ⟩ [3] ἄλλοτ' ἄλλο βλαστάνειν ἄτοπον, καὶ ταῦτ' ἀπὸ μιᾶς ὁρμῆς.
10 αἱ μὲν οὖν ἀπορίαι σχεδὸν αὗται καὶ τοιαῦται [4] εἴρηνται περὶ τούτων.[5]

11. 4 πρὸς δὲ τὸ τὴν αἰτίαν τοῖς εὐπορουμένοις [6] λέγειν πρῶτον ληπτέον, ὅπερ εἴρηται καὶ πρότερον, ὅτι πᾶν δένδρον ἀρχὰς πολλὰς ἔχει πρὸς τὴν βλάστησιν καὶ τὴν καρποτοκίαν· τοῦτο δ', ὥσπερ ἐλέχθη, 5 τῆς οὐσίας, ὅτι καὶ πολλαχόθεν ζῇ· διὸ καὶ βλαστητικόν. εἰ δ' οἱ [7] πρῶτοι τούτων ἀναγκαιότεροι, ἐκ τῶν ἀρχῶν δὲ τούτων οὔθ' ὅμοιοι πάντες οὔτ' ἴσοι οὔθ' ἅμα βλαστάνουσιν οἱ βλαστοί,

[1] πρῶται U: πρώϊαι Schneider; πορρώτεραι Wimmer.
[2] μηλέα Vasc.[2]: μελία U.
[3] ⟨καὶ⟩ Gaza (*et*), Vasc.[2]
[4] τοιαῦται u HP: τοιαῦτα U N a.
[5] Hindenlang would omit either αὗται καὶ τοιαῦται or εἴρηνται περὶ τούτων.
[6] εὐπορουμένοις U N H[ac]: ἀπορουμένοις H[css]P.
[7] εἰ δ' οἱ U N a: οἱ δὲ HP; εἰσὶ δ' οἱ Schneider.

---

[a] *Cf. CP* 5 1. 3–4.
[b] There are not two distinct annual periods of growth, as in most other trees, or two impulses, as with premature and normal sprouting, but there is a single activity extending uninterruptedly throughout the year.
[c] *CP* 1 3. 4: ". . . the side-shoots appear in all when a conflux of fluid accumulates in a certain spot and this on being warmed and concocted by the sun becomes as it were pregnant and brings forth offspring. In fact shoots are produced from the branches and other parts in the same way." *Cf. CP* 1 7. 4

ensues, just as premature sproutings come about in consequence in other trees.[a] But the citron and the like produce completed fruit and concoct it fully. This is why the divided performance of duties, one part sprouting now, another later, is odd, especially when a single impulse sets all this in motion.[b]

So the difficulties about these matters have been presented as being these (one may say) and of this description.

Sprouting at all Seasons: The Solution

11. 4 To give the reason that clears away the difficulties we must first take as premiss a point also made before,[c] that every tree has many starting-points for sprouting and fruiting. This (as was said) [d] is of the essence of a plant, that it also lives from a multitude of parts, which is why it can also sprout from them. It is true that the initial sprouts among them are of a character more determined by necessity; [e] but the sprouts that come from these are neither all of similar or equal size nor of simultaneous production, but are

and Aristotle, *On Length and Brevity of Life*, chap. vi (467 a 22): ". . . for everywhere in a plant it (*sc.* the starting-point) has both a root and a stem potentially."

[d] *HP* 1 1. 4: ". . . for a plant can sprout from any part, since it lives in each." *Cf.* Aristotle, *On the Parts of Animals*, iv. 6 (682 b 27–30): " Of necessity insects are segmented, for it is of their essence to have many starting-points, and in this they resemble plants. For like plants insects too can live when divided . . ."

[e] Root and stem must first germinate if there is to be a plant at all.

ἀλλ᾿ ὅταν ᾖ πλείων [1] συρροὴ καὶ ἰσχυροτέρα, κατὰ
10 ταῦτα θᾶττον καὶ πλεῖον· ἕκαστος γὰρ αὐτῶν
ὥσπερ φυτόν ἐστιν ἐν τῷ δένδρῳ, καθάπερ ἐν τῇ
γῇ. μὴ ἅμα δὲ τῆς βλαστήσεως οὔσης, μηδ᾿
ἴσης τῆς ἰσχύος, εὐλόγως οὐδὲ τῶν καρπῶν ἡ
11. 5 πέψις ἅμα γίνεται πάντων· ἐπεὶ καὶ ἡ θέσις διαφορὰν [2] ποιήσει τῶν μερῶν (οἷον ἢ πρὸς ἀνατολὰς ἢ
δύσεις, ἢ πρὸς ἄρκτον ἢ μεσημβρίαν)· ὥσπερ γὰρ [3]
καὶ ὅλων τῶν δένδρων, εἰ ἐν εὐδιεινῷ [4] τόπῳ (πρωϊ-
5 βλαστῆ γὰρ ταῦτα καὶ πρωΐκαρπα).

τοιαύτης δὲ τῆς φύσεως οὔσης τῆς κοινῆς,
ὅπου μὲν ὁ ἀὴρ μαλακὸς καὶ εὐδιεινός, ὀλίγον
χρόνον ⟨οὐ⟩ [5] βλαστητικὰ τὰ δένδρα γίνεται
(καθάπερ καὶ περὶ Αἴγυπτον)· ὅπου δὲ χειμέριος
10 καὶ σκληρός, πλείω.

11. 6 τὸ γὰρ ὅλον βραχεῖά τις ἂν γένοιτο διάλειψις [6]
⟨εἰ⟩ [7] ἥ τ᾿ ἐκ τῆς γῆς τροφὴ δαψιλὴς εἴη καὶ ὁ
ἀὴρ εὐκραής· ἐπεὶ καὶ τὰ μὴ φυλλορροοῦντα [8]
παρά τισιν (οἷον συκῆ καὶ ἄμπελος) διὰ τοῦτο οὐ
5 φυλλορροεῖ (καθάπερ εἴρηται), διότι διαρκὴς ἡ

[1] πλείων u HP: πλείω U N.
[2] διαφορὰν Ur: διαφθορὰν Uar.
[3] γὰρ U: γοῦν Wimmer.
[4] ἐν εὐδιεινῶι u (-νῷ HP): εν εὐδεινῶι U; ἐνευδινῷ N.
[5] ⟨οὐ⟩ Schneider (⟨μὴ⟩ later [vol. v, p. xxxvi]).
[6] διάλειψις HPc: -ηψις U N Pac(?).
[7] ⟨εἰ⟩ Gaza, Itali; ⟨εἴπερ⟩ Vasc.2
[8] φυλλοροοῦντα Uc from -ροῦν-.

---

[a] *Cf.* Aristotle, *On Youth and Age, Life and Death and Respiration*, chap. ii (468 b 9–10): "For such animals (*sc.* those

produced sooner and in greater extent with the greater extent and strength of the conflux, for each sprout is as it were a plant [a] growing in the tree as in the earth. Since their sprouting is not simultaneous, and again their strength not equal, it is reasonable that the concoction of fruit should also not be simultaneous in all of them. Indeed the position of the parts will make a difference in their sprouting, for instance if they face east or west, north or south, for it is as with the position of the whole tree when it is in a place where the weather is clear, trees so situated sprouting and fruiting early. 11. 5

Such being the nature common to all plants, we find that where the air is mild and clear there is only a brief interval when the trees do not sprout, as in Egypt,[b] but where the air is wintry and severe the interval is longer.

We can go further: there would be only the briefest interruption if the supply of food should be lavish and the air well-tempered; indeed the trees which do not shed their leaves in certain countries, such as the fig and vine,[c] retain them for the reason mentioned:[d] the supply of food is continuous. Now this 11. 6

that can live when divided) resemble many animals grown together."

[b] *Cf. HP* 3 5. 4: ". . . indeed in Egypt . . . the trees are practically always sprouting, or else the interval when they are not sprouting is brief."

[c] *Cf. HP* 1 3. 5: "So too with the distinction between deciduous and evergreen (*sc.* we must not take it too absolutely): thus at Elephantine it is said that not even the vine or the fig sheds its leaves;" *HP* 1 9. 5: "Some plants, not evergreen by their nature, are so because of the region where they grow, as we said of the plants at Elephantine and Memphis . . ."

[d] *CP* 1 10. 7.

τροφή. τουτὶ μὲν οὖν κοινὸν ἐπὶ πλειόνων ἐστίν· τὰ μὲν γὰρ διὰ τὴν φύσιν, τὰ δὲ διὰ τὸν τόπον, ἔχει τι ταὐτόν.[1]

διῃρημένων δὲ τῶν μὲν φυλλοβόλων, τῶν δὲ 10 ἀειφύλλων, αἰτίας δὲ οὔσης τοῖς ἀειφύλλοις τῆς διαρκείας τῆς τροφῆς ([δι'] [2] ὧν τὰ μὲν ⟨διὰ⟩ [3] τὴν ἰδίαν φύσιν, τὰ δὲ διὰ τὸν τόπον τοιαῦτα), τρίτον δὲ [4] καὶ ὥσπερ ἐφεξῆς τούτοις τὴν ἀειβλαστίαν θετέον, ὅτι τὸ ἀείφυλλον οὐκ ἀειβλαστές, 15 ἀλλ' ἐκεῖνα μὲν ὥσπερ διατηρεῖν μόνον δύνανται τὰ προϋπάρχοντα, ταῦτα δὲ καὶ προσεπιγεννᾶν 11. 7 ἕτερα [5] διὰ τὴν ἰδίαν δῆλον ὅτι φύσιν.[6] τοῦτο δὲ καὶ ἐν ἄλλοις μέχρι τινός· ἐπιβλαστάνει μὲν γὰρ τὰ μὲν ἅμα τοῖς ἄστροις, τὰ δὲ καὶ ἀορίστως, ὥσπερ καὶ ἄμπελος. ὃ δὴ τούτοις μέχρι τινός, ⟨καὶ⟩ [7] δι' ὅλου δέδωκεν ἡ φύσις, ὥστε καὶ καρποφυεῖν [8] καὶ 5 καρπογονεῖν.

οὔσης δὲ τοιαύτης τῆς οὐσίας, οὐδὲν ἄτοπον ἤδη τὰ μὲν τελεοῦν, τὰ δ' ἀνθεῖν, τὰ δὲ βλαστάνειν, τὰ δὲ μέλλειν, ἐπείπερ οὐδ' ἐν τοῖς ἄλλοις δένδρεσιν

[1] τί ταυτόν U: τοιαύτην Wimmer.
[2] [δι'] Gaza, Vasc.[2]
[3] ⟨διὰ⟩ a.
[4] δὲ U: δὴ Vasc.[2]
[5] ἕτερα u: ἑτεραν U (ἑτέραν N HP).
[6] φύσιν U[c]: φασιν U[ac].
[7] ⟨καὶ⟩ ego: ⟨ἐκείνοις⟩ Gaza (*illis*), Itali.
[8] καρποφυεῖν U (*cf.* η 119 τὰ μὲν φύει, ἄλλα δὲ πέσσει and the "potential fruit" of *CP* 1 12. 10): βλαστοφυεῖν Vasc.[2] (*semper germinent* Gaza).

continuity of supply is found in a greater number of trees than natural evergreens, since whereas some trees have an identical feature because of their nature, others have it because of the region.

Trees being divided into deciduous and evergreen, and the cause of being evergreen being the continuous supply of food (some plants being evergreen because of their distinctive nature, others because of the region), we must set up as a third character, coordinate with these, that of being ever-sprouting, because what is always in leaf[a] is not the same as what is always sprouting. Instead the former can only retain (as it were)[b] what it already has, whereas the latter can go further and generate fresh parts, evidently because of its own distinctive nature.[c] This power, up to a point, is also found in other trees. Thus whereas some put forth further sprouts at the rising of certain stars,[d] others do so at no fixed season, as does a vine.[e] What occurs in these only to a limited extent also has been given to a tree by its nature to do all the time, not only to begin the formation of fruit but to bear it.

11. 7

Such being the essence of the ever-sprouter, we now see that there is nothing odd in the simultaneous fruiting, flowering, sprouting and preparation for sprouting in the same tree. Indeed even in the rest the parts do not all go through each process simul-

[a] *Cf.* note *e* on *CP* 1 10. 7.

[b] *Cf. HP* 1 9. 7: " In evergreens the shedding and withering is gradual, for it is not the same leaves that persist always, but new ones are put out while the others wither."

[c] It is not due to the country, since there is no climate where the tree lacks this power.

[d] *Cf. HP* 3 5. 4, cited in note *c* on *CP* 1 10. 6.

[e] The " mad " vine: *cf. CP* 1 11. 3; 1 18. 4.

ἅπανθ' ἅμα τὰ μέρη· διαφέρει τε[1] τοῖς χρόνοις,
10 ὅτι τὰ μὲν ἄλλα παρ' ὀλίγον, τὸ δ'[2] ἐφεξῆς.

11. 8 ἀλλ' ἐκεῖνο μᾶλλον ζητητέον· τίς ἢ πόθεν ἡ κρᾶσις καὶ ἡ σύστασις τῶν τοιούτων· ἔοικεν δὲ παραπλήσιον, πλὴν χαλεπωτέρῳ,[3] καὶ τίς ἡ τῶν ἀειφύλλων, προσεπειπεῖν γὰρ δεῖ καὶ διὰ τί
5 τοσαύτην λαμβάνει καὶ δύναται πέττειν ὥστε γεννᾶν (ἡ γὰρ ἐπιβλάστησις καρποτοκίας γένεσις).[4] ἐπεὶ τό γε[5] πλείους ἅμα καρποὺς ἔχειν, τὸν μὲν τέλειον, τὸν δ' ἀτελῆ, τὸν δ' ὑποφυόμενον, συμβαίνει καὶ τῇ[6] ἀρκεύθῳ καὶ ἄλλοις ὧν βραδεῖά τε ἡ
10 ἔκπεψις καὶ δυσαπόπτωτος ὁ καρπός.

ἡ μὲν οὖν τούτων αἰτία μέχρι τούτων εἰρήσθω τὰ νῦν.

12. 1 πότερα[7] δ' ἡ βλάστη[8] καὶ αὔξησις ἅμα τῶν ἄνω τε γίνεται καὶ τῶν ὑπὸ γῆς, ἢ διῃρημέναι τοῖς χρόνοις;

§ 1. 1–11. Varro, *R.R.* i. 45. 2–3: sub terra et supra virgulta ne eodem tempore aeque crescunt. nam radices autumno aut hieme magis sub terra quam supra alescunt, quod tectae terrae tepore propagantur, supra terram aere frigidiore tinguntur. . . . nam prius radices quam ex iis quod solet nasci crescunt.

[1] τὲ U: δὲ Schneider.
[2] τὸ δ' ego (τὰ δ' Schneider): τῶ U.
[3] χαλεπωτέρωι U (-ω N HP): χαλεπωτέρως a; χαλεπώτερον Schneider.
[4] γένεσις ego (ἕνεκα Wimmer): γεννησις U.
[5] γε Vasc.[2]: τε U.
[6] τῇ Vasc.[2] (*cf. HP* 3 12. 3): τῶι U.
[7] πότερα N HP (ποτέρα u): ποτερα U.
[8] βλάστη HP (βλαστὴ u N): βλαστῆ U.

[a] And not merely leaves.
[b] *Cf. HP* 3 3. 8: "It happens that this (*sc.* the second

taneously; and there is a difference in the time, which in the rest is a very short interval, but here there is an uninterrupted succession.

11. 8 But what we should investigate is rather another question: what is the nature and source of the tempering of qualities and of the formation of ever-sprouters? The question when raised about evergreens appears similar, but as an easier question to a harder. For in answering the first we must also tell why the tree takes in and can concoct so much food that it keeps producing fruit [a] (the further sprouting being an initiation of fruit production). As for having fruit in several stages at the same time, some fully formed, some on the way, some just beginning, this is also found in Phoenician cedar [b] and other trees that are slow in completing concoction and have fruit that does not readily drop.

Thus far at present for the cause of these matters.

Is Growth Simultaneous Above Ground and Below?

12. 1 Do the parts above ground and those below increase and grow at the same or at different times?

Phoenician cedar) is the only tree practically speaking to carry its fruit for two years;" *HP* 3 4. 1: ". . . indeed even of those (*sc.* wild trees) that are later in bearing fruit—which some say take a year to bear it—as the Phoenician cedar and the kermes-oak, the sprouting occurs in spring;" *HP* 3 4. 5–6: ". . . for the Phoenician cedar is held to keep its fruit for a year, since the new fruit overtakes the fruit of the year before. And some say that it does not even ripen that fruit, which is why the fruit is taken from the tree unripe and kept for some time; whereas it dries up if left on the tree. The Arcadians say that the kermes-oak also takes a year to perfect its fruit, for it begins to show the new fruit while it is ripening last year's . . ."

ὥσπερ τινές φασιν τὰς μὲν ῥίζας αὐξάνεσθαι
5 μετοπώρου καὶ χειμῶνος, τὰ δὲ στελέχη καὶ τοὺς ἀκρεμόνας ἔαρος καὶ θέρους καὶ μάλισθ'[1] ὑπὸ ⟨τὸ⟩ ἄστρον·[2] τοῦτο δὲ κατὰ λόγον οὕτως, συμβαίνειν[3] γὰρ τῇ[4] πρώτῃ γενέσει κατακολουθεῖν· ἐν ἐκείνῃ γὰρ τὴν ῥίζαν πρότερον ἢ τὸν βλαστὸν
10 ἀφιέναι πάντα, προϋπάρχειν γὰρ ἀναγκαῖον ᾧ τὴν

12. 2 τροφὴν ἐπάξεται. φανερὸν δὲ καὶ ἐν ταῖς μετοπωριναῖς φυτείαις· τότε γὰρ ῥιζοῦσθαι[5] (οὐ βλαστάνει δὲ τὰ φυτευόμενα, ἢ[6] ἐπὶ βραχύ τι, ⟨κατὰ⟩[7] τὰ τοῦ ἀέρος),[8] εἰ γὰρ ἦν ἄρριζα, ἐσήπετ' ἄν. ἀλλὰ
5 διὰ τοῦτο ἐπαινοῦσι ταύτην τὴν φυτείαν, ὅτι μᾶλλον τὰς ἀρχὰς ἰσχυροτέρας ποιεῖ δι' ὧν καὶ ἡ τοῦ στελέχους καὶ ἡ τῶν ἄλλων γένεσις·[9] ὡς τά γ' εὐθὺς ἀνατρέχοντα πρὸς τὴν βλάστησιν ἀσθενῆ καὶ ἄκαρπα γίνεται (καθάπερ ἐπὶ τῶν σπερμάτων οἱ Ἀδώνιδος
10 κῆποι).

12. 3 πιθανὸς δὲ κἂν[10] ταύτῃ δόξειεν[11] ὁ λόγος· ὅτι

[1] μάλισθ' M a: μάλιστα θ' U N; μάλιστα HP.
[2] ⟨τὸ⟩ ἄστρον a: ἄστρων U^{ar} (-ον U^{r} N HP).
[3] συμβαίνειν Schneider, punctuating after it (συμβαίνειν· ἐν Vasc.[2]): συμβαίνει. ἐν U.
[4] γὰρ τῆι U: τῇ γὰρ Schneider.
[5] ῥιζοῦσθαι N: ριζοῦσται U (ῥ- u); ῥιζοῦται HP.
[6] ἢ u: ἡ U.
[7] βραχύ τι ⟨κατὰ⟩ ego (βραχύ τι Scaliger): βραχύτητα U.
[8] τοῦ ἀέρος U: τοῦ ἔαρος Palmerius; ⟨πρὸ⟩ τοῦ ἔαρος Wimmer.
[9] γένεσις N HP: -νν- U.
[10] κἂν ego: καὶ U.
[11] δόξειεν ⟨ἂν⟩ Wimmer.

---

[a] *Cf. HP* 1 7. 1: "In all plants the roots are held to grow before the upper parts . . .;" *HP* 8 2. 2: "In all (*sc.* cereal

A. *The Case for Priority of the Root*

So some say that the roots grow in autumn and winter, but the trunk and branches in spring and summer, especially in the dog days; and that this is reasonable, since it accords with the original generation, for then all send out the root before the shoot,[a] since the plant must first have the means to bring in its food. This (they say) can also be seen in autumn planting, for at that time the cuttings strike root (if they did not, they would decompose),[b] but do not sprout (or sprout for only a short time, depending on the weather). In fact this is why autumn planting is recommended: it does more to make the starting-points [c] stronger by whose agency both the trunk and other parts are produced, since the cuttings that run up at once turn out weak in the end and bear no fruit, like the gardens of Adonis [d] in the case of grains.

12. 2

A further consideration might also make their contention seem plausible: in winter the parts above

12. 3

and legume seeds) the root comes out a little before the stalk. But it happens in the cereal seeds at least that the shoot sprouts first in the seed itself, and as it grows the seeds split open (for all these seeds too are in a way double, whereas all legume seeds are visibly two-valved and composite), but the root pushes out of the seed at once. But in legume seeds this does not happen (*sc.* the previous sprouting of the stalk inside the seed) because root and stalk are on the same side of the seed, and the root comes out a little earlier than the stalk." *Cf.* also Aristotle, *On the Generation of Animals*, ii. 6 (741 b 36–37): ". . . for seeds send out the roots before they do the shoots."

[b] From the rains of autumn and winter.

[c] The roots.

[d] Also mentioned in *HP* 6 7. 3.

τὰ μὲν ἄνω κωλύεται διὰ τὸν πέριξ ἀέρα ψυχρὸν ὄντα, τὰ δὲ κάτω, στεγαζόμενα τῇ γῇ, καὶ ἅμα συγκατακλειομένου [ὕπο] τοῦ θερμοῦ[1] διὰ τὴν
5 ἀντιπερίστασιν, ἔτι τε ⟨τῷ τὴν⟩ ὑγρότητα καὶ τρέφειν,[2] προσαύξεται· πάντα γὰρ αὐτοῖς ὑπάρχει δι' ὧν ἡ αὔξησις καὶ γένεσις.[3] σημεῖον δὲ καὶ τὸ ἐπὶ τοῦ σίτου συμβαῖνον, ὃς ὑπὸ τοῦ χειμῶνος πιλούμενος ῥιζοῦται μᾶλλον, ὃ δὴ καρκινοῦσθαι
10 λέγουσιν, ὡς τὴν ἀπὸ τῶν ἄνω δύναμιν καὶ τροφὴν εἰς τὰ κάτω τρεπομένην.

12. 4 ταύτῃ μὲν οὖν δόξειεν ἂν μερίζεσθαι τὸ τῆς αὐξήσεως.

τῇδε δὲ πάλιν οὐκ ἂν δόξειεν· ἔν τε γὰρ τῇ πρώτῃ γενέσει προτερεῖ μὲν ἡ ῥίζα τῶν βλαστῶν,

§ 3. 2–4. *Cf.* Varro *R.R.* i. 45. 2–3 (cited on *CP* 1 12. 1).

[1] συγκατακλειομένου τοῦ θερμοῦ ego (ἀπὸ τοῦ θερμοῦ συγκατακλειομένου Schneider): συγκατακλειόμενα ὕπο τοῦ θερμοῦ U.
[2] ἔτι τε ⟨τῷ τὴν⟩ ὑγρότητα καὶ τρέφειν ego (ἔτι τε ⟨δι'⟩ ὑγρότητα καὶ τροφὴν Schneider; τῇ τε ὑγρότητι καὶ τροφῇ Wimmer): ἔτι τε ὑγρότητα καὶ τρέφην U; ἔτι τὲ ὑγρότητα καὶ τροφὴν u.
[3] ⟨ἡ⟩ γένεσις N HP.

---

[a] This is the theory of *antiperístasis* or "reciprocal displacement." Plato used it to avoid a vacuum and to account for the movement of inanimate things (*cf. Timaeus*, 59 A 1–8, 79 A 5–C 1) : A displaces B, B displaces C, and so forth until Z occupies the place left by A. Since like prefers like, the result of the shuffle is often that bodies of the same kind are massed together without intermixture. So the cold, when it prevails, presses out the hot until the hot is all in a mass, with no cold left in between. Aristotle often applies the theory to the concentrating effect of cold on heat or heat on cold: *cf. Meteorologica*, i. 10 (347 b 5–7), i. 12 (348 b 2, 349 a 8), ii. 4 (361 a 1–3),

ground are checked by the cold of the surrounding air, whereas the parts below keep growing, sheltered as they are by the earth; then too with their heat shut in, displaced by the cold,[a] and furthermore because the water [b] also feeds them, for they have at their disposal all the means of growth and generation.[c] A proof of this growth is what happens in grain, which roots better when it is compressed by the winter, a thing which is called "crabbing," [d] with the implication that power and food is diverted from the upper parts to the lower.

12. 4 All this, then, would make it appear that growth is carried on by taking turns.[e]

B. *The Case for Simultaneous Growth*

What follows would on the contrary make this appear not so.

The root, it is true, precedes the shoots when they are originally generated. But it does not precede

iv. 5 (382 b 8–10) [" for the cold is sometimes said both to burn and to heat, not as the hot does, but because it collects the hot and displaces it reciprocally "]; *On Sleep and Waking*, chap. iii (457 b 1–2, 458 a 25–30). Theophrastus often supplements the arguments of others with the theory, or appeals to it himself: *cf. CP* 1 13. 5; 2 6. 1; 2 8. 1; 2 9. 8; 6 7. 8; 6 8. 8; 6 18. 11–12.

[b] Greek winters are rainy.

[c] Shelter, heat and food.

[d] That is, tillering: *cf. HP* 1 6. 3; *CP* 3 21. 5; 3 23. 5. In the present passage a reason for the name is hinted at: the movement turns from the usual direction, as a crab walks sideways. But perhaps the word was suggested by the roots put out from the seed, like the legs extending from the body of a crab.

[e] This is very marked in some plants. In citrus trees the root system grows when the soil is colder than the air, and the top when the air gets warmer.

5 οὐ μὴν τοσοῦτον ὥστε χρόνου γίνεσθαι πλῆθος, ἀλλὰ βραχύ τι πάντων,[1] ὥσπερ καὶ ἐπὶ τῶν ζῴων ἡ καρδία καὶ τὰ περὶ τὴν καρδίαν· ἁπλῶς γὰρ ὡς εἰπεῖν ἡ φύσις οὐθέν, καθάπερ ἡ τέχνη, ποιεῖ κατὰ μέρος, ἀλλὰ πάντ᾽[2] ἀθρόα καταβάλλεται, 10 συντελεῖ[3] δὲ ἑτέρων ἕτερα πρότερον.

εἰ δὲ καὶ ἐν τῇ πρώτῃ γενέσει τοῦτ᾽ ἐπί τινων[4] ἀναγκαῖόν ἐστιν, ἀλλ᾽ οὔτι γε ἐν τῇ τροφῇ καὶ αὐξήσει, τὰ μὲν πρότερον τρέφεσθαι, τὰ δ᾽ ὕστερον, ὥσπερ οὐδ᾽ ἐπὶ τῶν ζῴων, ἀλλ᾽ ἅμα πως 15 μάλιστα πάντων,[5] καὶ ὁ ὅλος ὄγκος ὥσπερ[6] κατὰ συνέχειαν καὶ τρέφεται καὶ ἐπιδίδωσιν· ὃ καὶ κατὰ τὴν αὔξησίν ἐστι φανερόν.

12. 5 ἐπεὶ καὶ ἄτοπον εἰ[7] τὸ θρεπτικόν, ὃ δὴ διαπλάττει καὶ δίδωσιν τροφάς, διαιρεῖται κατὰ μέρη τὴν ἐνέργειαν, ἢ πάλιν εἴ τι[8] τῶν σωματικῶν τὸ ἐνεργοῦν, οἷον πνεῦμα ἢ πῦρ, οὐδὲ γὰρ ταῦτα

[1] πάντων U: πάντως Wimmer.
[2] πάντ᾽ u: πᾶν U.
[3] συντε | τελεῖ U.
[4] ἐπί τινων u HP: ἐπι τίνων U (ἐπὶ τίνων N).
[5] πάντων HP: πάντως U N.
[6] ὥσπερ Schneider: ὥστε U.
[7] εἰ U[r]: ἐπι U[ar]; ἐστὶ N; ἐστι HP.
[8] εἴ τι Gaza (*si quid*), Schneider: ἐπι U.

---

[a] A matter of weeks in some plants and even longer.

[b] The heart is produced before the rest: *cf.* Aristotle, *On the Generation of Animals*, ii. 1 (735 a 13–25); ii. 4 (739 b 33–740 a 23); ii. 6 (743 b 18–26); *On Youth and Age, Life and Death and Respiration*, chap. iii (468 b 28).

[c] That is, the blood-vessels, of which the heart is a part: *cf.* Aristotle, *History of Animals*, iii. 3 (513 a 24–25); *On the Parts of Animals*, iii. 4 (665 b 33–34).

[d] The root: *cf.* *CP* 1 12. 1.

them so greatly that any length of time intervenes, but in all by a very short interval,[a] as in animals with the heart [b] and its appurtenances.[c] For to put it in a word, nature does not, like art, make anything piecemeal, but lays down all the foundations together, although it finishes some things before others.

And even if this priority is necessary in the original generation of some parts,[d] it does not hold of feeding [e] and growth—that some parts are fed first, others later—any more than in animals either. Instead all parts (one might say) do so at more or less the same time in all, and the entire bulk not only feeds but also develops continuously as it were. This is also evident in the case of growth.[f]

12. 5 Indeed it would be strange if the nutritive faculty, which forms the plant and feeds what it has formed,[g] should exercise its activity by turning from one part to another; or again, supposing that what carries out this activity is something corporeal, as *pneuma* [h] or fire,[i] that this corporeal thing should do so, for it is

[e] "Feeding" (τροφή) is also "nurture," and implies more distinctly than our word "food" that what is fed has been previously engendered and is now being reared.

[f] Every part of a growing thing is seen to have grown: *cf.* Aristotle, *On Generation and Corruption*, i. 5 (321 a 2–3, 19–20, 321 b 32–322 a 33).

[g] The nutritive faculty is also the generative: *cf.* Aristotle, *On the Generation of Animals*, ii. 1 (735 a 15–20), ii. 5 (740 b 25–741 a 2); *On the Soul*, ii. 4 (416 a 19).

[h] For *pneuma*, literally "breath" or "wind," *cf.* A. L. Peck in his edition of Aristotle's *De Generatione Animalium* in the L.C.L., Introduction, p. liii and Appendix B, pp. 576–593.

[i] *Cf.* Aristotle, *On Sense*, chap. iv (441 b 27–442 a 2): the food is cause of increase and decline in growth insofar as it is hot or cold.

5 εἰκός· ἀλλ' ὅταν ἅμα ταῖς ὥραις κινηθῶσιν, ὁμοίως δι' ὅλων διήκειν τῶν φυτῶν. ἓν γάρ τι τὸ γεννῶν, οὐχ ὥσπερ Ἐμπεδοκλῆς διαιρεῖ καὶ μερίζει τὴν μὲν γῆν εἰς τὰς ῥίζας, τὸν δ' αἰθέρα εἰς τοὺς βλαστούς, ὡς ἑκάτερον ἑκατέρῳ [1] χωριζόμε-
10 νον, ἀλλ' ἐκ μιᾶς ὕλης καὶ ὑφ' ἑνὸς αἰτίου γεννῶντος, ὡς ἡ τῶν ὅλων σύστασις μικρὰν ἀεὶ περιλαμβάνει πρόσδεξιν [2] καὶ κατὰ τὰς τροφὰς καὶ κατὰ τὰς αὐξήσεις, ὥστε δ' ὅλαις ὥραις [3] χωρίζειν μέγα κομιδῇ [4] τὸ διάστημα τῶν χρόνων.

12. 6 ἡ δ' ὑπὸ τοῦ ψύχους κώλυσις τῆς βλαστήσεως ἀληθὴς μέν, οὐκέτι δὲ ποιεῖ μερισμὸν φυσικόν· ἀλλ' ὡσπερεὶ συμβαῖνόν τι[νι] τὸ προσαυξεστέρας [5] τὰς ῥίζας, ὡς [6] καὶ καθηκούσης ἐνίοτέ [τε] [7] γε
5 γίνεσθαι [8] τῆς ὥρας ὅταν ἀντικόψῃ [ὁ] [9] χειμών· ἡ μὲν γὰρ βλάστησις κατέχεται, τὸ δὲ τῶν ῥιζῶν οὐ κωλύεται διὰ τὴν ἀλέαν ὅταν ἤδη προορμώμεναι [10] τύχωσιν εἰς ἔκφυσιν καὶ [11] αὔξησιν [ἔρχεται]. [12] ἀλλὰ τοῦτο οὐ τῆς [13] φύσεως θετέον, ὃ

§5. 7–11. Empedocles, Frag. A70 Diels-Kranz, *Die Fragmente der Vorsokratiker*, vol. I[10], p. 296. 27–29.

[1] ἑκατέρωι U: ἑκατέρου Gaza (*ab altero*), Schneider.

[2] μικρὰν—πρόσδεξιν ego (μικρὰν ἀεί τινα λαμβάνει ἔφεξιν [ἐπίσχεσιν Hindenlang] Wimmer): μικρά. εἴπερ ἐλάμβανε πρὸς ἕξιν U.

[3] ὥραις Vasc.[2]: χώραις U.

[4] μέγα κομιδῇ HP: μετακομιδῇ U; μετὰ κομιδῇ N.

[5] ὡσπερεὶ—προσαυξεστέρας ego (ὡσπερεὶ συμβαίνοντι ἔοικε τὸ προσαύξεσθαι Schneider; ὥσπερ ἐν συμβαίνοντι τὸ προαύξεσθαι Wimmer): ὥσπερ εἰ συμβαίνοντι τινι προσαυξεστέρας U.

[6] ὡς ego: ὁ U.

[7] [τε] U[r] N HP.

[8] γε γίνεσθαι U N: γίγνεσθαι HP; γίγνεται Vasc.[2]

unlikely that these should operate in this way either. What is likely instead is that all these, when stirred to activity with the coming of the seasons, should pervade the whole plant equally. For what generates the plant is a single unit, and not divided as Empedocles divides it, letting earth work with the roots and aether [a] with the shoots, the generator being separate for each. No; the parts come from a single matter and are generated by a single cause. It is true that the formation of the whole plant involves, both in feeding and in growth, a constant small [b] accretion, but that this is such as to separate feeding and growth here from feeding and growth there by entire seasons is to make the intervals between accretions add up to far too long a time.

12. 6 True enough, the upper parts are checked from sprouting by the cold, but this produces a division of growth that is no longer natural. Instead the circumstance that here the roots grow more than the rest is as it were an accident, so that it even happens occasionally in the growing season, when a cold spell checks the plant: sprouting is stopped, but because of the warmth of the earth the roots are not held back once they happen to have already received the impulse to come out and grow. But we must not account as belonging to the plant's nature an occasional

[a] In Empedocles a synonym of fire.
[b] And so applying only to one part.

---

[9] [ὁ] ego.
[10] προορμώμεναι u: προορώμεναι U N HP.
[11] ⟨δὲ⟩ καὶ Schneider.
[12] [ἔρχεται (so U N HP)] ego (a variant of -έχεται in line 6); ἔρχονται Schneider; ἔρχεσθαι Wimmer.
[13] οὐ τῆς Gaza (*non*), Basle ed. of 1541: οὕτως U.

10 διακωλύειν ποτὲ συμβαίνει τι τῶν ἐκτός, ἀλλ' ὅσα τῇ ὁρμῇ γίνεται τῇ αὐτῆς.

12. 7 ὡραίων δὲ χειμώνων καὶ εὐδιῶν γινομένων ἅμα τά [1] τε ἄνω καὶ τὰ κάτω λαμβάνει τὰς αὐξήσεις, ἐπεὶ ὅπου ⟨ὁ⟩ ἀὴρ [2] εὐβλαστὴς καὶ μαλακὸς βραχύν [3] τινα χρόνον (ὥσπερ εἴρηται) διαλεί-
5 πουσιν αἱ βλαστήσεις, οὐ μεριζομένων τοὺς χρόνους. τοὺς δὲ χειμῶνας οὐ μόνον τὰς βλάστας, ⟨ἀλλὰ⟩ [4] καὶ τὰς ῥίζας, εὔλογον κατέχειν, εἴπερ ἀπὸ τῆς τοῦ ἡλίου θερμότητος καὶ αἱ τούτων αὐξήσεις καὶ γενέσεις, οὐκ εἰς πλεῖον καταλαμβά-
10 νουσαι [5] βάθος [6] ὧν [7] ὁ ἥλιος ἐφικνεῖται (πλὴν ἐάν που τόπος εὐδίοδος ᾖ καὶ μανὸς καὶ κενός). μὴ γὰρ τούτου [8] συμβαίνοντος [9] ἀπορήσειεν ἄν [10] τις εὐλόγως διὰ τί ποτ' οὐκ αὐξάνονται τοῦ χειμῶνος, τροφήν τε λαμβάνοντος [11] καὶ ἀποστέγοντος τοῦ
15 ψύχους.

τάχα δὲ καὶ τοῦτο κοινὸν ἀπόρημα καὶ ἐπὶ τῶν
12. 8 ἄλλων μερῶν· ἅπαν μὲν γὰρ τὸ ζῶν [12] τρέφεται, τὸ δ' ἐν ὁρμῇ τῆς αὐξήσεως ὂν καὶ αὔξεται· τὰ δὲ

6–8. Varro, *R.R.* i. 45. 3: neque radices longius procedunt nisi quo tepor venit solis. *Cf.* Pliny, *N. H.* 16. 129: quidam non altius descendere radices quam solis calor tepefaciat . . .

[1] τά U^r N HP: τάς U^ar.
[2] ⟨ὁ⟩ ἀὴρ Schneider: ἁὴρ U; ἀὴρ N HP.
[3] βραχύν u: βραχύ U.
[4] ⟨ἀλλὰ⟩ Gaza, Itali.
[5] καταλαμβάνουσαι u: -σιν U.
[6] βάθος U: βάθους Schneider.
[7] ὧν ego (ᾗ οὗ ἂν Schneider): ὃν U: οὗ u; ἂν N HP.
[8] τούτου u: τοῦτο U.
[9] συμβαίνοντος U^cc from -νουσι.
[10] ἄν U^r N HP: ἐν ἄν (?) U^ar.

impediment by something external, but only what comes from the impulse of that nature itself.

12. 7 But when cold weather and clear weather come in their season the parts above and parts below acquire their growth at the same time. Indeed in regions where the weather promotes sprouting, the sprouting is intermitted for only a brief interval (as we said),[a] and there is no separation of the times of growth.[b] It is reasonable not only that the shoots should be held back by winter, but that the roots should be held back as well, since growth and production of these too come from the heat of the sun, the roots going no deeper than the sun can reach (except where the soil offers easy passage and is loose and free from other plants).[c] For if this restriction to the sun's reach did not occur one could reasonably raise a problem: why do the roots not grow in winter, when the tree receives food and the cold seals in the heat?

12. 8 But perhaps the problem applies to the upper parts as well. For everything that lives feeds, and what has an impulse to grow grows as well; and plants,

[a] *CP* 1 11. 6.

[b] It is only in cold weather that the roots can grow without the rest, and here the cold weather is at a minimum.

[c] *Cf. CP* 3 3. 1 and *HP* 1 7. 1: ". . . no root goes down further than the sun reaches, for it is heat that generates. Nevertheless the following points contribute greatly to the depth of the root and even more to its length: the nature of the ground, when it is light and open-textured and yields easy passage . . . and what we see in cultivated plants, for when they have water they penetrate practically everywhere when the place is empty of other plants and there is nothing to oppose them."

---

[11] λαμβάνοντος U: -τα u.

[12] ζῶν Gaza (*quamdiu vivit*), Scaliger: ζῶιον U.

φυτὰ πανταχῇ, καὶ[1] τὰ νέα καὶ τὰ παλαιά, προορμᾷ πρὸς αὔξησιν.

5 εἰ μὴ ἄρα συμβαίνει τότε μὲν αὐξάνεσθαι τοὺς ὅλους ὄγκους, ἅμα δὲ τῇ ὥρᾳ διαγελώσῃ[2] τὰς ἐκβλαστήσεις γίνεσθαι· τοῦτο δὲ οἷον γένεσίς τις ἤδη· διὸ καὶ οὐκ ἄλογον ἐπ' αὐτοῖς[3] ὥσπερ κύοντα κατέχειν, ἢ συναθροίζοντα καὶ λαμβάνοντα 10 πρὸς τὴν κύησιν, ἐν οἷς ὄγκος γίνεται καὶ αὔξησις,

12. 9 εἶθ' ἅμα ταῖς ὥραις ἀποτίκτουσιν. ἔχουσιν γὰρ δή τινας οἱ κλάδοι καὶ οἱ ἀκρεμόνες ἐν ἑαυτοῖς ἀρχὰς ζωτικάς, αἳ διαθερμαινόμεναι τῇ ὥρᾳ προΐενται τοὺς βλαστοὺς καὶ μὴ ῥιζουμένων τῶν κάτω· 5 τοῦτο δὲ μάλιστα φανερὸν ἐν τοῖς ἀφαιρουμένοις φυτοῖς τῶν ἀμπέλων τε καὶ ἑτέρων καὶ συντιθεμένοις[4] ἐν πίθοις, προβλαστάνουσιν γὰρ οἱ βλαστοὶ τῆς ὥρας καθηκούσης· ὁτὲ δὲ καὶ τὰ πηγνύμενα κλήματα καὶ κράδαι καὶ χάρακες, ἐξ ὧν[5] ἄνωθεν 10 ἐβλάστησεν[6] ὅσον εἰς ἀρχήν, κάτωθεν δὲ οὐκ ἐρριζώθησαν, ὡς ἔχοντα μὲν ἐν ἑαυτοῖς ἤδη τὰς ἀρχὰς καὶ τὰς δυνάμεις, τροφῆς δὲ δεόμενα

[1] καὶ ego ([οὐκ] Gaza, Vasc.[2]): οὐκ U.
[2] διαγελώσῃ U[r] N (-η H)P: -ῃς U[ar].
[3] ἐπ' αὐτοῖς U: ἐν αὐτοῖς Gaza (*in se*), Schneider.
[4] συντιθεμένοις a: -ων U N HP.
[5] ἐξῶν U: ἐκ τῶν Heinsius (*parte sua superna* Gaza).
[6] ἐβλάστησεν U N HP: -αν u.

---

[a] For the distinction between generation or production (γένεσις) and growth *cf.* Aristotle, *On Generation and Corruption*, i. 5 (322 a 3–16).

[b] So the cutting of the vine is allowed to bleed: *cf. CP* 1 6.

both young and old, have in all their parts an impulse to growth.

Unless the answer is this: in winter a growth of the whole bulk takes place, but with the coming of mild weather the shoots are put forth. This production of shoots is no longer mere growth, but as it were a kind of generation.[a] So it is not unreasonable to suppose that in winter the upper parts, because they are (as it were) pregnant, hold back the shoots and keep them to themselves (or because they are collecting and taking in food for their eventual pregnancy), and that this involves an increase in bulk and so growth; after this, when the proper season arrives, they bring forth the shoots. 12. 9 For the twigs and branches contain within themselves certain starting-points of life, and these, warmed by the growing season, send out their shoots even when the lower parts have struck no root. This is most evident in cuttings taken from the vine and other trees and kept together in jars,[b] for the shoots come out part way when the season has arrived. So too occasionally the branches of vine or fig [c] and stakes of olive [d] set in the ground: from these there has been sprouting (enough for a start) above ground, although the pieces did not strike root below; this implies that whereas the pieces already contain within themselves the starting-points and the powers, they nevertheless require food after parturition, and

8. But it is kept from drying out by the jar and the presence of other cuttings in the bundle.

[c] *Cf. CP* 5 1. 4.

[d] *Cf. CP* 1 7. 4 and *HP* 2 1. 2: "Yet some say that it has happened that a stake of olive set in the ground as a prop for ivy grew with the ivy and became a tree . . ."

μετὰ τὸν τόκον, ἧς μὴ γινομένης καταξηραίνεται.
12. 10 διὸ καὶ οὐκ ἐοίκασιν κακῶς λέγειν οἱ φάσκοντες εὐθὺς ἀνθεῖν τοὺς νέους καρποὺς ἔτι τῶν ἑτέρων ἐπόντων· τοῦτο γὰρ δῆλον ὡς δυνάμει λέγουσιν.

τὰς [1] μὲν οὖν αὐξήσεις εἴη μὲν ἂν ἀμφοτέρως 5 συμβαίνειν, οὐ μὴν ἀλλὰ διὰ ταῦτά γε μᾶλλον ἄν τις ὅλων τῶν δένδρων ⟨ἢ⟩ [2] μεριζομένας ὑπολαμβάνοι γίνεσθαι.

13. 1 ζητήσειε δ' ἄν τις ἐκ τῶν μικρῷ πρότερον εἰρημένων [3] πότερον τὰ δένδρα κατὰ χειμῶνα κύει πρὸς καρπογονίαν, τοῦ δ' ἦρος ἀποτίκτει, καὶ τοῦθ' οἷον περίοδός ἐστιν χρόνοις ὡρισμένη,[4] καθάπερ 5 τοῖς ζῴοις, ἢ [5] διὰ τὴν ἔνδειαν τῆς τροφῆς συμβαίνει καὶ διὰ τὴν ψυχρότητα τοῦ ἀέρος.[6] εἰ γὰρ ἐν τοῖς εὐδιεινοῖς καὶ μαλακοῖς αἰεὶ βλαστάνουσιν, οὐκ ἂν εἴη [7] τεταγμένη [ἐν] [8] τοῖς χρόνοις

[1] τὰς Gaza, Vasc.²: ὡς U.
[2] ⟨ἢ⟩ Vasc.²
[3] εἰρημένων u: -ου U.
[4] ὡρισμένη Vasc.²: -νοις U.
[5] ἢ u: ἧ U.
[6] ἀέρος u: ἀέρου U.
[7] εἴη Schneider: εἴ U; ἧ u N; ἦν HP.
[8] [ἐν] Vasc.²

[a] Homer, *Odyssey* vii. 117–119 (cited in note *f* on *CP* 1 11. 1) and Empedocles (cited in *CP* 1 13. 2).
[b] *CP* 1 12. 4–9.
[c] *CP* 1 12. 8.
[d] Literally, " pregnant for fruit-generation." The fruiting

without it dry out. (This is why it appears that those[a] who speak of the new fruit as already in flower while the old is still on the tree have not spoken ill, since they evidently mean that the flowering is potential.) 12. 10

In conclusion, growth could occur in both ways. These last considerations[b] nevertheless would make one suppose that it belongs to the tree in its entirety rather than at separate times to the lower and upper parts.

Sprouting: Is There a Fixed Period of Gestation?

13. 1 In connexion with what was said a short while before[c] one might enquire whether during winter trees are pregnant with their produce[d] and bring forth in spring, and that this constitutes a cycle with fixed times of gestation and delivery, as in animals,[e] or whether what happens is due to their insufficient supply of food[f] and the coldness of the air.[g] For if in regions of clear weather and mild climate the trees sprout continually, their pregnancy would not be fixed in the times of its occurrence, or else it would not be fixed by the familiar seasons in such a way that

shoot is called καρπός ("fruit") even before it flowers or bears.

[e] Aristotle compares crop production in a plant to multiple pregnancy in an animal: *On the Generation of Animals*, i. 18 (723 b 9–11): "Furthermore, since some animals produce many young from a single union (plants in fact do this without exception, for it is evident that they bear their whole annual crop from a single impulse) . . ."

[f] In summer, when the rains cease.

[g] In winter.

ἡ κύησις, εἰ δὲ μή, οὔτι[1] γε ταύταις ταῖς ὥραις ὡς
10 ὅλον ἅμα βλαστάνειν πάλιν καὶ εἴ τι τὸ ὀργῶν
ἀποτίκτειν.

13. 2 καὶ εἴ γε[2] συνεχῶς ὁ ἀὴρ ἀκολουθοίη τούτοις
ἴσως οὐδὲ τὰ παρὰ τῶν ποιητῶν λεγόμενα δόξειεν
ἂν ἀλόγως[3] ἔχειν, οὐδ' ὡς Ἐμπεδοκλῆς ἀείφυλλα

καὶ ἐμπεδόκαρπά

5 φησιν θάλλειν

καρπῶν ἀφθονίῃσι[4] κατ' ἠέρα[5] πάντ' ἐνιαυτόν

ὑποτιθέμενός τινα τοῦ ἀέρος κρᾶσιν, τὴν ἠρινήν,
κοινήν. ἐκεῖνο δ' ἄν τις ἴσως ἐν τούτοις ἀπορή-
σειεν· πότερα καὶ πεπάνσεις ὅμοιαι τῶν καρπῶν
10 ἢ ἐνδεέστεραι γίνονται, ἀσθενεστέρου ὄντος τοῦ
θερμοῦ καὶ τῆς ὥρας ὑγροτέρας; τοῦτο μὲν οὖν
ὡς καθ' ὑπόθεσιν θεωρείσθω.

13. 3 ἐν δὲ τῇ νῦν περιόδῳ τῶν ὡρῶν ἔοικε τὰ δένδρα,
κενωθέντα τοῦ θέρους ἐκ τῆς βλαστήσεως καὶ τῆς
καρπογονίας, ἀντιπληροῦσθαι πάλιν, εἶτ'[6] ἐκ
ταύτης τῆς ἀντιπεριστάσεως ἀποτίκτειν καὶ βλα-
5 στάνειν κατὰ[7] τοὺς ἱκνουμένους καιρούς, ἔχοντά
πως τὴν κύησιν καὶ ⟨ἐν⟩[8] τοῖς μέρεσιν καὶ ἐν τοῖς
ὅλοις.

§ 2. 3–6. Empedocles Frag. B77–78, Diels-Kranz, *Die Fragmente der Vorsokratiker*, vol. i[10] p. 339; ⟨δένδρεα δ'⟩ ἐμπεδόφυλλα καὶ ἐμπεδόκαρπα τέθηλεν καρπῶν ἀφθονίῃσι κατ' ἠέρα πάντ' ἐνιαυτόν.

[1] οὔτί U[cc] (ί from ε).
[2] καὶ εἴ γε ego (εἰ δὲ καὶ Vasc.[2]): και εἴτε U.
[3] ἂν ἀλόγως Gaza (*absurdum sit*), Itali: ἀναλόγως U.
[4] ἀφθονίῃσι U[r]: -ιν U[ar].
[5] κατ' ἠέρα Vasc.[2]: κατῆρα U.
[6] εἶτ' u HP: εἴτ' U N.

the whole tree sprouts simultaneously once more and the parts that have the impulse to do so bring forth fruit.

13. 2 And supposing our own trees favoured by uninterrupted mildness in the air, perhaps even what the poets [a] say would not sound unreasonable, or even Empedocles'[b] words about trees evergreen

> with never-failing fruit
> Bearing profuse year-long, so mild the air

when he supposes a certain vernal tempering of qualities in the air common to all seasons. But with regard to our own trees we might find another point difficult: What of the ripening of the fruit? Is the ripening under those circumstances equal to the ripening now? Or is it worse, since the heat then is weaker than that of our summer and there is more rain? But so much for this matter, which is to be considered as hypothetical.

13. 3 In the present round of seasons, on the other hand, it appears that trees are emptied in summer in consequence of their sprouting and production of fruit, and then are replenished with food again, and as a result of this counter-displacement bring forth fruit and sprout at their proper times, and that this pregnancy is present in a way both in the parts and in the entire tree.[c]

[a] Homer, *Odyssey* vii. 114–121, cited in note *f* on *CP* 1 11. 1.
[b] See Testimonium 3–6.
[c] *Cf. CP* 1 12. 10.

---

[7] κατὰ Gaza, Vasc.[2]: καὶ U.
[8] ⟨ἐν⟩ HP.

εὐλόγως δὲ καὶ αἱ φυτεῖαι καὶ ⟨αἱ⟩ [1] ἐπιβλαστήσεις γίνονται κατὰ πλείους ὥρας. δοκοῦσιν δὲ
10 ἐναντία πως [2] εἶναι μετόπωρόν τε καὶ ἔαρ καὶ Κυνὸς ἐπιτολή (ἡ γὰρ περὶ [3] τροπὰς βραχεῖα, ἀλλὰ μετ' Ἀρκτοῦρον ἐπιβλαστάνει πλείω καὶ ὑπὸ Κύνα).

13. 4 τὸ μὲν οὖν ἔαρ οὐδὲ θαυμάζεται, ζωτικωτάτη γὰρ ἡ ὥρα καὶ μάλιστα γόνιμος, ὑγρά τε οὖσα καὶ θερμή· τὸ δὲ θέρος ὥσπερ [οἱ] [4] ἐναντίον, ξηρά τε γὰρ καὶ ἐμπυρωτάτη, καὶ μάλισθ' [5] ὑπὸ τὸ ἄστρον.
5 ἔτι δὲ τὸ μετόπωρον οὐ μόνον ξηρὸν ἀλλὰ καὶ ψυχρόν, ἤδη μεταβάλλον, ἐπιλαμβανούσης τῆς ὥρας, ἐναντιωτάτη δὲ βλαστήσει ψυχρότης καὶ ξηρότης. ἀλογία μὲν γὰρ [6] δή τις φαίνεται διὰ τούτων.

10 οὐ μὴν οὔτε πρὸς τὰς βλαστήσεις ἐναντία ταῦτ' ἐστὶν οὔτε αὖ ἀσύμφωνα κατὰ τὰς δυνάμεις, ἀλλ'
13. 5 ἔχοντά τινα ὁμοιότητα. δεῖ γὰρ δὴ τὴν ὥραν ὑγρότητά τινα καὶ θερμότητα ἔχειν, ὥσπερ καὶ τὸ ἔαρ. αὕτη μὲν ὁμολογουμένη μάλιστα πρὸς βλάστησιν, ἐν ἀμφοῖν δὲ τοῦτο συμβαίνει, καὶ ἐν τῇ τοῦ

[1] ⟨αἱ⟩ HP.
[2] ἐναντία πως HP: ἐναντίαν ὡς U N.
[3] περὶ Vasc.[2]: επι U.
[4] ὥσπερ [οἱ] HP: ὥσπερ οἱ U; ὡσπερεὶ u; ὡς περὶ N.
[5] μάλισθ' ego (taking οἱ in line 3 as a misread θ'): μάλιστα U.
[6] [γὰρ] Gaza, Schneider.

---

[a] For the association of planting with the additional annual sprouting *cf*. *CP* 1 6. 3; *CP* 3 2. 6–3 3. 4.

[b] Spring is not a season of additional annual sprouting, but of planting and of the first annual sprouting.

Sprouting: Explanation of the Three Seasons

It is moreover reasonable that planting and the additional annual sprouting [a] should occur at more than one season. Yet autumn, spring [b] and the dog days are held to be somehow contrary to one another (as for the season at the summer solstice,[c] it is very brief, and more trees have additional sproutings after the rising of Arcturus and in the dog days).

13. 4 That spring should be such a season is not even felt as a problem, for this season is the greatest furtherer of life and procreation, being both wet and warm. Summer on the other hand is an opposite (as it were), a dry and most torrid season, above all in the dog days. So too autumn is opposite: it is not only dry but already turning to cold as the season advances, and coldness and dryness are most unfavourable to sprouting. It is these considerations that make the problem felt.

Nevertheless these other two seasons are neither unfavourable to sprouting nor again at variance with spring in their effects, but bear a certain similarity to it. 13. 5 For the season must possess some fluid and warmth, like spring. Spring is the season admittedly most favourable to sprouting, yet wetness and warmth are found in both of the other seasons, in the dog days

[c] *Cf. HP* 1 9. 7: "In evergreens the loss and desiccation of the leaves is gradual; for the same leaves do not always remain. Instead there is an additional sprouting of some while others wither. This happens mainly around the summer solstice." *Cf.* also *CP* 2 19. 2.

5 *Κυνὸς ἐπιτολῇ ⟨καὶ⟩*[1] *μετ' Ἀρκτοῦρον· ὑπὸ γὰρ αὐτὸ τὸ ἄστρον, καίπερ ὄντος ἐμπύρου τοῦ ἀέρος, ὅμως καὶ νότια πνεῖ καὶ νέφη συνίσταται καὶ αὐτὰ τὰ δένδρα διυγραίνεται φανερῶς καὶ ὑπὸ τὸν φλοιὸν αὐτῶν διαδίδοταί τις ὑγρότης, ὅθεν καὶ ῥοαὶ*[2] *καὶ*[3] 10 *κατὰ τοῦτον τὸν καιρόν, εἴτ'*[4] *οὖν συνελαυνομένου τοῦ ὑγροῦ καὶ ἀντιπεριστάσεως γινομένης, εἴτε δι'* 13. 6 *ἄλλην αἰτίαν· πλὴν συμβαίνει γε τοῦτο καὶ τοῖς ἀνθρώποις· διὸ καὶ κοιλίαι*[5] *μάλιστα λύονται, καὶ πυρετοὶ πολλοὶ γίνονται, καθυγραινομένων τῶν σωμάτων. δοκεῖ δὲ καὶ ἡ γῆ τότε καθυγράνθαι* 5 *μᾶλλον· ὅθεν καὶ ὑδάτων ἀναδόσεις*[6] *καὶ ἕτεραι*[7] *μεταβολαὶ γίνονται πλείους. ἀλλὰ δι' ἣν μὲν αἰτίαν ἕκαστα συμβαίνει τούτων ἕτερος λόγος· ὅτι δὲ ἐξυγραινομένων τῶν φυτῶν καὶ τοῦ ἔξωθεν ἀέρος οὐκ ἀντιπίπτοντος ἡ ἐπιβλάστησις οὐκ ἄλογος* 10 *φανερὸν ἐκ τῶν εἰρημένων.*

13. 7 *τὸ δὲ μετόπωρον οὐχ, ὥσπερ ἐλέχθη, ξηρὸν καὶ ψυχρόν ἐστιν, ἀλλὰ μᾶλλον θερμόν, ἅμα δὲ*[8] *ἐν*[9] *τῇ τοῦ ἄστρου μεταβολῇ*[10] *[ἐν]*[11] *τῷ ἀέρι γίνεται, διὸ καὶ ὥσπερ μῖξίν τινα συμβαίνει γίνεσθαι τοῦ ὑγροῦ*

[1] ⟨καὶ⟩ u: U N HP omit.
[2] ῥοαὶ Itali: ῥόαι U.
[3] καὶ U N: HP omit; τισι Wimmer.
[4] εἴτ' N HP: ἦτ' U.
[5] καὶ κοιλίαι H (καὶ αἱ κοιλίαι Schneider): αἱ κοιλιαι καὶ U (αἱ κοιλίαι καὶ u N); αἱ κοιλίαι P.
[6] ἀναδόσεις U^r (-δώ- U^{ar}): -δύ- N HP.
[7] ἕτεραι u: αἰτέραι U.
[8] δὲ U: γὰρ Wimmer.
[9] [ἐν] Schneider.
[10] μεταβολῇ N (*simul ac syderis facta mutatio sit: aer humidus*

and after the rising of Arcturus. For in the actual dog days, although the air is torrid, yet south winds blow,[a] clouds form and the trees themselves become noticeably fluid and a certain fluidity is transmitted under the bark. In consequence at this time too there is in certain trees a flow of fluid,[b] whether due to the concentration of the fluid by reciprocal displacement [c] or to some other cause. At all events this also occurs in man, and this is why the bowels are loosest at this time and there is a great incidence of fevers, since the body becomes fluid. It is held that the earth too is then fuller of fluid; so waters burst forth and a number of other changes occur. But to give a cause to each of these occurrences belongs to another discussion. What has been said, however, makes it evident that further annual sprouting is not unreasonable when the plants become fluid and the air outside offers no opposition. 13. 6

Autumn is not (as was said) [d] dry and cold. It is rather hot than cold, and with the season of Arcturus a change occurs in the air, which is why a certain mixture (as it were) occurs of the fluid with the hot, 13. 7

[a] *Cf.* Theophrastus, *On Winds*, chap. viii. 48 and [Aristotle], *Problems*, xxvi. 12 (941 a 37): "Why does the south wind blow in the dog days . . ." *Cf.* also the calendar excerpted in the Introduction (p. xlix), where Eudoxus mentions the morning rising of Sirius under Cancer 27 (July 17), and under Cancer 31 (July 21) says that the south wind blows.

[b] *Cf.* *HP* 9 1. 6: "Frankincense and myrrh trees are reported incised during the dog days and on the hottest days of his period; so too with the Syrian balsam."

[c] *Cf.* note *a* on *CP* 1 12. 3.

[d] *CP* 1 13. 4.

satis redditur Gaza; μεταβολῇ ⟨ὑγρότης⟩ Itali): μεταβολῆι U (-ῇ HP).

11 [ἐν] ego.

5 καὶ τοῦ θερμοῦ, καθάπερ καὶ ἐν τῷ ἦρι προσγίνεται τὸ θερμὸν ἀπὸ τῆς ὥρας· ἐνταῦθα δέ, τοῦ θερμοῦ προϋπάρχοντος, ὑγρότης ἐπιγίνεται διὰ τὴν πύκνωσιν καὶ κατάψυξιν τοῦ ἀέρος.

13. 8 εὐλόγως δὲ καὶ τὰ μὴ κάρπιμα, καὶ ὅλως τὰ νέα τῶν [1] πρεσβυτέρων, ἐπιβλαστικώτερα, καὶ τὰ ἐν ταῖς ὑγραῖς καὶ χειμερίοις χώραις.[2] τὰ μὲν γὰρ κάρπιμα καὶ πρεσβύτερα ξηρότερα (τὰ μὲν εἰς 5 τοὺς καρποὺς καταναλωκότα, τὰ δὲ καὶ τῇ φύσει τοιαῦτα)· τὰ δὲ ἄκαρπα καὶ νέα καὶ ὑγρότητα ἔχει καὶ θερμότητα [ἔχει] [3] πλείω. πάλιν δέ, ὅπου μὲν οἱ τόποι χειμέριοι καὶ ὑγροὶ τυγχάνουσιν καὶ ὁ ἀὴρ εὔπνους, ἐνταῦθα καὶ τὰ μετοπωρινὰ 10 γίνεται μακρὰ καὶ ὑγρὰ καὶ καλά· πολλάκις δὲ καὶ ὕδατα θερινὰ (κατά γε τὰς πλείστας). ὥσθ' ὑγροῦ τοῦ ἀέρος ὄντος, ἐπιγινομένης [4] ἑτέρας,[5] εὐβλαστότερα γίνεται καὶ εὐαξῆ μᾶλλον.

13. 9 ἐπεὶ καὶ τὰ δοκοῦντα δικαρπεῖν μηλεῶν [6] τέ τινα

[1] τῶν ⟨καρπίμων καὶ⟩ Vasc.[2]
[2] χώραις Gaza: ὥραις U.
[3] [ἔχει] HP.
[4] ἐπιγινομένης U^{t}: -οις U^{c} (οι ss.); -ου N HP.
[5] ἑτέρας ego (*aeris ea ipsa humiditate plenius germinant* Gaza; θέρμης Schneider): ἑτέρως U; ἕτερως N; ἑτέρου H^{t}P; ἑτέρα H^{css}.
[6] μηλεῶν HP: μηλιῶν U N.

---

[a] To the wetness already present from winter.

[b] *Cf.* Aristotle, *Meteorologica*, i. 12 (348 b 28–29): ". . . in spring it (*sc.* the air) is still fluid, and in autumn it is already turning fluid."

[c] The summer is not the dry and torrid Mediterranean summer, but that of more northern climates. "Winter" in Greek suggests rain almost as much as it does cold.

just as in spring, where heat is added by the season.[a] But in the present case it is heat that is already present, and the fluid is added by the condensation and chilling of the air.[b]

13. 8 It is also reasonable that trees bearing no fruit, and indeed all young trees as compared to older ones, should do more of this further annual sprouting, and so too trees in humid and winter-like[c] climates. For those with fruit and the older ones are drier, the former having spent their fluid on the fruit, the others being dry by their nature; whereas the ones without fruit and the young have more of both fluid and warmth. Again, where the country has a winter-like and humid climate and the winds are temperate, the autumns turn out to be long and humid and fine; often there are also summer rains (at least in most of these places). So that the air being humid,[d] and further fluid being added,[e] the trees are better sprouters and grow taller.[f]

13. 9 In fact the kinds of apple and pear tree[g] that are

[d] The summer is humid and there is commonly rain.

[e] With the autumn rains.

[f] *Cf. HP* 3 5. 4: "The sproutings that take place in the dog days and at the rising of Arcturus, after the spring sprouting, are common to practically all trees, although more evident in the cultivated, among these above all in the fig, vine, pomegranate and in general all that are good feeders and grow in soil providing abundant food. This is why the most abundant sprouting at the rising of Arcturus is said to occur in Thessaly and Macedonia; for here it also happens that the autumn turns out fine and long, and thus the mildness of the weather contributes."

[g] *Cf. HP* 1 14. 1 (of trees bearing from this year's shoots, from last year's and from both): ". . . from both last year's and this year's shoots bear certain twice-bearing apples and other fruit trees . . ."

γένη καὶ ἀπίων ἐν τούτοις μάλιστα γίνεται τοῖς τόποις, ἅτε παρεμπιπτούσης ἐπὶ πολὺ τῆς ὥρας· καὶ ταχὺ τῶν πρότερον [1] ἀφαιρουμένων, τὸ [2] καὶ
5 ⟨μὴ⟩ [3] πρωϊκαρποῦν [4] ὅμως διφορεῖ,[5] θᾶττον γὰρ πάλιν ἀναπληροῦνται ⟨καὶ κυΐσκονται⟩.[6] διὸ καὶ χρῶνταί τινες αὐτῷ [7] τῶν πρὸς τὴν ἀγορὰν βλεπόντων [καὶ κυΐσκονταί τινες αυτῶν προς τὴν ὥραν βλεπόντων].[8] οὐ μὴν ἀλλ' οὐδέν γε (ὡς
10 εἰπεῖν) τῶν δικαρπούντων [9] ὅμοιον γίνεται τοῖς ἐξ ἀρχῆς, ἀλλ' ὡς ὁρμὴν μόνον τοῦ φυτοῦ λαμβάνοντος, οὐ συνεκτελέσαντος δὲ τοῦ ἡλίου καὶ τοῦ

13. 10 ἀέρος. καὶ ἐφ' ἑτέρων ἔτι μᾶλλον συμβαίνει· προφαίνει γὰρ καρπὸν καὶ ἡ ῥόα καὶ ἡ μύρρινος, ἀλλὰ μέχρι τοῦ προδεῖξαι μόνον.

ἐπιβλαστικώτατα [10] δ' οὖν ὡς εἰπεῖν τὰ μάλιστα
5 εὐβλαστῆ τῇ φύσει, πλὴν εἴ τι διὰ ξηρότητα κεκώλυται,[11] καθάπερ καὶ ἡ ἀμυγδαλῆ καὶ εἴ τινων ὑπόγυος ἡ ἀφαίρεσις τῶν καρπῶν πρὸς τὴν τῆς ὥρας ἐπιβλάστησιν.

[1] τῶν πρότερον ego: τὸν πρότερον U; τῶν προτέρων u.

[2] τὸ ego: ὁ U; ὃ u.

[3] ⟨μὴ⟩ ego.

[4] πρωϊκαρποῦν HP: προϊκαρπεῖν U N.

[5] διφορεῖ Vasc.[2]: διαφέρει U.

[6] ἀναπληροῦται (sic) ⟨καὶ κυΐσκονται⟩ Schneider: ἀναπληροῦται U.

[7] αὐτῷ HP: αὐτῶν U N.

[8] [καὶ—βλεπόντων] Wimmer.

[9] δικαρπούντων u: διακαρποῦντων U.

[10] ἐπιβλαστικώτατα Schneider: ἔπειτα βλαστικώτατα. βλαστηκώτατα U.

[11] κεκώλυται ego: καὶ κω(κο- N)λύεται U N; κωλύεται HP.

held to bear twice a year grow mainly in these parts, since a long fruiting season intervenes.[a] And if the first crop is harvested without delay even a tree that does not bear early will yield a second, since the quick harvest lets the tree fill up again more rapidly and become pregnant, and this is why some persons harvest early with an eye to the market. Yet hardly any twice-bearer yields a second crop as good as the first, but only the sort of fruit that is found when the tree has only made a start without the help of the sun and air to bring the product to completion. This happens still more in other trees; so both the pomegranate and the myrtle[b] give promise of a second crop but go no further. 13. 10

At all events those trees have the best second annual sprouting (one might say) that by their nature sprout most readily, except for cases where the tree is prevented from doing so by its dryness, as with the almond[c] or where the fruit is harvested too late for the tree to take advantage of the second growing season.

[a] Between the heat of summer and the cold weather.

[b] *Cf. HP* 1 14. 1: " Trees also differ in fruiting in this respect: some bear from this year's shoots, some from last year's, some from both. Bearers from this year's shoots are the fig and vine; from last year's olive, pomegranate, apple, almond, pear, myrtle and practically all of this description (*sc.* that have dry shoots)—if it should happen that one of them gets pregnant and flowers from this year's shoots, this too taking place in a few, as in the myrtle, and especially one might say when they sprout after the rising of Arcturus, the trees are unable to perfect the fruit, which perishes half-formed . . ."

[c] *Cf. CP* 1 3. 2: the branches of the almond are too dry to serve as cuttings.

13.11 ἐὰν οὖν μακρὸν γίνηται τὸ μετόπωρον, οὐκ ἄλογον καὶ ῥόδα γίνεσθαι καὶ ἄλλ' ἄττα τῶν τοιούτων (ὥσπερ καὶ περὶ Δῖόν[1] φασιν τῆς Μακεδονίας), οὐκ ἰσχυρᾶς[2] γε δεόμενα τῆς 5 πέψεως· χρόνον δὲ λαβόντα ἱκανὸν ἐξέφηνεν τὸ ἄνθος. ὅλως δέ, ὃ πλεονάκις εἴρηται, μαλακοῦ καὶ ὑγροῦ τοῦ ἀέρος, καὶ τὸ σύνολον εὐκρᾶτος,[3] αἰεὶ δυνατὸν βλαστάνειν οὐ πάντα, ἀλλ' ἔνια τῶν δένδρων, τὰ δ' ἐλάττω, τούτων ἔτι μᾶλλον. ὃ καὶ 10 νῦν ἐπί τινων στεφανωμάτων συμβαίνει, τόπους

13.12 ἐχόντων εὐσκεπεῖς καὶ προσείλους. διατελεῖν γὰρ ἀνθοῦντα δοκεῖ καὶ ἡ οἰνάνθη καὶ τὸ ἴον τὸ μέλαν καὶ ἄλλ' ἄττα, θεραπείας δέ τινος προσγενομένης ἔτι μᾶλλον. ὅτι δὲ μεγάλην ῥοπὴν ὁ τόπος παρ-5 έχεται πρόσειλος ὢν καὶ εὐσκεπής, καὶ αὐτὰ τὰ δένδρα μαρτυρεῖ· φύεται γὰρ ἐν τούτοις ἔνια κατὰ τὴν ἄλλην χώραν οὐ φυόμενα, καὶ καρποφορεῖ τῶν ἄλλων οὐ φερόντων, καὶ προανθεῖ καὶ προβλαστάνει πρότερον τῶν λοιπῶν.

10 ὑπὲρ μὲν οὖν τούτων ἱκανῶς εἰρήσθω.

[1] περὶ δίον HP: περίδιον U N.
[2] ἰσχυρᾶς (*vehementiore* Gaza), Vasc.[2]: ισχυρῶς U.
[3] εὐκρᾶτος ego: εὐκράτου U.

---

[a] That is, of the lesser plants.
[b] *CP* 1 11. 6; 1 12. 7; 1 13. 2.
[c] *Cf. HP* 6 8. 2: ". . . whereas the violet, as we said (*HP* 6 6. 2), remains throughout the year, if tended. So too does dropwort . . . if one pinches off the flowers and does not allow them to go to seed, and if moreover the plant has a sunny position."
[d] *Cf. HP* 3 4. 1 (of the sprouting of wild trees): "Trees of the same kind differ among themselves in sprouting earlier

Second Blooming in Lesser Plants

13. 11 Consequently if the autumn is a long one it is not unreasonable that roses too and certain other crops of the sort [a] should be produced again, as is reported to happen at Dium in Macedonia, needing as they do no great power for their concoction: when the plant gets time enough, it produces its flower. And in general (as we have said several times before) [b] if the air is mild and humid, and in a word well-tempered, it is possible for sprouting to occur constantly, if not in all trees, yet in a few, and still more in smaller plants. In fact this occurs at present in certain coronary plants in sheltered and sunny places. 13. 12 So dropwort, violet and some others are held to keep producing flowers, and to do so even more when tended in a certain way.[c] The great importance of a sunny and sheltered locality is attested even by the trees,[d] some growing in such spots but not elsewhere in the region, some bearing here but not elsewhere, and some flowering and sprouting here before the rest of their kind.

Let this suffice for the discussion of second crops in the lesser plants.

and later according to the locality. As they say in Macedonia the first to sprout are the trees in the marshes, second come those in the plains, and last those on the mountains;" *HP* 3 3. 5: "The nature of the locality also makes a great difference in bearing or failing to bear, as with the *persea* and the date-palm. The *persea* bears fruit only in Egypt and certain neighboring places, but in Rhodes only gets as far as flowering. The date-palm is remarkably fruitful in Babylonia, but in Greece does not even ripen its fruit, and in some places does not even show it."

14. 1 τῷ[1] δ' ἐπὶ τῶν διφορούντων[2] δένδρων ὅμοιόν τινα τρόπον ἐστὶν τὸ[3] ἐπὶ τῶν προβάτων γινόμενον·[4] ἐκεῖνα γὰρ εὐτοκήσαντα καὶ εὐγονοῦντα πάλιν ὁρμᾷ πρὸς κύησιν ἐκποιούσης[5] ἔτι τῆς
5 ὥρας. καὶ τὰ δένδρα παραιρεθέντων τῶν πρώτων καρπῶν γονεύει πάλιν ἑτέρους, δεῖται δ' ἴσως χώρας τε εὐτρόφου τὰ τοιαῦτα καὶ τῆς θεραπείας[6] πλείονος, ἢ[7] ἀμφοῖν, καὶ μάλισθ' (ὡς εἰπεῖν) τῆς τοῦ ἀέρος κράσεως ὅπως λάβῃ χρόνον ἱκανὸν εἰς
14. 2 τὴν κύησιν. διὰ τοῦτο γὰρ οὐδ' ἐάν τις ἀφέλῃ τὸν καρπὸν ἢ ⟨τὸ⟩[8] ἄνθος, δύνανται πάλιν ἕτερα γεννᾶν, διὰ τὸ μὴ λαμβάνειν τὸν τῆς κυήσεως χρόνον· οὐ γὰρ οἷόν τ' ἄνευ τοῦ κυῆσαι γεννᾶν, ἐξανηλωμένου
5 τοῦ προϋπάρχοντος· ἅμα δ' ὥσπερ πηροῦσθαι συμβαίνει τὴν ἀρχὴν διὰ τὴν ἕλκωσιν, ὥστε μὴ βλαστάνειν ἀπὸ τούτου, καινῆς ⟨δ'⟩[9] ἄλλης γινομένης οὐκ ἐκποιεῖ ⟨τὸ⟩[10] τῆς ὥρας.

τούτων μὲν οὖν οὕτως ταύτας ὑποληπτέον τὰς
10 αἰτίας.

§ 2. 1–5. Varro, *R.R.* i. 44. 4: itaque si florem acerbumve pirum aliudve quid decerpseris, in eodem loco eodem anno nihil renascitur, quod praegnationis idem bis habere non potest.

[1] τῷ ego: το U.
[2] διφορούντων u: διαφοροῦντων U; διαφερόντων N HP.
[3] τὸ ego: τῶ U.
[4] γινόμενον U: γινομένω u N^c(γί- N^ac) HP: γενομένῳ a.
[5] ἐκποιούσης N (ἐκ πιούσης H)P: ἐμποιούσης U.
[6] καὶ τῆς θεραπείας ego: τῆι θεραπείᾳ U; τῆς τε θεραπειας (-είας HP) u; τῆς θεραπείας N.
[7] [ἢ] Scaliger.
[8] ⟨τὸ⟩ Schneider.

Twice-Bearing Trees Continued; Conclusion

The twice-bearing trees in a way resemble sheep: when these have yeaned well and are well provided with generative power they are impelled to gestate a second time when the season still allows it.[a] So the trees, when their first crop is removed, proceed to bear a second. But all such twice-bearers perhaps require both a country that supports them well and a greater amount of tendance, or both, and above all (one may say) an equable tempering in the air so that they may get sufficient time for their gestation. This last requirement explains why plants are unable to generate a new crop even when the fruit or flower has been removed: they get no time for gestation. For it is impossible to generate without previous gestation, once the store of fluid has been exhausted. And at the same time the starting-point is maimed as it were by the attendant wounding, so that the plant does not generate from this source, and the season does not allow sprouting from a newly formed starting-point.

14. 1

14. 2

We are then to take the causes of these matters in the way indicated.

[a] *Cf.* Aristotle, *History of Animals*, vi. 19 (573 b 20–22): "Both sheep and goats have a period of gestation of five months. So in some places that are sunny and where the animals prosper and have abundant food, they bear twice."

---

[9] ⟨δ'⟩ Vasc.[2]
[10] ⟨τό⟩ Vasc.[2]

14. 3 αἱ δ᾽ ἀνθήσεις τῶν καρπῶν οὐ κατὰ τὰς βλαστήσεις γίνονται· βλαστάνει γὰρ τά γε πολλὰ σύνεγγυς αὑτοῖς[1] κατὰ μίαν (ὡς εἰπεῖν) ὥραν. τοῦ δὲ πεπαίνειν τοὺς καρποὺς πολλοῖς χρόνοις 5 ὕστερον ἐκείνην τὴν αἰτίαν ὑποληπτέον, ὅτι ⟨τὰ⟩[2] μὲν τῶν φύλλων καὶ βλαστῶν εὐκινητότερα[3] καὶ ῥᾴονα,[4] σωματικωτέραν ἔχοντα καὶ περιττωματικὴν τὴν ὕλην, οἱ δὲ καρποὶ καθαρωτέραν, καὶ μάλιστα δὴ τοὺς χυλοὺς αὐτούς· αὕτη[5] δ᾽ ἡ τῶν 10 καρπῶν πέπανσις, εἰς ἣν πλείονος δεῖται καὶ δυνάμεως καὶ κατεργασίας.

14. 4 ἔτι δὲ μεγάλας εὐθὺς διαφορὰς ἔχουσιν οἱ καρποὶ τῷ ξυλώδεις ἢ γεώδεις ἢ ξηροὶ ἢ λιπαροὶ τὴν φύσιν εἶναι· δυσκατεργαστότεροι γὰρ οἱ τοιοῦτοι, διὸ καὶ προανθοῦντ᾽ ἔνια τὸν[6] 5 καρπὸν ἔχει πολὺν χρόνον, ὥσπερ ἀμυγδαλῆ,[7] δυσαπόσπαστος[8] γὰρ ὁ ξυλώδης· τὸ δ᾽ ἄνθος ὠθεῖ[9] διὰ τὴν πρότερον λεχθεῖσαν αἰτίαν, εἴπερ βούλονται[10] πάντα ταῦτα συμμετρίαν τινὰ ἔχειν καὶ τάξιν. ἀλλὰ καὶ αὐτῶν τῶν ἀνθῶν ὅσα μετὰ σωματικῶν

[1] αὑτοῖς Vasc.[2]: αὐτοῖς U.
[2] ⟨τὰ⟩ Schneider.
[3] εὐκινητότερα N HP: εὐκινητοτέρα U.
[4] ῥᾴονα Vasc.[2]: ῥαον ἂν U.
[5] αὕτη Vasc.[2]: αὐτὴ U.
[6] τὸν u: τῶν U.
[7] ἀμυγδαλῆ HP: -άλη U N.
[8] δυσαπόσπαστος U[ar] (-από- U[r]): δύσπεπτος Schneider; δυσκατάπεπτος Wimmer.
[9] ὠθεῖ u: ὤσθεῖ U.
[10] βούλονται U[c]: βάλονται U[ac].

[a] That is fruiting, reckoned (in contrast with sprouting) as beginning with flowering.

Cf. HP 3 4. 3 (of wild trees): " The flowering times answer more or less to the times of sprouting, yet there is some

The Times of Sprouting Compared With Those of Fruiting

The times of flowering[a] of the fruit do not answer to the times of sprouting, for most trees sprout at times close to one another at (one might say) a single season.[b] We must take the cause of their ripening the fruit much later to be a different one: that whereas the leaves and shoots are more readily set growing and easier to form, having as they do matter with more body and a character of being residuary, the fruit on the other hand has purer matter, above all the juice itself, and this refinement of the juice is the ripening of the fruit, and for this refinement the tree requires both more power and more elaborate preparation. Further the fruit to begin with varies widely according as it is woody, earthy, dry or oily in its nature, because in the likes of these it is harder to refine the juice. This is why some trees, although they flower early, keep their fruit for a long time, as the almond (woody fruit being hard to detach); but it is quick to flower for the reason mentioned before,[c] since all these matters involve a nice adjustment of resources and priorities.[d] Again of the flowers themselves those associated with bodily bulk come out

14. 3

14. 4

variation. The variation is greater and more widespread in the case of the maturing of the fruit."

[b] *Cf. HP* 3 4. 1 (of wild trees): "The sprouting of some occurs at the same time as that of the cultivated forms, of others slightly later, and of others considerably later; but in all it occurs in the season of spring."

[c] *CP* 1 14. 3 (producing the leaves and shoots takes less power and is a less complicated process).

[d] That is, the earliness and ease of production of the flower leaves more time and power for the development of the fruit.

10 ὄγκων [1] ὀψιαίτερόν τι γίνεται, καθάπερ τὸ τῆς ῥόας, ἐν γὰρ τῷ κυτίνῳ [2] τὸ [3] ἄνθος.

15. 1 τὰ δ' ἀργὰ τῶν εἰργασμένων πρωϊβλαστότερα (καθάπερ ἄμπελος μηλέα ἐλαία συκῆ τὰ [δ'] [4] ἄλλα) διὰ τὸ κατέχειν ἔνια τὴν θερμότητα μᾶλλον μὴ ἀνασκαπτομένης τῆς γῆς μηδὲ γυμνουμένων [5] 5 τῶν ῥιζῶν, αὕτη γὰρ ἡ κινοῦσα· καὶ διὰ τὸ μηδεμίαν ἐν τοῖς ἄνω γίνεσθαι πληγὴν κλωμένων ἢ καρπολογουμένων,[6] ἀφελκούμενα [7] γὰρ πόνον τε [γὰρ] [8] παρέχει καὶ καταψύχει καὶ εἰς ὀλίγον συστέλλει (τοῖς δὲ πολλοῖς διαδίδοται καὶ τὰ [9] 10 μικρά, διὸ ταῦτα [10] ταῖς ὥραις ὑπακούει)· καὶ ἔτι δὴ καὶ μάλισθ' (ὡς εἰπεῖν) διὰ τὸ ἀκλάστων ὄντων καὶ ἀκαθάρτων ἐν πολλοῖς εἶναι καὶ κατὰ [11] σμικρὰ τὰς γονίμους [12] ἀρχάς, ὧν ἕκαστον διὰ τὴν ὀλιγότητα σμικρᾶς δεῖται καὶ κινήσεως καὶ εὐθὺς

15. 2 ποιεῖται τὰς ἐκβλαστήσεις. ὃ καὶ ἐπὶ τῶν ἀπίων ξυμβαίνει· καὶ γὰρ ἐκεῖνα εὐβλαστότερα τῶν ἡμέρων ἐστὶν διὰ τὰς αὐτὰς αἰτίας· εἰς πλείω γὰρ καὶ κατὰ μικρὰ μεμερισμένων τῶν ἀρχῶν, εὐκίνητα

[1] ὄγκων u N (ἄγκων H)P: ὄγνων U.
[2] κυτίνωι u: κτινωι U (κτίνω N [-ῳ H]P).
[3] τὸ u: τῶι U.
[4] [δ'] Wimmer (τ' Vasc.²)
[5] γυμνουμένων Scaliger: -ης U.
[6] καρπολογουμένων U (the βλαστολογία is meant, for which see *CP* 3 16. 1–2): καρφολογουμένων Scaliger.
[7] ἀφελκούμενα HP: -αι U N.
[8] [γὰρ] HP.
[9] καὶ τὰ U: κατὰ Vasc.²
[10] ταῦτα Scaliger: ταῦταις U.
[11] κατὰ Vasc.² (κατὰ τὰ Itali): τὰ U.
[12] γονίμους Gaza, Vasc.²: μονίμους U.

somewhat later, as the flower of the pomegranate (the flower being inside the pot-like structure).[a]

Why Untended Trees Sprout Earlier

Trees left untended sprout earlier than the ones that are tended, as vine, apple, olive, fig and the rest. This is because 15. 1

(1) some retain their heat better when the ground is not dug up and the roots are not exposed, for heat is the agent of sprouting.

(2) the parts above ground escape the blows sustained in pruning and in thinning the fruiting branches, for the wounds cause distress and the removal chills the tree and reduces the number of its parts (whereas when the parts are numerous, even small portions of food are distributed to them, which is why these trees respond to the seasons).

(3) further and most important (one might say), when nothing is broken off or pruned the generative starting-points are spread in smaller size over a greater number of parts, and each such part, owing to its smallness, needs but a small stimulus to set it sprouting at once. (This is also found in the pear: 15. 2 the untended trees sprout better than the cultivated for the same reason: the starting-points are divided among a greater number of parts in smaller portions, and a part with such a portion is therefore

[a] *Cf. HP* 1 13. 5 for a description (corrupt) of the pot-like envelope.

The ovary, which bears the ovules, and the floral cup, which bears the stamens, petals and sepals, are adnate. The part of the floral cup above the ovary is tubular and throat-like. The epigynous flower as a result is a " pot-like " cup.

5 γίνεται τῷ περιέχοντι πρὸς βλάστησιν.
ἔτι δὲ καὶ ἡ ξηρότης συμβάλλεται· καὶ γὰρ διὰ τοῦτ' ἐλάττων[1] ἡ ὑγρότης, τὸ δὲ ἔλαττον εὐκινητότερον.
καλλίων[2] μὲν οὖν καὶ ἀθροωτέρα τῶν εἰργασμέ-
10 νων ἡ βλάστησις καὶ ἡ καρπογονία, προτέρα δὲ ἐκείνη.

15. 3 αὐτὸ δὲ τοῦτ' ἄν τις ἀπορήσειεν, διὰ τί τὰ ἄγρια, τῶν ἡμέρων ἰσχυρότερα ὄντα, τοὺς καρποὺς οὐ[3] πεπαίνει· κατὰ μὲν[4] γὰρ τὰς δυνάμεις ἐχρῆν καὶ τὰς πέψεις εἶναι.
5 μία μὲν δή τις αἰτία τὸ πλῆθος τῶν καρπῶν. οὐ γὰρ τοσοῦτον ὑπερίσχουσιν[5] τῷ ἰσχύειν ὅσῳ πλείω τοῦ συμμέτρου τὸν καρπὸν ἴσχουσιν· ἅμα[6] δὲ πλῆθος πολὺ καὶ[7] οὐκέτι γίνεται πέψις, διὸ καὶ οἱ γεωργοὶ παραιροῦσιν ὅταν ὦσι πλείω.
10 ἑτέρα[8] δ' ὅτι πυκνότερα καὶ ξηρότερα καὶ εἰς ἑαυτὰ μᾶλλον ἕλκοντα τὸ ὑγρόν, ἡ δὲ τροφὴ καὶ ἡ

[1] ἐλάττων U^c: ἐλάττον U^ac.
[2] καλλίων u: -ίω U.
[3] οὐ N HP: ὄντα U^ar (οὐ U^r).
[4] [μὲν] HP.
[5] ὑπερίσχουσιν Schneider: ὑπερισχυουσιν U^ar (-σι U^r).
[6] ἅμα U: ἂν Wimmer.
[7] καὶ U: κἀκείνων Wimmer.
[8] ἑτέρα u N aP^c: ἐτερα U; ἕτερα HP^ac(?).

---

[a] That is, to make it edible or at least to change its colour and flavour and consistency.

[b] *Cf. HP* 3 2. 1: "Peculiar to wild trees in comparison with the cultivated are late fruiting, strength and abundance of

easily set moving toward sprouting by the surrounding air.)

(4) dryness furthermore also contributes, for this too leads to fluid in smaller amounts, and the smaller amount is more easily set in motion.

So sprouting and fruiting is finer in tended trees and more simultaneous in its occurrence in the tree, but in the untended it is earlier.

Why Wild Trees Fail to Ripen their Fruit

15. 3 One might take this separate point and raise the difficulty: why do wild trees, though stronger than the cultivated, fail to ripen [a] their fruit? For concoction should answer to power.

(1) One cause is the abundance of their fruit.[b] For the superior strength of wild trees is more than offset by their superabundance of fruit; and along with a heavy yield goes a failure to concoct it all, which is why growers remove some fruiting parts when there are too many of them.[c]

(2) Another cause is that wild trees are denser, drier, and more apt to draw the fluid to the main body

fruit (in the sense of promising a greater yield); for they ripen the fruit later, and to speak comprehensively, are for the most part later in flowering and sprouting; and they are stronger in their nature; and more fruit is promised, although less is concocted, if not by all, at least in comparison to cultivated trees of the same kind, as in the wild olive and wild pear as compared to the cultivated olive and cultivated pear;" *HP* 1 4. 1: " So wild trees are held to bear more abundant fruit, as the wild pear and wild olive, whereas the cultivated trees bear finer fruit . . ."

[c] So in the cultivated vine the fruiting shoots are thinned: *CP* 3 16. 1–2.

πέψις ἐνδόσει τῆς ὑγρότητος, ἣν οὐ ῥᾴδιον ἀντισπωμένην [1] λαμβάνειν.

15. 4 ἁπλῶς δ' οὐ τὰ ἰσχυρότερα καὶ τροφιμώτερα, καθάπερ οὐδ' ἐπὶ τῶν ζῴων, ἀλλ' ἑτέρα τις καθ' ἑαυτὴν πρὸς καρπογονίαν ἰσχὺς καὶ δύναμις. μανὸν γὰρ καὶ εὐδίοδον καὶ ὑγρὸν εἶναι δεῖ τὸ 5 καρποτοκῆσον, ἡ δὲ πυκνότης ἐναντίον, ὥσπερ καὶ ἐπὶ τῶν γυναικῶν καὶ ἐπὶ τῶν ἄλλων ζῴων· ὃ καὶ ἡ γεωργία βούλεται ποιεῖν, ἀφαιροῦσά τε τὰ [2] περιττά, καὶ τροφὴν παρέχουσα καὶ εὔειλα [3] καὶ εὔπνοα ποιοῦσα.

10 προφαίνει μὲν οὖν πλείω καρπὸν διὰ τὸ πρότερον εἰρημένον, τοῦτον δὲ οὐκ ἐκπέττει διὰ ταύτας τὰς αἰτίας.

16. 1 ἡ δὲ πέψις ἐστὶν ἐν τῷ περικαρπίῳ· τοῦτο δὲ δεῖ γίνεσθαι καὶ λαβεῖν χυλὸν ἁρμόττοντα πρὸς τὴν ἡμετέραν φύσιν.

ἴσως δ' αὐτὸ τοῦτο πρότερον εὖ ἔχει διελεῖν, ὅτι 5 πέψις ἐστὶν ἡ μὲν οὖν τῶν περικαρπίων,[4] ἡ δ' αὐτῶν τῶν καρπῶν, καὶ ἡ μὲν πρὸς τὰς ἡμετέρας

1 ἀντισπωμένην u: ἄν τις σπωμένην (-ωμενην U) N HP.
2 τὲ τὰ HP: τά τε U N.
3 εὔειλα Dalecampius (εὔηλα Itali): εὔδηλα U.
4 περικαρπίων u N H^1m (καρπίμων H^t)P: περικαρδίων U.

a *Cf.* Aristotle, *On the Generation of Animals*, i. 18 (725 b 25–34): "In many animals and plants there is a difference in this (*sc.* in the production of seed) . . .; for some have much, some little, and some none at all, not from weakness, but in some the opposite is the case, for it is used up on the body, as in some men; for enjoying a fine constitution and becoming more fleshy or too fat, they do not emit seed to the same extent and have less desire for intercourse."

of the tree, whereas a tree feeds and concocts a part by imparting to it some of the fluidity, and it is not easy to obtain this when it is pulled the other way.

15. 4 To put it simply, the stronger group is not also the better rearer of young, any more than in animals;[a] instead, the strength and power that leads to fruit production is a distinct and separate one. For the tree that is to produce fruit must have an open texture, offer easy passage, and be fluid; the close texture, on the other hand, is unfavourable to generation, as it is in women and the animals; and husbandry has this aim when it removes superfluous parts, supplies food, and provides for the proper exposure to sun and wind.

So wild trees give promise of more fruit for the reason mentioned earlier,[b] but fail to concoct all of it for the reasons given now.

16. 1 But concoction is in the pericarpion;[c] and this must be produced and must acquire a savour that agrees with our human nature.

The Two Concoctions

Perhaps it is well to make a distinction about this last point. There is to be sure a concoction of the pericarpion, but there is another of the fruit proper;[d]

[b] *CP* 1 15. 1–2.

[c] The word *perikárpion*, literally "what surrounds the fruit," was probably coined by Aristotle, who like Theophrastus uses it mainly of the fleshy part (pulp) of seed-vessels, and opposes it to the "fruit" or seed proper.

[d] "Fruit" (*karpós*) is used by Theophrastus in three senses that concern us here: (1) of the whole structure containing the seed, including pericarpion and integuments; (2) of the structure within the pericarpion, when it is sometimes called "kernel" (*pyrēn*); and (3) of the true seed.

τροφάς, ἡ δὲ πρὸς γέννησιν καὶ διαμονὴν τῶν δένδρων, οἱ γὰρ καρποὶ καὶ τὰ σπέρματα τούτων χάριν. ἑκατέρα δέ πως ἐναντιοῦται πρὸς τὴν
10 ἑτέραν·[1] ἅμα γὰρ τὸ περικάρπιον[2] ὑγρότερον καὶ πλεῖον[3] καὶ ὁ καρπὸς ἐλάττων, καὶ ἅμα μείζων οὗτος[4] καὶ τὸ περικάρπιον ἔλαττον καὶ σκληρότερον καὶ δυσχυλότερον.

16. 2 πρὸς ὃ δὴ καὶ ἡ γεωργία μεμηχάνηται, κωλύουσα τὴν τούτων αὔξησιν καὶ τροφήν. ἅπαν γὰρ (ὡς εἰπεῖν) καὶ ἥμερον ἀγρίου καὶ γεωργούμενον ἀγεωργήτου καὶ κάλλιον εἰργασμένον τοῦ χεῖρον,[5]
5 μικροπυρηνότερον ἀνυγραινόμενόν[6] τε μᾶλλον καὶ τὴν τροφὴν περισπῶν εἰς τὸ περικάρπιον, ἔτι δὲ τοὺς χυλοὺς ἐκ⟨πε⟩παῖνον[7] εἰς συμμετρίαν τῆς ἡμετέρας χρείας.

αἱ μὲν οὖν πέψεις τοσοῦτον διεστᾶσιν, εἴπερ χρὴ
10 καὶ τὴν μὴ εἰωθυῖαν πέψιν λέγειν.

16. 3 τάχα δ᾽ ἄν τις ἀπορήσειεν ἐκεῖνο, καὶ ἀξιώσειεν ὃ τὸ[8] ἰσχυρότερον[9] κατακρατεῖ, τοῦτο καὶ τὰ ἀσθενέστερα κατακρατεῖν· τὸ δὲ σπέρμα πάντων ἰσχυρότατον, κοινὸν γὰρ τέλος πάντων τῶν φυτῶν

1 ἑτέραν u (ε- U^r): εκατέραν U^ar.
2 περικάρπιον Vasc.2: ἐπικαρπιον U.
3 πλεῖον u N: πλειω U; πλεῖστον HP.
4 οὗτος u: οὕτως U.
5 χεῖρον Gaza (*deterius*), Vasc.2: χείρονος U.
6 ἀνυγραινόμενόν Vasc.2 (*quoniam . . . humescant* Gaza): ἄν. ὑγραινόμενόν U.
7 ἐκπεπαῖνον u: ἐκπαῖνον U.
8 ὁ το U: ὅ, τὰ u.
9 ϊσχυρότερον U: ἰσχυρότερα u.

and the former concoction serves to provide man with food,[a] the latter serves the generation and perpetuation of the tree, this being what fruit and seed are for. Each of the two concoctions interferes in a way with the other: with greater fluidity and size in the pericarpion goes smaller fruit,[b] and with larger fruit goes a smaller, harder and more ill-flavoured pericarpion.

16. 2 It is to meet this last situation that husbandry has been devised, preventing the further growth and feeding of the fruit.[c] Compare cultivated to wild, tended to untended, better tended to tended worse, and in practically every case the former has smaller stones, is more fluid, and diverts the food more to the pericarpion; it moreover ripens the juice to the point where this is adjusted to man's requirements.

So there is this wide difference between the concoctions, if we may use the term " concoction " of the one not ordinarily so called.[d]

A Problem: Concoction of the Seed Should Involve Concoction of the Pericarpion

16. 3 Perhaps one might raise another problem, and set up the principle that what masters the stronger thing should also master the weaker; but the seed is strongest of all.[e] For it is the consummation common to

[a] So Plato lets plants be created by the lesser gods for our food: *Timaeus*, 77 C 6–7.

[b] The fruit proper or seed.

[c] Of the fruit proper (or seed).

[d] Concoction of the fruit (that is, the seed) as opposed to that of the pericarpion.

[e] That is, the tree that concocts or masters the " fruit " or seed, which is stronger, should also master or concoct the pericarpion, which is weaker.

5 ἐστιν, ἐπείπερ ἡ τοῦ ὁμοίου γένεσις τέλος, ἅμα δὲ καὶ ἐν τοῖς ζῴοις δοκεῖ τελειουμένης ἐπιγίνεσθαι τῆς φύσεως, ὅταν δὲ ἐλλείπῃ διὰ τὴν ἡλικίαν, ἢ παρακμάσῃ [1] διὰ τὸ γῆρας, ἐξαδυνατεῖ τὸ γεννᾶν.

16. 4 οὐκ ἀλόγως δ' ἂν οὔτε ἀπορήσειεν οὔτε ἀξιώσειεν· ἐν τελειότητι μὲν γάρ τινι τὸ σπερμοφυεῖν καὶ τῶν ζῴων ὡς ἂν [2] ἡλικίᾳ [3] λάβοι τις ἄν.

οὐ μήν γε οὐδὲ [4] τὰ ἰσχυρότατα σπερματικώτατα, 5 ἀλλὰ σχεδὸν ἐναντίως, ἑκατέρωθεν μεριζομένης τῆς τροφῆς καὶ δυνάμεως, ὃ δὴ καὶ ἐπὶ τῶν φυτῶν συμβαίνει,[5] κατὰ λόγον συμβαῖνον· οὕτω δὲ καὶ [6] τὴν ἀναλογίαν λαμβάνειν ὡς εἰς ὁπότερον [7] ἂν τούτων ὁρμήσῃ, θάτερον ἐλλιπέστερον ἔσται, 10 διαρκεῖν γὰρ οὐ δύναται πρὸς ἄμφω· τοῦτο γὰρ σχεδὸν ἐν ἅπασιν ὁμολογούμενον.[8]

[1] παρακμάσῃ Vasc.[2]: παρακμασιν U.
[2] ὡς ἂν U: ὅσων Wimmer.
[3] ἡλικίᾳ U: ἡλικίαν N HP.
[4] οὐδὲ Vasc.[2]: οὔτε U.
[5] συμβαίνει Gaza (*evenire*), Vasc.[2]: σημαίνει U.
[6] καὶ U: δεῖ Wimmer.
[7] εἰς ὁπότερον Gaza (*ad utram . . . partem*), Vasc.[2]: εἰ πότερον U; ὁπότερον u H; ὁπώτερον N P.
[8] ὁμολογούμενον U[c]: ὁμολούμενον U[ac].

---

[a] *Cf.* Aristotle, *On the Generation of Animals*, i. 4 (717 a 22): "Now most animals have, like plants, no other function but seed and fruit;" i. 23 (731 a 24–26): "For the essence of plants has no other task or activity to perform than the generation of the seed . . ."

[b] *Cf.* Plato, *Symposium*, 208 A 7–B 2: "For it is in this way that all that is mortal is preserved, not by remaining (like the divine) entirely the same, but by this: that what departs and

all plants,[a] since the end is the generation of like.[b] Then too in animals as well, the seed is held to be produced when their nature reaches its perfection, and when the animals are too young to have achieved it or too old to have retained it, they are unable to generate.[c]

16. 4 There is nothing unreasonable in either raising the problem or in setting up the principle; for it is true that in animals too the production of seed is found in a certain perfection, if we judge perfection by the time of life.

Solution

But this does not also make the strongest animals the most productive of seed; indeed the opposite is more nearly the case, since the food and power are devoted to either the one result or the other,[d] and this happens in plants too, and happens in a way parallel to what is found in animals. We are to take the parallel in the following sense: whatever of the two things the animals and plants set out to do involves deficiency in the other, since they are not equal to both. For the impossibility of achieving the two tasks in all is (one may say) agreed.

grows old leaves behind a young replacement like itself." *Cf.* also *CP* 1 16. 12 with note *b*.

[c] *Cf.* Aristotle, *On the Generation of Animals*, i. 18 (725 b 19–25): "Furthermore no seed (semen) is present either in childhood or in old age . . .; . . . in old age because the nature of the old man does not concoct enough, and among the young because of their growth; for everything is first used up in that . . ."

[d] *Cf.* Aristotle, *On the Generation of Animals*, i. 18 (725 b 25–34), translated in note *a* on *CP* 1 15. 4, and the rest of the passage (to 726 a 6).

16.5 ἐν δὲ τοῖς φυτοῖς τρεῖς τινές εἰσιν οἱ μερισμοί· πρὸς αὐτὸ τὸ δένδρον καὶ τὴν βλάστησιν, ὅπερ ἐναντίον τοῖς καρποῖς ἐὰν πλείων[1] γένηται τοῦ ξυμμέτρου, διὰ τοῦτο γὰρ ἀκαρπία· καὶ πάλιν ἐν 5 αὐτῷ[2] καὶ τοῖς περικαρπίοις, ἡ γὰρ εἰς τὸ ἕτερον ὁρμὴ κωλύει τὴν ἑτέραν. ὥσπερ οὖν ἐπ' ἐκείνων οὐκ ἄλογον, οὐδ' ἐπὶ τούτων, ἀλλ' ἔχει τὸ ἀνὰ λόγον· τὰ γὰρ ἄγρια καὶ εἰς τὴν τροφὴν[3] καὶ εἰς αὐ⟨τὰ⟩[4] τὴν ὑγρότητα ἄγοντα παραιρεῖται[5] 10 τῶν περικαρπίων, ὥστε μείζω γίνεσθαι τὸν καρπόν.[6]

16.6 τάχα δ' οὐδ' ἄν τῳ[7] δόξειεν ὅλως ἄτοπον εἶναι τὸ μᾶλλον ἐφικνεῖσθαι τὰ ἄγρια τῶν σπερμάτων, ὥσπερ ἄρρενα ὄντα, πυκνότερά τε καὶ ξηρότερα τὴν φύσιν· διαθηλύνουσι γὰρ αἱ κατεργασίαι καὶ 5 αἱ τροφαί. τοῦτο μὲν οὖν ὡς καθ' ὁμοιότητά τινα λεγέσθω, πορρωτέρω[8] κείμενον.

τῶν δὲ πεπάνσεων εἰς μὲν τὴν γένεσιν[9] αὕτη

[1] πλείων U^c: πλείω U^ac; πλεῖον N HP.

[2] ἐν αὐτᾷ [sic] Heinsius (ἐν αὐτοῖς Vasc.²; *ad fructum* Gaza): ἐν ἑαυτῶι U.

[3] τὴν τροφὴν U: τὸν καρπὸν Gaza (*fructui*), Vasc.²

[4] αὐτὰ (αὐτὰ ego) τὴν Vasc.² (*pulpae* Gaza): αὐτὴν U.

[5] παραιρεῖται Schneider: παρατρία τὲ (τε U^r) U^ar.

[6] τὸν καρπὸν u: τὸν καρπῶν U.

[7] τῳ ego (*cf. CP* 6 5. 3; που Schneider): ποι U.

[8] πορρωτέρω N HP: πορρωτετρω U; πορρωτέρων u.

[9] γένεσιν N HP: γέννεσιν U.

16. 5 In plants however the distributions [a] are (one may say) three in number: the food can serve the tree proper and its vegetative growth (and this distribution, if excessive, is prejudicial to the fruit, leading as it does to failure to bear); and again the distribution can occur in the fruit proper and in its pericarpion, the movement of the food in the one direction checking its movement in the other. Then just as in the first case [b] there is nothing unreasonable in such interference, so too in the second; [c] instead the one case is parallel to the other. For the wild trees, devoting the fluid both to their own feeding and to the fruit proper, take it away from the pericarpia, with the result that the fruit proper increases at the expense of the latter.

16. 6 Perhaps someone might even think that there is no problem at all in this greater success of wild trees with their seeds,[d] wild trees being as it were male and in their nature closer in texture and drier, for cultivation and good feeding have an effeminating effect. This remark, however, is to be taken as resting on a certain resemblance, and the parallel is pretty remote.

Of the two ripenings this of the seed is the more important for reproduction, that of the pericarpion

[a] The following diagram may serve as illustration:

	in animals	(1) for the body	
		(2) for the seed	
distribution of food			
	in plants	(1) for the body	
		(2) for the fruit	(a) the fruit proper
			(b) the pericarpion

[b] Of the tree itself and its fruit.

[c] Of the fruit proper and its pericarpion.

[d] Than with their pericarpia.

κυριωτέρα, πρὸς δὲ τὴν ἡμετέραν χρείαν ἡ τῶν περικαρπίων. ἐν ποτέρᾳ δὲ δεῖ θέσθαι[1] τὸ
10 τελειότερον ἄλλος λόγος· ἐπεὶ οὕτω γε καὶ ὧν τοῖς φύλλοις μόνον χρώμεθα καὶ ὧν[2] ταῖς ῥίζαις, ὥσπερ τῶν λαχάνων, αὕτη κυριωτέρα πέψις ἔσται. καίτοι γε τοῦτο[3,4] τέλος, ἐν τοῖς σπέρμασιν, οἷς ἡμεῖς οὐδὲν χρώμεθα πρὸς τὴν τροφήν.

16. 7 ἔστιν δέ τις καὶ οὗτος[5] ὁ λόγος, ὡς διὰ ψυχρότητα τῶν ἀγρίων οὐ δυναμένων πέττειν, τοὺς δὲ πυρῆνας ἐκ τῆς ξυλώδους καὶ περιττωματικῆς γίνεσθαι τροφῆς, ὥσπερ τὰ[6] ἐν τοῖς ζῴοις.

5 αὐτὸ μὲν τοῦτ' ἴσως οὐ κακῶς, εἰ ὁ πυρὴν ἐκ τοῦ γεώδους καὶ ξυλώδους, ἀλλὰ τὸ σπέρμα οὐκέθ' ὁμοίως, ἀλλ' ἐκ τῆς[7] καθαρωτάτης, ὅπερ ἐν τούτῳ. πέφυκεν δὲ τὰ θερμὰ μάλιστα σπερμοφυεῖν· ᾗ[8] θερμότερα ἂν εἴη. καίτο⟨ι⟩ γε[9] τῆς
10 τῶν ἡμέρων θερμότητος ἐκεῖνο φέρεται σημεῖον, ἡ τῶν ὀπῶν[10] δύναμις·[11] ὁ μὲν γὰρ τῆς συκῆς

[1] δεῖ θέσθαι Itali: δεῖσθαι U.
[2] ὧν—ὧν Moldenhawer: ἐν—ἐν U.
[3] τοῦτο U: τούτων Vasc.[2] (*eorum* Gaza).
[4] γε τοῦτο U: τοῦτό γε Hindenlang.
[5] καὶ οὗτος U[cm]: U[t] omits.
[6] τὰ ⟨ὀστᾶ⟩ Gaza (*ossa*).
[7] τῆς Gaza: γῆς U.
[8] ᾗ Gaza (*quo*), Itali: εἰ U.
[9] καίτοι γε Wimmer: καὶ τό γε U HP; καὶ τὸ N.
[10] ὀπῶν u: ὅπλων U.
[11] δύναμις u: δυναμεις U.

---

[a] Compare the proof that the seed is strongest (*CP* 1 16. 3), since the perfection of the producer of the seed is contemporary

the more important for human requirements. To which of the two ripenings we are to assign the greater achievement by the tree of its goal[a] is another question. Indeed if we assign it to the ripening of the pericarpion we should have to say that in plants whose leaves (or again whose roots) we use alone, as vegetables, the concoction of these parts is the more important; and yet the goal lies here, in their seeds, which we do not use for food at all.

Cold (and Heat) as Explanations

16. 7 There is also this other explanation to the effect that wild trees are unable to concoct because of coldness,[b] and the stones come from woody and residuary food, like the hard parts in animals.[c]

Its Difficulties

Now this last point, considered by itself, is perhaps not badly taken, that the stone comes from the earthy and woody part. But this could no longer be said if the seed is meant, which comes from the purest food, and the seed is inside the stone; and it is hot plants that are naturally the greatest producers of seed, and this would make wild trees hotter than the cultivated. (Yet in proof of the heat of cultivated trees another piece of evidence is adduced: the potency of the sap. Thus the sap of the fig curdles

with its production. Theophrastus hints that the goal of wild trees and plants is not the perfection of the pericarpion.

[b] *Cf. CP* 1 21. 7.

[c] The author of this explanation is probably Menestor: *cf. CP* 1 21. 6.

τὸ γάλα πήγνυσιν, ὁ δὲ τοῦ ἐρινεοῦ οὐ πήγνυσιν ἢ κακῶς.

16. 8 πρὸς αὐτὸ δὲ τοῦτο πάλιν ἀντίκειταί τις ἑτέρα καθόλου πίστις ὑπὲρ τῆς θερμότητος, ὅτι τὰ ἄγρια μᾶλλον ἐν τοῖς ψυχροῖς δύνανται διαμένειν, καὶ ὅλως δὲ διὰ θερμότητα ἡ ἰσχύς.
5 ἀλλ' ὑπὲρ μὲν τούτων τάχ' ἂν ὕστερον εἴη λεκτέον· πλείων γὰρ ὁ λόγος καὶ ἔχων [1] τινὰ ἀπορίαν, ποῖα θερμὰ καὶ ψυχρὰ καὶ τοῖς ποίοις [2] διοριστέον, καὶ προσέτι τῶν ποίων [3] τὸ αἴτιον, εἴτε μόνων [4] εἴτε μεθ' ἑτέρων.

16. 9 καί τινων δὲ ἡ πέψις δόξειεν ἂν ἔχειν διαίρεσιν καὶ ἁπλῶς καὶ πρὸς ἡμᾶς, οἷον ὅσα ταῖς δριμύτησιν, καὶ ὅσα τοῖς φαρμακώδεσιν καὶ ὀπώδεσιν [5] χυλοῖς, ἃ δὴ καὶ ἐπαινοῦσιν καὶ μάλιστα χρῶνται
5 τοῖς τοιούτοις. ταῦτα μὲν οὖν ὥσπερ ἰδιότης τις φύσεως, πρὸς ἣν δῆλον ὅτι καὶ αἱ τροφαὶ καὶ αἱ κατεργασίαι τείνουσιν, ἢ τοὐναντίον [6] αἱ ἀργίαι

[1] ἔχων u HP: ἔχον U N.
[2] τοῖς ποίοις u HP (ποίοις Wimmer): τοῖς ποιοῖς U N.
[3] τῶν ποίων u HP (τὸ ποῖον Wimmer): τῶν ποιῶν U N.
[4] μόνων ego: μόνον U.
[5] καὶ ὀπώδεσι Wimmer: καὶ σκώδεσιν U; N HP omit.
[6] ἢ τοὐναντίον u: ἧτ' οὖν ἀντιον U.

---

[a] For heat as the operative factor in the curdling of milk by fig-juice *cf.* Aristotle, *On the Generation of Animals*, iv. 4 (772 a 23–25).

[b] *Cf. HP* 3 2. 4 (of the distinction of wild trees [and plants] from cultivated): " Furthermore they are distinguished by a greater liking for cold and for mountain country, for this

milk,[a] whereas that of the wild fig either fails to do this or does it badly.)

16. 8 This last proof, that cultivated trees are hot, is met by another proof of a general character establishing the heat of wild trees: that they are better able to survive in cold regions [b] and indeed that their whole strength is due to heat.

(But these matters must perhaps be discussed later,[c] since we have here a question of some length and involving a certain difficulty: what kind of plants are hot and what kind are cold, and by what characters are the two groups to be distinguished; and furthermore to what characters, whether taken alone or accompanied by others, is the cause of this failure to concoct to be assigned?)

The Cases Where the Two Concoctions Are the Same

16. 9 Again the concoction of some plants would appear to be distinguished from that of the rest in being both concoction pure and simple and concoction for our use, as in the plants that are serviceable because of their pungency or their medicinal and rennet-like juices, products that are in high esteem and in great demand. Now these characters are a kind of distinctiveness belonging to the plant's nature, and to this distinctive nature in the one case the rearing and

point too is considered in determining the wildness of trees and of plants in general, whether taken by itself or as incidental to other distinctions."

[c] In the discussion of hot and cold plants in *CP* 1 21. 4–1 22. 7. *Cf.* also *CP* 1 21. 5–6 and the related view of Clidemus (*CP* 1 10. 3).

(καθάπερ τῷ σιλφίῳ καὶ τῇ καππάρει καὶ εἴ τι ἄλλο [τί] [1] φεύγει τὴν ἐργασίαν, ἢ εἴ τι πάλιν αὖ
10 διώκει τὴν ξηρὰν καὶ ὑγρὰν καὶ χειμερινήν).

16. 10 ἐξ αὐτοῦ δὲ τούτου τάχ᾿ ἂν τις πάλιν ἀπορήσειεν κοινήν τινα ἀπορίαν καὶ καθόλου, πότερα [2] τὴν φύσιν ἐκ τῶν αὐτομάτων μᾶλλον θεωρητέον ἢ ἐκ τῶν κατὰ τὰς ἐργασίας, καὶ ἐν ποτέροις τὸ κατὰ
5 φύσιν. (σχεδὸν δὲ τούτῳ ταὐτόν,[3] μᾶλλον δὲ μέρος τούτου, καὶ πότερον ἐκ τῶν ἀγρίων ἢ ἐκ τῶν ἡμέρων.)

ἡ μὲν γὰρ φύσις ἐν αὑτῇ τὰς ἀρχὰς ἔχει, καὶ λέγομεν τὸ ⟨μὲν⟩ [4] κατὰ φύσιν (τὸ δ᾿ ἐκ τῶν
10 αὐτομάτων τοιοῦτον), τὸ δ᾿ ἔξωθεν, ἄλλως τε [5] καὶ
16. 11 κατὰ τέχνην, ἀφ᾿ ἑτέρας γὰρ ἀρχῆς. οὐδ᾿ ἐν τοῖς ζῴοις ὅσα πλάττεται ἢ καταναγκάζεται πρὸς μικρό-

[1] [τί] HP.
[2] πότερα u: ποτέρα U N (-ως HP).
[3] ταυτὸν U[c] from ταυτὸ.
[4] ⟨μὲν⟩ Wimmer.
[5] ἄλλως τε u: ἀλλ᾿ ὥστε U.

---

[a] *Cf.* *CP* 2 1. 1; 3 1. 1: agriculture endeavours to help a plant achieve its nature.
[b] *Cf.* *HP* 1 3. 6; 3 2. 1; *CP* 3 1. 1.
[c] *Cf.* *CP* 3 1. 3–6.
[d] *Cf.* *HP* 4 5. 1 (some shrubs are fonder of cold regions) " as centaury and wormwood, and furthermore those with medicinal powers in their roots or juices, as hellebore, squirting cucumber, scammony and nearly all whose roots are taken."
[e] In a sense what comes from the external environment (when this is not altered by man) is also natural: *cf.* *CP* 2 1. 1.
[f] *Cf.* Aristotle, *On the Generation of Animals*, i. 18 (724 a 31–

cultivation of the plant is evidently directed,[a] and in the other on the contrary the omission of cultivation (as with silphium, caper and others that dislike tendance,[b] and so again with any that seek out dry [c] or wet and wintry country).[d]

A General Problem: Is Nature and the Natural to be Seen in What Grows Unaided or in What is Under Cultivation?

16. 10 But starting from this last point one could perhaps raise a further problem, this time one that applies to all plants and is of general scope: are we to study the nature of a plant in those that grow without human aid or in those growing under various forms of cultivation, and which of the two kinds of growth is natural? (Much the same as this, or rather a part of it, is the question whether we are to study the nature of a given kind from its wild or cultivated form.)

Unaided Growth is Natural

For the nature contains the starting-points in itself, and we speak here of the " natural " (and what we see in plants that grow unaided by man is of this description), contrasting it to what is of external causation, especially when it is due to art,[e] for the starting-point is different.[f] 16. 11 And in animals too one must not count as natural those cases where mould-

35): " Of such things (*sc.* where B is from A because A is the starting-point of motion) the starting-point is in some in the things themselves . . ., but in some is outside (as the arts are outside their products and the lamp outside the burning house)."

τητα καὶ μέγεθος καὶ τὸ⟨ν⟩[1] ὅλον τύπον τῆς μορφῆς, οὐ θετέα ταῦτα κατὰ φύσιν· ἡ δ᾽[2] ἀεὶ
5 πρός τε[3] τὸ βέλτιστον ὁρμᾷ[4] καὶ τοῦθ᾽ ὥσπερ ὁμολογούμενόν ἐστιν.

ταύτῃ δὲ τὰ ἐκ[5] τῆς θεραπείας· ἅμα γὰρ καὶ τελείωσις γίνεται τῆς φύσεως ὅταν ὧν[6] ἐλλιπὴς τυγχάνει,[7] ταῦτα προσλάβῃ διὰ τέχνης (οἷον
10 τροφῆς τε ποιότητα καὶ ἀφθονίαν καὶ τῶν ἐμποδιζόντων καὶ τῶν κωλυόντων ἀφαίρεσιν), ἃ παρέχουσιν δῆλον ὅτι καὶ οἱ οἰκεῖοι τόποι πρὸς ἕκαστον, ἐν οἷς δή φαμεν δεῖν θεωρεῖν τὰς φύσεις

16. 12 αὐτῶν. ἀλλ᾽ ἐκεῖνοι[8] μὲν ἀπὸ τῶν ἔξωθεν μόνον παρέχουσιν, οἷον ἀέρος καὶ πνεύματος καὶ ἐδάφους καὶ τροφῆς, ἡ δὲ γεωργία καὶ ἐν αὐτοῖς[9] μετακινεῖ καὶ μετατίθησιν·[10] ὥστ᾽ εἴπερ[11] καὶ ἐκεῖνό[12] γε
5 προσαπαιτεῖ πρὸς τὸ βέλτιον, καὶ ταῦτα προσδέχοιτ᾽ ἂν ὡς ἂν οἰκεῖα· προσαπαιτεῖν[13] δ᾽ αὐτὴν

[1] τὸν u HP: τὸ U N.
[2] ἡ δ᾽ N: ἡ δ᾽ U; ἥδ᾽ u HP; ᾗ δ᾽ Zeller.
[3] [τε] Schneider.
[4] ὁρμᾶι u: -αὶ U.
[5] τὰ ἐκ U: καὶ τὰ Wimmer.
[6] ὧν Wimmer: οὖν U.
[7] τυγχάνει U^{ac}: -η U^{c}.
[8] ἐκεῖνοι Gaza, Vasc.[2]: ἐκεῖνο U.
[9] αὐτοῖς U^{c}: αὑτοῖς U^{ac}.
[10] μετατίθησιν u: μετατιθέασιν U.
[11] εἴπερ U^{c}: εἰ περὶ U^{ac}.
[12] ἐκεῖνο U: ἐκεῖνα Gaza, Vasc.[2]
[13] προσαπαιτεῖν Gaza (*desyderent*), Itali: προσαπαιτεῖ (πρὸσά- U^{ar}). ἡ U^{r} N; προσαπαιτῇ· ἡ HP.

---

[a] *Cf. CP* 5 6. 7. Theophrastus is thinking of serpents: *cf.* Aristotle, *On the Parts of Animals*, iv. 1 (676 b 6–10): "Because

ing or forcing produces small or large size or a general physical outline.[a] The nature instead always sets out to achieve what is best,[b] and about this (one may say) there is agreement.

Cultivation is Natural

But what proceeds from husbandry does this too. For the nature of the plant is also fulfilled when that nature obtains through human art what it happens to lack, such as food of the right kind and in plentiful supply and the removal of impediments and hindrances, all of which evidently is also provided by the regions appropriate to a given plant, the regions in fact where we assert that the natures of plants should be studied.[c] But the appropriate region only provides external help, such as weather, wind, soil and food, whereas husbandry also introduces different movements and arrangements within the plant itself.[d] So if the nature of a plant demands that external aid for the achievement of what is better, it would also accept these internal modifica-

16. 12

of the shape of their body, which is long and narrow, serpents also have viscera that are long and dissimilar to those of other animals, the shapes having been (as it were) moulded in a frame on account of the restricted space; " Aristotle, *History of Animals*, ii. 17 (508 a 14–17): "They (*sc.* serpents) have the rest of their internal parts the same as the lizards do, except that their viscera are all narrow and long because of the animals' narrowness and length."

[b] *Cf.* Aristotle, *On Generation and Passing Away*, ii. 10 (336 b 27–28): ". . . we assert that in everything nature always aims at what is better; " Theophrastus, *On the Senses*, chap. vi. 32; *CP* 6 4. 2.

[c] *Cf. CP* 2 7. 1; also 1 9. 3, 2 16. 7–8, 3 1. 6, 3 6. 7, *HP* 4 4. 1.

[d] By directing the movement of the food to the pericarpion (*CP* 1 16. 2).

καὶ ζητεῖν εὔλογον, ἄλλως τε[1] καὶ ἐκ τούτων ἠρτημένην καὶ ἐν τούτοις ἔχουσαν τὰς ἀρχάς· ἐπεὶ κἀκεῖνο τοῖς αὐτομάτοις ἄτοπον συμβαίνει καὶ
10 ὥσπερ παρὰ φύσιν, τὸ ἐκ τῶν σπερμάτων χείρω γίνεσθαι καὶ ὅλως μὲν[2] ἐξίστασθαι τοῦ γένους· οὐδὲ γὰρ δὴ τοῦτο κατὰ φύσιν, ἀλλ' εἰς[3] τὸ ὅμοιον ἀπογεννᾶν.

16. 13 αἱ μὲν οὖν ἀπορίαι σχεδὸν αὗταί τε καὶ τοιαῦται.

φαίνεται δὲ καὶ ἐκ τούτων[4] πρότερον εἶναι δῆλον ὅτι[5] διαιρετέον τὰς φύσεις ὥσπερ καὶ τὰς πέψεις λέγομεν· τοῖς μὲν γὰρ ἡ αὐτόματος ἡ οἰκειοτέρα,
5 τοῖς δ' ἡ τῆς θεραπείας καὶ γεωργίας, ἔνια δ' ἀμφοτέρως, ἐξ ὧν καὶ θεωρητέον,[6] ὥσπερ καὶ ἡ φύσις διῄρηται ἡμέροις καὶ ἀγρίοις, ὁμοίως ἔν τε ζῴοις καὶ φυτοῖς· ἑκατέροις γάρ ἐστιν [φύλλα][7] φυσικὰ καὶ οἰκεῖα, καὶ πρὸς σωτηρίαν, καὶ πρὸς

[1] ἄλλως τε u: ἀλλ' ὥστε U.
[2] μὲν U N: μὴ HP; Schneider deletes.
[3] εἰς U: ἀεὶ Gaza (*semper*).
[4] τούτων U: τῶν Wimmer.
[5] πρότερον εἶναι δῆλον. ὅτι U: εἶναι δῆλον, ὅτι πρότερον Schneider.
[6] θεωρητέον U: διαιρετέον Schneider.
[7] [φύλλα] ego (αἴτια Schneider; πολλὰ Wimmer).

---

[a] *Cf. CP* 1 9. 1.

[b] *Cf. CP* 1 16. 3 with note *b* and Aristotle, *On the Soul*, ii. 4 (415 a 26–b 2): ". . . for the most natural of their functions in things that live, when complete and not cripples, or when not produced by spontaneous generation, is to create something else like itself, an animal an animal, a plant a plant, so that they may partake in what way they can of the eternal and divine; for all aim at this, and for its sake they do all that they do naturally;" *Politics*, i. 2 (1252 a 27–30) [of the coupling of

tions as appropriate to itself; and it is reasonable that it should demand and seek them, especially since it depends on what is internal and has its starting-points there. In fact in trees that grow without human aid there is this strange and (as it were) unnatural result: produced from seed they deteriorate and even undergo a complete mutation of variety,[a] for this degeneration too is nothing natural, what is natural being instead to achieve similarity in reproduction.[b]

16.13 These, then, are the problems (one may say) and such is their character.

The Solution: Two Kinds of Nature

This discussion too [c] makes it appear evident that we must make a prior distinction of the natures just as we say that we must do with the concoctions: [d] so for some plants their nature as it develops unaided by man is more appropriate, for others their nature as developed by care and cultivation, and a few do well in both ways; [e] and we must rest our study on this distinction, just as their natures are distinguished for the domesticated and wild, in animals and plants alike, for each of the two groups has things that are natural and suited to it, conducive not only to pre-

male and female]: ". . . and this is not the result of choice, but as in the other animals and in plants it is natural to aim at leaving behind another like oneself . . ."

[c] That is, the argument about nature in *CP* 1 16. 10–12. The conclusion also rests on the discussion of the two ripenings in *CP* 1 16. 1–6.

[d] *CP* 1 16. 9.

[e] So with pungent and medicinal plants, where sometimes cultivation, and sometimes its omission, favours concoction: *CP* 1 16. 9.

10 διαμονήν, καὶ πρὸς αὔξησιν καὶ βλάστησιν, καὶ πρὸς τὴν τῶν καρπῶν γέννησιν. ἴσως δὲ καὶ ἐν αὐτοῖς [1] τοῖς καρπίμοις πάλιν ⟨ἄν⟩ [2] τις διέλοι τὰ μὲν εἰς τὸ αὐτόματον ἀφιείς, τὰ δὲ εἰς ἐπιμέλειαν καὶ κατεργασίαν μόνον.
15 ἀλλὰ γὰρ τούτων μὲν ἐνταῦθ' ὁ διορισμός·

17. 1 περὶ δὲ τῆς πέψεως, ὅθεν ὁ λόγος ἐξέβη, πάλιν τὰ ἐπίλοιπα λεκτέον.

ὡς γὰρ ἐπὶ τὸ πᾶν ὧν [3] μὲν ὁ καρπὸς ὑγρὸς καὶ γυμνὸς ἢ λεπτὸν ἔχων περὶ αὐτὸν [4] κέλυφος, 5 ταῦτα μὲν πρωΐκαρπα, καθάπερ ἄμπελος καὶ συκῆ, μάλιστα δὲ συκάμινος· αὕτη γὰρ γυμνὸν ἔχει τὸν καρπόν, ὥσθ' ὅσον [5] ἂν ὁ ἥλιος ἐπιβῇ [6] ταχὺ προηλλοίωσεν, βραχείας δεόμενον θερμότητος· ἅμα δὲ καὶ ἐν αὐτῷ συνεργάζεται δύναμις 10 ἰσχυρὰ καὶ ἀθρόος [7] ἐπιοῦσα, καθάπερ καὶ πρὸς τὴν
17. 2 βλάστησιν. ἡ γὰρ ὀψιότης ἐποίησεν ἀθροισμόν, ὅθεν

§ 2. 1–4. *Cf.* Pliny, *N.H.* 16. 102: serotino quaedam germinatu florent maturantque celeriter, sicuti morus, quae novissima urbanarum germinat nec nisi exacto frigore, ob id dicta sapientissima arborum. sed cum coepit, in tantum universa germinatio erumpit, ut una nocte peragatur etiam cum strepitu.

[1] αὐτοῖς u: ἑαυτοις U.
[2] ⟨ἄν⟩ HP.
[3] ὧν u: ὃν U.
[4] αὐτὸν P (αὐτὸν H): αὐτο U; αὐτὸ u; αὐτὸ N.
[5] ὅσον U^r N HP: ὅσων U^ar.
[6] ἐπιβῇ ego (ἐπιθίγῃ Schneider; ἐπιθέρῃ Coray; ἐπέλθῃ Wimmer): ἐπιθῆι U.
[7] ἀθρόος ego: ἀθρόως U.

---

[a] As opposed to the pungent and medicinal plants of *CP* 1 16. 9, where the distinction between those that do better when

servation and survival and to growth and sprouting, but also to the generation of the fruit. Perhaps within the group of fruitful plants as well [a] one might make the distinction again, letting some bear unaided, the rest only under care and cultivation.

In these matters, then, this is the line to be drawn.

Concoction Concluded: Early Bearers: Mulberry, Fig and Vine

17. 1 But we must return to complete the account of concoction from the point where we digressed.[b]

Broadly speaking all trees bear early whose fruit is (1) fluid and (2) naked or with a thin covering, as the vine and fig and most of all the mulberry. For the mulberry has naked fruit, and so the sun quickly begins the alteration of as much as it reaches, naked fruit requiring but little heat; and accompanying this is a force cooperating within the fruit that is strong and comes on all at once, just as when it leads to the sprouting. 17. 2 For its lateness in sprouting [c] brings

uncultivated and those that improve with cultivation is a recognized one.

Cf. also *HP* 1 3. 6: "It is right to speak of cultivated or tame and of wild not only with reference to these characters (*sc.* growing at all or growing better under cultivation) but also with reference to the tamest of all creatures; and man is that creature which alone or above all others is tame."

[b] The digression began at *CP* 1 16. 10.

[c] For the late sprouting *cf.* *HP* 1 9. 7.

αἵ τε βλαστήσεις ἀθρόοι καὶ μετὰ φορᾶς γίνονται νεανικῆς, ὥστε καὶ ψόφον ποιεῖν (καθάπερ τινές φασιν), καὶ αἱ πέψεις ταχεῖαι·[1] παρόμοιον γὰρ τὸ
5 συμβαῖνον ὥσπερ καὶ τοῖς σίτοις ὑπὸ τῶν χειμώνων κατεχομένοις,[2] ἀνεθέντες γὰρ οὕτω[3] ταχείας ποιοῦνται τὰς αὐξήσεις ὥστε μὴ πολὺ καθυστερεῖν
17. 3 ἢ[4] μὴ κατὰ λόγον τοῖς εὐδιεινοῖς. [ἔτει δε και ταυτηι συμβαίνει το μὴ ἀθροοι. ἀλλὰ κατὰ μερος πέττειν. διο καὶ πολυν διαμένει χρόνον καὶ ἐπι πλεῖον δὴ τοῦτοις ἡ ὑγρότης λεπτῆι και
5 ὑδατωδης.][5]

ἡ μὲν οὖν τῆς συκαμίνου διὰ τοῦτο πρώϊος (ὡς δὲ Μενέστωρ φησίν, ἡ μὲν βλάστησις αὐτῆς ὀψία διὰ τὴν ψυχρότητα τοῦ τόπου,[6] ἡ δὲ πέψις ταχεῖα διὰ τὴν ἀσθένειαν).
10 οἱ δὲ τῆς ἀμπέλου καὶ τῆς συκῆς ὀψιαίτεροι, ὅτι καὶ τὰ κελύφη περίκειται καὶ ὑγρότης πλείων καὶ παχυτέρα. καὶ τὸ μὲν σῦκον μεῖζον τῷ ὄγκῳ, καὶ ἀθροωτέρα πως ἡ πέψις αὐτῶν· οἱ δὲ βότρυς[7] καὶ τῷ πλήθει πλείους ὡς πρὸς τὴν δύναμιν καὶ ἐν
15 τούτοις ἀθρόος[8] ὁ καρπὸς καὶ οὐ διειλημμένος· ἔτι δὲ ὑπόσκιος καὶ οὐχ ὁμοίως ὑπαίθριος, καὶ ἡ ὑγρότης πολλή (φύσει γὰρ τὸ δένδρον φίλυδρόν ἐστιν). ⟨ἔτι δὲ καὶ ταύτῃ συμβαίνει τὸ μὴ ἀθρόον[9] ἀλλὰ κατὰ μέρος πέττειν· διὸ καὶ πολὺν

§ 3. 7. Menestor Frag. 4, Diels-Kranz, *Die Fragmente der Vorsokratiker*, vol. i[10], p. 375. 22–24.

[1] ταχεῖαι Schneider: ταχεις U.
[2] κατεχομένοις u: -νων U.
[3] οὕτω Gaza (*adeo*), Basle ed. of 1541: οὗτοι U.
[4] ἢ Schneider: εἰ U.

about an accumulation of fluid, so that (1) sprouting occurs all at once and is attended with so impetuous a movement that (as some say) it makes a noise, and (2) concoction is rapid, for the case is like that of cereals held back by cold weather,[a] for once released from the restraint they grow so rapidly that they do not fall much behind cereals that have grown in fair weather, or else the lag is not proportionate to the initial delay. So the mulberry ripens early for this reason. (But according to Menestor it sprouts late because of the coldness of the region, but ripens rapidly because the tree is weak.)

17. 3

The fruits of the vine and fig ripen later than this because they have coverings and their fluid is greater in amount and thicker.[b] As for the fig, its fruits are larger and are ripened more nearly at the same time; grape-clusters on the other hand are not only more numerous relative to the power of the tree, but in the clusters the fruit is bunched and not isolated; furthermore the fruit is shaded[c] and not so much in the open and has a great deal of fluid, the tree being by nature fond of water. Then there is the further point that the vine ripens its fruit not all at once but successively, and this is why the later fruit remains

[a] *Cf. CP* 2 1. 4.
[b] That is, the fluid that the ripe fruit is to contain.
[c] By the leaves. In the fig the leaf is behind the fruit (*CP* 5 2. 2).

---

5 [ἔτει—ὑδατώδης] transferred by Schneider to lines 18–21.
6 τόπου U: ὀποῦ Bruns (but the tree is hot: *HP* 5 3. 4).
7 βότρυς U: -υες u.
8 ἀθρόος N HP: -ως U.
9 ἀθρόον u: ἄθροοι (?) U.

20 διαμένει χρόνον, καὶ ἐπὶ πλεῖον δὴ τούτοις ἡ ὑγρότης λεπτὴ καὶ ὑδατώδης⟩·[1] ἅπαντα γὰρ τὰ γλυκέα βραχυτέρας ποιεῖ⟨ται⟩[2] τὰς πεπάνσεις, ἐὰν δὲ δή τις καὶ πρὸς τὰς ἡμετέρας χρείας ποιῇ⟨ται⟩[3] τὰς ἀφαιρέσεις, ἔτι μᾶλλον.

17. 4 αὐτῶν μὲν οὖν τούτων σχεδὸν ἐν τούτοις αἱ αἰτίαι τοῦ πρότερον καὶ ὕστερον.

οὐ μὴν καθ' ὅλων τῶν δένδρων ἐστὶν τὸ πρώϊον· πολλαὶ γὰρ διαφοραὶ καὶ ἀμπέλων καὶ συκῶν, ὥστ' 5 ἐνίων πόρρω πάνυ πεπαίνεσθαι, διόπερ ἴσως τὰς καθόλου λεκτέον αἰτίας. πρωΐκαρπα μὲν ὅσα μήτε κάθυγρα[4] μήτε ψυχρὰ τοῖς ὀποῖς, ἔτι δὲ γυμνὰ ἢ λεπτοῖς ὑμέσι περιεχόμενα, καὶ τὴν πέψιν ἔχοντα τῶν χυλῶν ὑδαρῆ καὶ μὴ παχεῖαν.

17. 5 ὀψίκαρπα δὲ τὰ ἐναντία τούτων· ὅσα κάθυγρα καὶ ψυχρά, καὶ[5] τοῖς καρποῖς ἢ τοῖς περικαρπίοις ξυλώδη καὶ σκληρά, καὶ ὧν αἱ περιοχαὶ τοιαῦται, καὶ ἅμα πλείους· ἔτι δὲ ὧν οἱ χυλοὶ καὶ πρὸς τὰς 5 πέψεις λιπαροί, καὶ ἄλλην τινὰ ἔχοντες παραπλησίαν δύναμιν (ἐὰν ξηροί τε καὶ ὀλίγην ἔχοντες ἅμα καὶ τοιαύτην ὑγρότητα πρὸς τὴν πέψιν)·[6] ἅπαντα γὰρ ταῦτα κωλυτικὰ τῶν πέψεων.

[1] ⟨ἔτι—ὑδατώδης⟩ transferred by Schneider from lines 1–5.
[2] ποιεῖται ego: ποιεῖ U.
[3] ποιῆται ego: ποιεῖ U (-ῇ N HP).
[4] κάθυγρα Schneider (*cf. CP* 1 17. 5): κάθυδρα U.
[5] καὶ Gaza (*et*), Schneider: ἢ U.
[6] [ἐὰν—πέψιν] Schneider.

[a] The mulberry, fig and vine.

a long time on the tree and its fluid stays thin and watery longer. For all trees with sweet fruit ripen sooner, and the time for the later fruit is further shortened if the earlier fruit is picked as soon as it is ready for our consumption.

17. 4 So considering these trees alone,[a] the causes of relative earliness and lateness lie (one may say) in the points mentioned.

General Characterization of Early and Late Bearers

But early ripening does not apply in each of these trees to the entire kind, for there are many varieties both of vine and of fig, so that in some the ripening occurs very late indeed, and this is why we should perhaps give the general causes: early fruiting are all that are (1) neither very fluid (2) nor with cold sap, and that further have fruit that is (3) naked or (4) wrapped in thin membranes or that have (5) juice which on ripening is watery and not thick.

17. 5 The opposites of these fruit late: (1) those full of fluid and (2) cold; (3) those with fruit or pericarpia that are woody and hard; (4) those with envelopes of this description, and then too with several of them; (5) moreover those with juice that to be concocted must be oily and those where it must have some similar power [b] (if the juice is dry [c] and for concoction must have fluid not only small in quantity but of this description); for all this hinders concoction.

[b] So viscosity requires more concoction: *cf.* *CP* 1 17. 6; 1 22. 5; 4 8. 2; 4 15. 1.

[c] For " dry " juice *cf.* *CP* 6 6. 5.

17. 6 ἐκ δὲ τῶν καθ' ἕκαστα θεωροῦσιν σύμφωνος ὁ λόγος τῶν γιγνομένων.

τῶν γὰρ ἀειφύλλων ἅπανθ' (ὡς εἰπεῖν) ὀψίκαρπα. ξυλώδεις δὲ οἱ καρποὶ καὶ τὰ περικάρπια, καθάπερ [1]
5 πεύκης πίτυος κυπαρίττου (τούτων δὲ [2] ξηροί)· ἢ λιπαροὶ ἢ γλισχρότητά τινα ἔχοντες, ὥσπερ ὁ τῆς κέδρου καὶ τῆς ἰξίας, διὸ καὶ οὐκ ὄντες μεγάλοι δυσκατέργαστοι τῷ εἶναι τοιοῦτοι, καὶ ἅμα διὰ τὴν πυκνότητα μικρὰ καὶ ἡ ἐπιρροὴ καὶ ἡ ἐπίσπα-
10 σις [3] ὅλη.

17. 7 τῶν δὲ μὴ ἀειφύλλων ὅσα κάθυγρά τε καὶ ψυχρά, καὶ ὅσα γεώδη· καὶ γὰρ ἡ ψυχρότης καὶ τὸ πλῆθος δυσέργαστον, καὶ τὸ γεῶδες καὶ ἡ ξηρότης (ὥσπερ τῶν ἀχράδων καὶ τῶν βαλάνων).[4] ὅσα
5 δὲ καὶ κέκραται [5] ⟨καὶ⟩ [6] πρὸς τούτοις [καὶ] [7] ἐν θερμασίᾳ τυγχάνει καὶ μανά, ταῦτα καὶ πρωϊβλαστῆ καὶ πρωΐκαρπα, συμμετρίαν ἔχοντα τῆς μίξεως καὶ ἐν αὑτοῖς [8] καὶ πρὸς τὸ περιέχον.

17. 8 χρὴ δὲ λαμβάνειν ἕκαστα (ὥσπερ τῶν εἰρημένων) ἐὰν μή τις ᾖ κώλυσις· οὐδὲν γὰρ ἐν τούτῳ [9] κύριον ἐπενεγκεῖν οὔτε πρωϊκαρπίαν οὔτε ὀψικαρπίαν, οὐδέ γε πλείω [10] πάντα, ἐὰν ἕτερ' ἅττα ἐναντιῶ-
5 ται.[11] λέγω δ' οἷον ἔνια γυμνόκαρπα μὲν ὄψια δέ,

[1] καθάπερ HP: καθὰ U N.
[2] τούτων δὲ U: ἄλλων δὲ Schneider (*alii* Gaza); τῶν δ' αὖ Wimmer.
[3] ἐπίσπασις Gaza (*accessus*), Heinsius: ἐπίστασις U.
[4] βαλάνων u: βαλανῶν U.
[5] καὶ κέκραται ego: καὶ κρατε U; κέκραται u.
[6] ⟨καὶ⟩ ego.
[7] [καὶ] ego.
[8] αὑτοῖς Wimmer: αυ- U (αὐ- u).

When we consider the particular kinds of trees (or plants) the explanation of their ripening early or late agrees with this general formulation. 17. 6

Thus practically all evergreens fruit late. Their fruit or pericarpion is woody, as in pine, Aleppo pine, cypress. It is dry in these; in others it is oily or has a certain viscosity, as in cedar and mistletoe. This character makes the fruit, although not large, hard to prepare; then too the close texture of the tree not only makes the influx of food to the fruit a small one but also the whole intake.[a]

Of non-evergreens those are late fruiting whose fruit is very fluid and cold, and those whose fruit is earthy; for not only coldness and the large quantity are hard to elaborate, but also earthiness and dryness, as in the fruit of the wild pear and in acorns.[b] But where the fruit is tempered in its qualities and is in addition exposed to warmth and not bunched, the tree both sprouts and fruits early, having a mixture of qualities both in itself and in relation to the surrounding air that avoids excess and defect. 17. 7

We must take each character (as in the case of the trees mentioned) with the proviso that there is no impediment. For no single character can in the meantime determine either early or late fruiting, nor yet any combination of several of them, if certain other characters oppose. I mean for example that 17. 8

[a] Hence no doubt the constant supply of food in evergreens.

[b] *Cf. HP* 3 4. 4: " alder, hazel and a certain kind of wild pear produce their fruit in autumn; oak and chestnut later still; . . . and the late wild pear in winter . . ."

---

9 *οὐδὲν γὰρ ἐν τοῦτωι* U: *οὐδὲ γὰρ ἓν τούτων* Wimmer.

10 *πλείω* ⟨*ἢ*⟩ Wimmer.

11 *ἐναντιῶται* Schneider: *ἐναντιώταται* U^{ar} (-*τα* U^{r}).

καθάπερ μῖλαξ καὶ ἄλλ' ἄττα βοτρυώδη (τὰ δὲ πρὸς τῷ ὀψίῳ καὶ κατὰ μέρος πεπαίνεται, καθάπερ ὁ βάτος). τούτων γὰρ [ὅτι][1] τὰ μέν, ὅτι[2] ψυχρὰ [τὰ] τῇ[3] φύσει, ὀψίκαρπα καὶ ὀψιβλαστῆ
10 (ὡς[4] δὲ γυμνὰ καὶ ἀκέλυφα, οὐχ ἱκανά, περικαταλαμβανόμενα[5] τῇ ὥρᾳ)· τὰ δ', ὅτι ξηρὰ τῇ φύσει· πᾶν δὲ τὸ ξηρὸν ἰκμάδος δεῖται καὶ πρὸς
17. 9 τροφὴν καὶ πρὸς πέψιν. ὧν[6] δὲ ὁ βλαστὸς ἀμφοτέρων μετέχει τῶν ὡρῶν, ἐπὶ πολὺν χρόνον παρεκτείνων.

ἀλλ' ὥς γε[7] τύπῳ εἰπεῖν ταύτας ὑποληπτέον
5 εἶναι τὰς αἰτίας. ἐπεὶ καθ' ἡλικίαν ὀψικαρπότερα, καὶ μὴ κατὰ τὰς ἐνιαυσίους ὥρας,[8] οἷον τὰ νέα τῶν φυτῶν διὰ πλῆθος ὑγρότητος καὶ τὸ ὅλον τροφῆς ὀψίκαρπα· τὰ δ' αὖ πάλιν ὡς[9] ὀψιφόρα[10] πόρρω τῆς ἡλικίας ὄντα, καθάπερ τὸ[11] ἐν Αἰγύπτῳ
10 λεγόμενον δένδρον ὃ ἑκατοστῷ[12] ἔτει μυθολογοῦσιν φέρειν καρπόν. ἡ δ' αὖ συκάμινος οὐδὲ πέττειν δύναται δι' εὐτροφίαν καὶ πλῆθος ὑγρότητος μὴ ἐπικνισθέντων καὶ ἐπαλειφθέντων ἐλαίῳ τῶν καρπῶν. οὐ μόνον δὲ τὸ πλῆθος ὀψικαρπεῖν[13]

[1] [ὅτι] HP.
[2] [ὅτι] HP.
[3] [τὰ] τῇ Schneider: τὰ τῇ U N aP; τὰ H.
[4] ὡς ego: ὅσα U.
[5] περικαταλαμβανόμενα Schneider (περιλαμβανόμενα Wimmer): περιλαμβανομένων U.
[6] ὧν Schneider (τῶν Hindenlang): ὡς U.
[7] ἀλλ' ὥς γε Scaliger (*verum quantum* Gaza): ἄλλως τε U N; ἀλλ' ὥστε HP.
[8] ὥρας Gaza (*tempore*), Itali: χώρας U.
[9] [ὡς] Schneider.
[10] ὀψίφορα Itali: ὅτιφορα U.
[11] τὸ u N HP: τῷ U.

›me plants have naked fruit but are nevertheless .te, as smilax and some others that bear clusters ınd of these plants some are not only late but ripen ıe fruit successively, like bramble). For of these .te plants with naked fruit some are late in fruiting ıd in sprouting because they are naturally cold (and ıeir character of having fruit that is naked and ithout covering overtaxes their powers when the ːason turns cold); others again are late because they ·e naturally dry, and everything dry requires moisıre both for feeding and for ripening.[a] In some 17. 9 ſ these the growing of the fruit takes so long that it xtends into the next recurrent season.[b]

But roughly speaking we must take these to be the ıuses. Indeed there are late fruiting trees in which ıe lateness is a matter of their period of life, and not ' the annual seasons, such as young trees, which uit late in this sense because of their abundance ' fluid and in general of food; again, on the other ınd, there are those which count as late bearing ecause they fruit when far advanced in age, as the ee reported[c] in Egypt of which the story is told ıat it bears fruit in its hundredth year. The g-mulberry again cannot, owing to its rich feeding ıd abundance of fluid, ripen its fruit at all unless the uit is scratched and smeared with oil.[d] Too much

[a] And so must wait for the rains of autumn.

[b] *Cf.* the passages cited in note *b* on *CP* 1 11. 8.

[c] Not mentioned elsewhere.

[d] *Cf. HP* 4 2. 1 (of the fig-mulberry): "It cannot ripen the ıit unless the fruit is scratched, and they do this with iron ıws, and the fruit that has been scratched ripens on the ırth day . . ."

ı2 ἑκατοστῷ Gaza (*centensimo*), Itali: ἑκάστωι U.

.3 ὀψικαρπεῖν Heinsius: ὀψιοκαρπεῖν U.

15 ποιεῖ τῆς τροφῆς, ἀλλὰ καὶ ἀκαρπεῖν ἔνια, καθά
ἐπί τε τῶν ἀμπέλων εἴρηται καὶ ἐπὶ τῶν ἀμ
δαλῶν καὶ ὅλως τῶν διατετραινομένων[1] καὶ τ
πληγαῖς[2] [τιτραινομένων καὶ][3] κολαζομέν
17. 10 ἅπαντα γὰρ ὅταν τοῦτο πάθωσιν, τῆς ὑγρότη
ἀπερασθείσης[4] τὰ μὲν ἐξ ἀκάρπων κάρπιμα,
δὲ καλλικαρπότερα καὶ ἐγχυλότερα γίνεται.
δὲ ἀμυγδαλῆς ἐάν τις, ἐκκόψας[5] τὸν πάτταλ
5 ἐπικαθάρῃ[6] τὴν ἐπιρροὴν τῆς ὑγρότητος ἐπὶ
δύο ἢ τρία, καὶ γλυκεῖαν ἐκ[7] πικρᾶς γίγνεσ
φασιν. δοκεῖ δὲ καὶ ἡ συκῆ ῥιζοτομηθεῖσα
κατασχασθεῖσα[8] εὔφορός τε ἐξ ἀφόρου γίνεσ
καὶ πολυκαρπεῖν μᾶλλον. σχεδὸν δὲ καὶ ⟨τὸ
10 περὶ τὰς ἀμπέλους τὰς τραγώσας ὅμοιόν ἐσ

[1] διατετραινομένων ego (*cf.* τετρένεται, τετρανη, τετρᾶναι U at *HP* 5 4. 5): -τιτ- U.
[2] ταῖς πληγαῖς U: τοῖς παττάλοις Wimmer.
[3] [τιτραινομένων καὶ] Wimmer.
[4] ἀπερασθείσης Schneider: ἀπελαθείσης U.
[5] ἐκκόψας U: ἐγκόψας Schneider.
[6] ἐπικαθάρηι U: ἀποκαθάρῃ Schneider.
[7] ἐκ N HP: ἐκκ U.
[8] κατασχασθεῖσα Schneider (*cf. HP* 2 7. 6: κατασχῶσι Const tinus: καταχοῦσι U): κατασχισθεῖσα U.
[9] ⟨τὸ⟩ Schneider.

[a] So with the "mad vine" (*CP* 1 11. 3; *cf. CP* 1 18. 4) the "goaty vine" (*HP* 2 7. 6; *CP* 1 5. 5; *cf. CP* 1 17. which may be the same. For the almond the reference is *CP* 1 9. 1. For the insertion of a peg into the almond *HP* 2 2. 11 and 2 7. 6–7. In *HP* 2 7. 6–7 this operation, ca "chastising," is said to be performed on the pear and on sc

ood not only causes late fruiting but in some trees
:ven causes failure to bear, as we said [a] of vines and
f almonds and in general of all trees that have holes
lriven in them and that are chastened by blows.[b]
'or when this is done to them, all of them, with the 17. 10
luid drained off, either bear when they had failed to
ear before or bear finer and more succulent fruit.
n the almond, if one knocks out the peg and purges
he influx of fluid over a period of two or three years
he tree is even said to change from bitter to sweet.[c]
[he fig too, when roots are cut and the trunk scari-
ied, is held to change from a non-bearer to a bearer
nd from bearing less to bearing more.[d] The
reatment of the "goaty" vine is also (one may say)
imilar: here too we must reduce the amount of fluid

thers besides the almond; and that when it is performed in
rcadia on the sorb it is called "correcting."

[b] *Cf*. *CP* 2 14. 4.

[c] *Cf*. *HP* 2 2. 11: "By tendance the pomegranate and
lmond change; . . . the almond when one inserts a peg
nd removes over a period of time the exudation that forms,
vhile keeping up the usual tendance;" *HP* 2 7. 6–7 (of trees
hat turn to leafy growth and fail to bear): "In the almond
hey even drive in an iron peg and after making a hole replace
he peg with a wooden one and cover the spot with earth; and
his some call 'chastising' the tree, as if it were getting out of
and. 7. . . . They say that an almond tree will even change
rom the bitter variety to the sweet if one digs around the
runk and after making a hole in it about a span deep allows
he exudation that collects from all sides in the hole to flow
way."

[d] *Cf*. *HP* 2 7. 6 (of trees that turn to leafy growth and fail
o bear): "In the fig in addition to cutting roots all around the
ircumference they sprinkle ashes and make slits in the trunk
nd say that the tree does more bearing."

καὶ γὰρ τούτων ἀφελεῖν δεῖ καὶ ἀντισπάσαι τὴν ε
τὴν βλάστησιν ὁρμὴν ὅπως καρποτοκῶσιν.

18. 1 ἐν τῷ αὐτῷ δέ πως γένει τῆς αἰτίας ἐστὶ καὶ
μὴ τὴν ἀρίστην καὶ πίειραν καὶ βαθύγειοι
ἀρίστην εἶναι τοῖς δένδροις, ἀλλὰ τὴν δευτέραν, τ
δὲ σίτῳ ἐκείνην. ἐν μὲν γὰρ τῇ,[2] κατὰ βάθ
5 ἰούσης τῆς ῥίζης, καὶ τῆς χώρας εὐτρεφοῦς
πλείω τῆς συμμέτρου τροφὴν ἐπισπῶνται· ἐν
τῇ λεπτογείῳ καὶ μὴ βαθείᾳ τὰς ῥίζας ἀναγκαῖ
ἐπιπολαιοτέρας εἶναι, καὶ τροφὴν [4] ἐλάττω κ
σύμμετρον. ἐπεὶ καὶ ἐν ταῖς ψαφαραῖς [5] καὶ τα
10 πετρώδεσι [6] δύνανται λαμβάνειν ἱκανήν, καθιέν
καὶ βιαζόμενα ταῖς ῥίζαις, ἔτι δὲ καὶ καταψύχε
τὰς ῥίζας ᾗ [7] γε πετρώδης δοκεῖ μᾶλλον, ὅπ
ἐπιζητεῖ τὰ δένδρα.

18. 2 ὁ δὲ σῖτος ἐν μὲν ταῖς ἀγαθαῖς συλλαμβάν
μὲν [8] πλείω διὰ τὸ μὴ κατὰ βάθους εἶναι [9] τ
ῥίζας,[10] ἐν δὲ ταῖς μοχθηραῖς καὶ καταξηραίνετ
διὰ τὸ μὴ ἔχειν πολλάς,[11] καταψύξεως δὲ δεῖται δ
5 τὸ μὴ κατὰ βάθους [12] εἶναι· ὅταν δὲ ἐπομβρί
γένωνται καὶ [13] πολὺς ὁ διασῳζόμενος, χείρων

[1] βαθύγειον u: -γιον U.
[2] τῇ ⟨ἀρίστῃ⟩ Wimmer.
[3] εὐτρεφοῦς U: εὐτραφούς u^{ac} (-οῦς u^{css}); ἀτρεφοὺς (ἀτρεφοῦς HP).
[4] ⟨τὴν⟩ τροφὴν Schneider.
[5] ψαφαραῖς U HP: ψαθαραῖς u (θ ss.); ψαφεραῖς (misreadi the θ as ε) N.
[6] πετρώδεσι u: περώδεσι U.
[7] ᾗ u: εἴ U N HP.
[8] συλλαμβάνει μὲν u: συλαμβάνειν ἒν U.

and pull against the movement toward leafy growth so that the trees may bear.[a]

18. 1 Under what we may call the same class of cause comes this: the best land, land with soil that is rich and deep, is not the best for trees, but the second-best is here the best, the first being best for cereals.[b] For in the best land, since tree roots go deep and the soil feeds them well, the trees attract more food than is good for them, whereas in light soil that has no depth the roots are of necessity shallower and the food is less and of the right amount. Indeed even in crumbly or rocky soil a tree is able to get enough food by forcing its roots deep; furthermore rocky soil is believed to cool the roots,[c] a thing sought by trees.

18. 2 Grain [d] on the other hand, because its roots are not deep, takes in more food when the soil is good, but in poor soil it even dries out, not having many roots, and needs cooling for them, since they are not deep; but when there are rains and much of the crop

[a] *Cf. HP* 2 7. 6 (of the vine): "If some tree fails to bear fruit and turns instead to leafy growth, they slit the part of the trunk that is underground and insert a stone to make it split open, and say that after this it bears."
[b] *Cf. CP* 2 4. 2, 3, 10.
[c] *Cf. CP* 3 4. 3; 3 17. 3.
[d] Wheat is meant: *cf. CP* 3 21. 2.

---

[9] *μη κατα βάθους εἶναι* U: *κατὰ βάθους ἰέναι* Schneider (after Gaza).
[10] *ῥίζας* ⟨. . .⟩ Wimmer.
[11] *πολλάς* u: *πολλούς* U.
[12] *βάθους* U^c: *βάθος* U^ac.
[13] *καὶ* U: *οὐ* Schneider.
[14] ⟨*καὶ*⟩ *χείρων* Schneider.

ἐστίν. ἐπεὶ καὶ οἱ τῶν δένδρων καρποὶ διὰ τὸ κρατεῖσθαι τῷ πλήθει τῆς τροφῆς ἐξίστανται τῶν γενῶν (ὥσπερ ἐλέχθη πρότερον).

18. 3 ἄτοπον δ' ἂν δόξειεν τὸ τῶν ὁμογενῶν ἔνια τὰ μὲν εἶναι πρώϊα, τὰ δὲ ὄψια, καθάπερ συκαῖ τέ τινες καὶ ἄμπελοι καὶ μηλέαι καὶ ἄπιοι καὶ τἆλλα· τῶν γὰρ ζῴων οὐδὲν τοιοῦτον πλὴν κυνός, ἀλλὰ 5 παρισόχρονα [1] κατὰ τὰς κυήσεις καὶ τὰς ἐκτροφάς, ἀλλὰ μόνον παραλλάττει ταῖς ὥραις κατὰ τοὺς τόκους [2] καὶ μάλιστα τά γε ξυνανθρωπευόμενα.

τάχα μὲν οὖν καὶ αἱ βλαστήσεις ἐνίων ὕστεραι, καὶ παραδιδόασιν [3] τὸ ἀνὰ λόγον· οὐ μὴν ἀλλ' ἐφ' 10 ὧν τοῦτο μέν [4] ἐστιν, ἔοικεν ὥσπερ ἐν ὁμωνυμίᾳ 18. 4 γίνεσθαι τὸ ἀπόρημα. ἔστιν γὰρ εὐθὺ τῇ φύσει

[1] ἀλλὰ παρισόχρονα Schneider (*pari tempore* Gaza; ἀλλ' ἰσόχρονα Itali): ἀλλὰ περισσόχρονα U.
[2] τόκους ego: τόπους U.
[3] καὶ παραδιδόασι U^r (-δώασιν U^ar) N HP: ἀποδιδόασι Schneider.
[4] μέν U: μή Schneider (*desit* Gaza).

---

[a] *Cf. CP* 2 16. 2–3 for the change of wheat to darnel owing to rainy weather. *Cf. HP* 8 6. 6–7: "On the whole drought is better for grain than rainy weather; for rain, besides being unfavourable in other ways, even destroys the seeds themselves, and if it does not do this, causes a luxuriant growth of weeds, so that the grain is stifled by them and starved for food."
[b] "Fruit" is here (as often) virtually synonymous with "seed;" *cf.* note *b* on *CP* 1 19. 1.
[c] *CP* 1 9. 1.
[d] Theophrastus returns to a point made at *CP* 1 17. 4; *cf.* also *CP* 4 11. 2.

survives,[a] what survives is inferior. Indeed the fruit [b] of trees also departs from its kind (as we said earlier) [c] by being overpowered by too much food.

A Problem: Early and Late Varieties of the Same Tree

It might seem odd that within the same kind certain varieties should be early, others late (as with some figs, vines, apples, pears and so on).[d] In animals the like is found only in the dog,[e] animals otherwise taking the same time for pregnancy and developing the embryo, varying only with the season of birth, as do above all the animals that live with man.[f] 18. 3

In some of the varieties of trees, to be sure, it may be that the sprouting is also late, and so corresponds to the late bearing. Nevertheless in the kinds of tree where the difference does occur the difficulty appears to rest on a mere community of name (as it were). For just as the distinction that in animals as well as plants makes some tame and some wild comes 18. 4

[e] *Cf.* Aristotle, *History of Animals*, vi. 20 (574 a 20–30): "The Laconian bitch is pregnant for one sixth of a year . . . Some bitches are pregnant for one fifth of a year . . ., some for one fourth of a year . . ." It might appear that Theophrastus is ignoring the difference between seven, eight, nine and ten month pregnancies in man (*cf.* Aristotle, *History of Animals*, vii. 4 [584 a 34–b 1]), but these do not differ with different varieties of man.

[f] *Cf.* Aristotle, *History of Animals*, v. 8 (542 a 26–30) [of the times of copulation]: ". . . but man most of all does this at all seasons, and so do many animals living with man, because of the warmth and good feeding, among those whose pregnancy is of short duration . . ."

καὶ[1] τοῖς ζῴοις καὶ τοῖς φυτοῖς, ὥσπερ τὰ ἥμερα καὶ τὰ ἄγρια, καὶ τὰ[2] πολύκαρπα, τὰ δ' ὀλιγόκαρπα, τὰ δ' ὅλως ἄκαρπα. τὸ[3] γὰρ τῶν
5 ἐρινεῶν ἕτερον γένος, οὐ δυνάμενον πέττειν οὐδὲ δικνεῖσθαι πρὸς τὴν τελείωσιν. ἕτερον δὲ[4] καὶ τὸ τῶν ἀμπέλων τῶν μαινομένων καλουμένων, αἳ οὐ μόνον βλαστάνουσιν[5] ἀλλὰ καὶ πέττουσιν καὶ ἀνθοῦσιν καὶ βοτρυοῦνται, καὶ οὐ δύνανται
10 τελειοῦν. ὡσαύτως δὲ καὶ τῶν ῥοῶν,[6] καὶ εἴ τι ἄλλο μέχρι τοῦ ἄνθους ἀφικνεῖται μόνον. ἐν γὰρ τῇ ἰδίᾳ φύσει τὰς διαφορὰς ἕκαστα τούτων ἔχοντα, δικαίως ἕτερ' ἂν[7] λέγοιντο κατὰ τὸ εἶδος.

18. 5 ὅσα δὲ δύνανται πεπαίνειν καὶ βλαστάνειν καὶ ἀνθεῖν κατ' ἄλλα καὶ ἄλλα μέρη, καθάπερ ἡ Μηδικὴ μηλέα, ταῦτα μείζω τινὰ ἔχει καὶ ἰδιωτέραν δύναμιν ἐν ἑαυτοῖς, εἴπερ αἰεὶ διὰ τέλους
5 τοῦτο δρᾷ· παρόμοιον γὰρ τὸ συμβαῖνον ὥσπερ ἐν τῷ ἀέρι πρότερον ἐλέχθη τῷ μαλακῷ καὶ εὐκράτῳ[8]

[1] καὶ ⟨ἐν⟩ Schneider.
[2] τὰ ⟨μὲν⟩ Schneider.
[3] τὸ Schneider: τὰ U HP; ἀλλὰ N.
[4] δὲ u: δι U.
[5] βλαστάνουσιν u N HP: βλαστανουσῶν U.
[6] ῥοῶν U^ac: ῥόων U^c.
[7] ἕτερ' ἂν Wimmer: ἕτερα ἂν HP^c(ἕτερα P^ac); ἕτερον U (ἕ- u N).
[8] εὐκράτωι u: εὐβράτωι U.

[a] *Cf. HP* 1 3. 6: "And at the same time plants appear to have a certain natural distinction beginning immediately with the two groups of wild and tame . . .;" *HP* 3 2. 2: "for whatever does not accept domestication, as among animals,

directly from a difference in their nature,[a] so too with the distinction that makes some trees bearers of much fruit, others of little, and others of none. For wild figs are a different kind of tree from figs, a kind that lacks the power to concoct [b] and reach full development. So too the so-called mad vines [c] are a different tree from the vine: they not only sprout but also initiate concoction and flower and form clusters, but they lack the power to complete the fruit. Similarly with those pomegranates [d] and other trees that only get as far as the flower. For since each of these has the character that marks it off in its own distinctive nature, it would rightly be called distinct in kind from the other varieties.

Ever-Fruiters

18. 5

As for those trees that are able to ripen (and sprout and flower) in one part after another, like the citron,[e] they have within themselves some greater and more distinctive power, if they keep this up throughout the year. For what happens here is similar to what we said earlier [f] took place in that mild and well-

is wild by its nature;" *HP* 1 14. 3–4: "Of just about all trees and plants there happen to be a number of sub-kinds in each kind, for practically no kind is simple. Rather the distinction in kind between those called tame and wild is the most evident and great, as between fig and wild fig, olive and wild olive, pear and wild pear . . ."

[b] *Cf. CP* 2 9. 6; 2 9. 14; 4 4. 3.

[c] *Cf. CP* 1 11. 3.

[d] *Cf. HP* 1 13. 4: "Indeed certain varieties of both vine and pomegranate are unable to produce completely developed fruit, generation proceeding only as far as the flower."

[e] *Cf. CP* 1 11. 1; *HP* 4 4. 3.

[f] *CP* 1 13. 1–2; *cf. CP* 1 11. 6.

καθ' ὃν ἡ καρποφορία καὶ ἡ βλάστησις· πλὴν ἐκείνων μὲν ὁ ἀὴρ αἴτιος, διὸ καὶ πᾶσι κοινόν,[1] ἐνταῦθα δὲ ἡ τοῦ δένδρου φύσις καὶ δύναμις, 10 εὔκρατος οὖσα πρὸς ἁπάσας τὰς ὥρας.

19. 1 ἐπεὶ δ' ἕτερον τὸ περικάρπιον, τοῦτο γὰρ πρὸς τὴν χρῆσιν ἡμῶν, ἀνάγκη[2] μὲν τούτων[3] ὅρον τινὰ καὶ[4] τῶν χρόνων[5] εἶναι, τὸν δὲ ὑπερβάλλοντα λυμαίνεσθαι χειμώνων τε καὶ ὑδάτων καταλαμβα- 5 νόντων· τῶν δὲ καρπῶν[6] μὴ εἶναι, συγκαταρρέουσι γὰρ τούτοις ⟨ἢ⟩[7] προεκπηδῶσιν,[8] ὥσπερ οἱ τῆς πιττώδους πεύκης καὶ ὅλως τῶν κωνοφόρων, προσηρτημένων[9] γὰρ ἔτι τῶν κώνων ἐκπηδᾷ τὰ κάρυα καὶ καταλείπονται κενοί. ταὐτὸ δὲ συμ- 10 βαίνει τοῦτο καὶ ἐπὶ τῶν κυπαρίττων, ἀλλ' ἔνθα μὲν τὸ σπέρμα καρυῶδες, ἔνθα δὲ ὑμενῶδες καὶ ἀμενηνόν.[10]

19. 2 ὅσα μὲν οὖν ξυλώδεσιν ἢ δερματικοῖς τισιν περιέχεται, καθάπερ τά τε κάρυα καὶ βάλανοι, ταῦτα μὲν περιστεγόμενα διατηρεῖται

[1] πᾶσι κοινὸν Wimmer (κοινὸν Link): πᾶσι κοινός U.
[2] ἀνάγκη U^r N HP: -ην U^ar.
[3] τούτων u^c N HP (τοῦτων U): τοῦτον u^ac.
[4] [καὶ] a.
[5] τῶν χρόνων ego (χρόνον Schneider): τὸν χρόνον U.
[6] τῶν δὲ καρπῶν Gaza, Itali: τὸν δε καρπὸν U.
[7] ⟨ἢ⟩ ego (*alia* Gaza).
[8] προεκπηδῶσι Gaza, Scaliger: πρὸσεκπιδῶσιν U (προσεκπηδῶσιν u).
[9] προσηρτημένων u: -νω U.
[10] ἀμενηνὸν u: -νῶν U.

---

[a] *Cf.* Empedocles' κατ' ἠέρα (*CP* 1 13. 2).
[b] That is, the seed or germ, as opposed to the pericarpion (literally, " what surrounds the fruit ").

tempered air to which [a] fruiting and sprouting were due, except that there the air was the cause, which was why the effect was the same in all, whereas here the cause is the nature and power of the tree, and it is this that is well-tempered to all the different seasons.

The Pericarpion has a Time Limit that does not Apply to the Seed

19. 1 Since the pericarpion is distinct from the fruit proper, the pericarpion serving our needs, it necessarily has a time limit, any time in excess of this limit injuring the pericarpia, since they are overtaken by the winter cold and rains; but the limit necessarily does not apply to the fruit.[b] For this is either shed with the pericarpion or first drops out of it, as in the pitch-pine and cone-bearers in general, where the nuts drop out while the cones are still on the tree and leave them empty; the same also happens in the cypress (in the cone-bearers however the seed is nut-like, whereas here it is membranous and thin).[c] 19. 2 Now all seeds in woody or leathery shells, as nuts and acorns,[d] are preserved by this shelter until the

[c] *Cf. CP* 1 5. 4; 4 4. 3.

[d] For these seeds and the following *cf. HP* 1 11. 3–4: " The seeds themselves are in some fleshy from the start [that is, as we proceed from outside in], as all that are nut-like and acorn-like; in others a stone encloses the fleshy part, as in olive, bay and others; in others there is only a stone or they are stone-like and as it were dry . . . This is most evident in the seed of the date-palm, for it has not even an internal hollow but is solid throughout . . . 4. They also differ in that some are massed together, whereas others are separated and in rows . . . And of those massed together some are enclosed by a single container, like those of pomegranate, pear, apple, vine and fig . . ."

πρὸς τὴν τῆς βλαστήσεως ὥραν· ὧν δὲ σαρκώδη
5 τὰ περικάρπια, ταῦτα δὴ [1] σηπομένων καὶ περιρρεόντων αὐτὰ καθ' αὑτὰ σῴζεται, τὰ μὲν ὄντα [2] ξυλώδη, καθάπερ τὸ γίγαρτον καὶ ὁ τοῦ φοίνικος καὶ ὁ τῆς ἐλαίας [3] πυρήν, τὰ δὲ ἐν ὑμέσι καὶ χιτῶσι περιεχόμενα πλείοσιν, τὰ δὲ καὶ ἀλλήλοις
10 πως συνημμένα καὶ κοινὴν περιοχὴν ἔχοντα, καθάπερ καὶ τὰ τῶν ἀπίων καὶ μήλων,[4] ἅπαντα γὰρ ταῦθ' ὡς εἰπεῖν ἐν μείζοσι περικαρπίοις ἵνα πλείω διαμένῃ χρόνον.

19. 3 τῶν μὲν οὖν καρπῶν αὕτη φυλακὴ καὶ σωτηρία πρὸς τὴν γένεσιν· ἡ δὲ [περὶ] [5] τῶν περικαρπίων ὅμως [6] ὅρον τινὰ ἔχει (καθάπερ ἐλέχθη) πρὸς τὴν χρείαν. ἐπεὶ [7] καὶ οἱ χυλοὶ χρονιζομένων,[8] καὶ
5 ἀνυγραινόμενοι χείρους γίνονται, τῶν ⟨δὲ⟩ [9] οὐδὲ πλείους οὐδὲν [10] ἀπό τινος ὥρας, ὥσπερ οὐδὲ τῶν ἐλαῶν ἀπ' Ἀρκτούρου· μέχρι τούτου γὰρ τὸ ἔλαιον ἐγγίνεσθαι δοκεῖ, κατὰ δὲ τοῦτον [11] τῆς σαρκὸς ἡ αὔξησις, καὶ ἐάν γε δὴ πλείω ποιῇ ὕδατα,[12] καὶ
10 χεῖρον γίνεσθαι τὸ ἔλαιον, ἀμόργην [13] λαμβάνον

[1] δὴ U N HP: δὲ a.
[2] μὲν ὄντα Basle ed. of 1541: μένοντα U.
[3] ἐλαίας Schneider: ἴτεας U.
[4] μήλων u: μηλῶν U.
[5] [περὶ] HP.
[6] ὅμως ego (παραμονὴ Wimmer): ὁμοια U (ὁμοία u N); ὁμοίως HP.
[7] ἐπεὶ u: ἐπὶ U.
[8] χρονιζομένων ego (τῶν μὲν χρονιζόμενοι Schneider): χρονιζόμενοι U.
[9] ⟨δὲ⟩ HP.
[10] οὐδὲν U^{ar}: οὐδὲ U^{r} N HP.
[11] κατὰ δὲ τοῦτον ego (μετὰ δὲ τοῦτο Schneider; ἀπὸ δὲ τούτου

rrival of their sprouting season. But those in leshy pericarpia are preserved without the peri-arpia, which decompose and fall away, some of the eeds being woody, like the grape pit and stone of he date-palm and olive, others are wrapped in several nembranes and coats, some moreover are attached n a way to one another and also enclosed in a com-non container,[a] as again with pears and apples (for ractically all this last group are in larger pericarpia, o that they may survive longer).

19. 3

This, then, is how the fruit proper is kept and pre-erved for reproduction. But the preservation of he pericarpion has nevertheless a certain time limit as was said)[b] that is fixed by its service to man. ndeed even the juice, when the pericarpion is etained for long, not only gets watery [c] and deteri-rates, but in some pericarpia it does not even in-rease after a certain season, as in olives after the ising of Arcturus.[d] For up to that time (it is held) he oil is produced in them, but at the time of Arc-urus what increases is the flesh, and indeed if there s more rain than usual the oil is said to deteriorate, cquiring more watery sediment, this being often

[a] That is, pome fruits, which are composed of seed sur-ounded by pericarp, and pericarp surrounded by fleshy floral up.

[b] *CP* 1 19. 1.

[c] When the rains begin.

[d] *Cf. CP* 6 8. 1–5.

Vimmer): κατα δε τοῦτου U (κατὰ δὲ τούτου u N); κατὰ δὲ οῦτο HP.

12 [ὕδατα] Hindenlang, understanding αὔξησιν with πλείω.

13 ἀμόργην u N (ἀμοργὴν HP): αμωλργην U[ar]; αμωργην U[r].

πλείω, πολλάκις δὲ[1] καὶ σηπομένου τοῦ καρπο
19. 4 περιμένουσι δὲ τὴν πέπανσιν καὶ οὐκ εὐθὺς ἀφα
ροῦσιν, ὅτι καὶ ἡ κατεργασία καὶ ἡ ἀφαίρεσ
χαλεπωτέρα, καὶ ἔτι τὰ δένδρα λυμαίνοιτ'
ῥαβδιζόμενα. φαίνεται δ'[2] οὖν (εἴπερ τοῦ
5 ἀληθές) ἡ τοῦ θερμοῦ φύσις δημιουργεῖν
ἔλαιον καὶ τὴν λιπαρότητα τοῦ χυλοῦ, συμμετρί
ἔχουσα πρὸς τὸ ὑποκείμενον· ἡ δὲ πλείων
ὥσπερ ἀλλοτρία καὶ ἐπίθετος πρὸς περιττότητος
χώραν,[5] οἷον ἀντισπῶσα μᾶλλον εἰς τὴν σάρ
19. 5 ⟨τὴν⟩[6] τοῦ χυλοῦ δύναμιν. ὃ καὶ τοῦ θέρο
καὶ τοῦ χειμῶνος συμβαίνει, γιγνομένων ὑδάτ
ἐκ Διός, καὶ βρεχομένων τοῖς ναματιαίοις[7]
ὁρμῇ τῆς αὐξήσεως οὔσης. ἐκσαρκοῦνται γ
5 καὶ ἀπολλύασιν τὸ ἔλαιον διὰ τὴν πολυτροφία
ἂν μὴ μετὰ ταῦτα αἰθρίαι γινόμεναι καταξ
ράνωσιν, οὕτως δὲ σῴζεται καὶ πληθύει μᾶλλο
ὅπερ ἤδη καὶ πρότερον πολλάκις γέγονεν κ
τὸ τελευταῖον [ἤδη][8] ἐπ' ἄρχοντος Νικοδώρο
10 διὸ καὶ ἡ ῥύσις[9] ἐγένετο καλλίων, ἐκ γὰρ τ
ἡμίσεων ἡ αὐτή.

τὰς μὲν οὖν πέψεις ὅτι πρὸς τὴν χρείαν τ
ἡμετέραν εὑρίσκομεν ἐκ τούτων, καὶ ἐκ τ
πρότερον δῆλον ἐν οἷς ὑπὲρ τῶν ἀγρίων εἴπομεν.

[1] [δὲ] Schneider.
[2] [δ'] Schneider.
[3] πλείων u HP: πλεῖον U (πλείον N).
[4] περιττότητος ego: περιττώματος U.
[5] χώραν U^c: χάριν U^ac.
[6] ⟨τὴν⟩ HP.
[7] ναματιαίοις Schneider: ναματίοις U.
[8] [ἤδη] Schneider; δὴ Keil.
[9] ῥύσις U^r N HP: ῥύησις U^ar.

attended by decomposition of the fruit. Producers however wait for the olives to ripen and do not harvest them as soon as the oil is ready because it is harder before this not only to prepare the oil but also to harvest the olives, and the trees moreover would suffer from the attendant cudgelling. However that may be, it appears (if the belief is true) that it is the nature of heat [a] that manufactures the oil and produces the fattiness of the juice, when the amount of the heat is right for the material on which it operates; and that more heat is (so to say) an inappropriate and extraneous addition, to be accounted an extravagance, since it (as it were) diverts from its task the power that was busy with the juice and forces it to attend more to the flesh. This diversion also occurs in summer and in winter, when the rains fall [b] and when the trees are watered with ground water [c] when growth is under way. For the fruit turns to flesh and loses its oil owing to the abundant feeding (unless clear weather follows and dries the water out, in which case the oil is saved and increases in amount. This has happened many times before, most recently in the archonship of Nicodorus.[d] This is why the yield from the oil-presses was finer, half as many olives yielding the same quantity of oil).

19. 4

19. 5

And so we find from all this that concoction [e] serves the requirements of man. This is also evident from the earlier [f] discussion where we dealt with the fruit of wild trees.

[a] Both external and internal heat.
[b] In (early) winter.
[c] In summer.
[d] 314–313 B.C.
[e] Of the pericarpion.
[f] *CP* 1 16. 1.

20. 1 φέρει δὲ τοὺς καρποὺς τὰ μὲν ἐκ τῶν ἔνων, τὰ δὲ ἐκ τῶν νέων βλαστῶν, διεστῶτα ταῖς φύσεσιν εὐθὺς κατὰ τὰς κράσεις· ὅσα μὲν [1] ξηρὰ καὶ πυκνὰ καὶ ξυλώδη, ταῦτα μὲν ἐκ τῶν ἔνων, ἅτε μικρᾶς καὶ 5 βραχείας οὔσης τῆς ἐπιρροῆς, ἅμα δὲ καὶ ὁ βλαστὸς ἀσθενής, ὥστε μήτε μετενεγκεῖν δύνασθαι μήτε κατασχεῖν (ὥσπερ ὁ τῆς ἐλαίας). ὅσα δὲ ὑγρὰ καὶ μανὰ καὶ τὸ ὅλον εὐτραφῆ, ταῦτα ἐκ τῶν ⟨νέων⟩,[2] ἀθρόος [3] γὰρ [4] ἡ ὁρμὴ καὶ πολλή, διὸ 10 ἄμφω δύναται ποιεῖν ἅμα, τόν τε βλαστὸν καὶ τὸν καρπόν. οὐ μὴν ἀλλὰ [5] καὶ ἐκ [6] τῶν ἔνων φέροντ' ἀποβλάστημά τι [7] ποιεῖται μικρόν, καὶ οὐκ εὐθὺς ἐκ τοῦ ξυλώδους ὁ καρπός· οὐδὲ γὰρ οὐδὲ πέφυκεν, ἂν μή τι παράλογον.[8]

20. 2 ἰδιωτάτη δὲ καὶ πρὸς τὰ ἄλλα καὶ πρὸς αὐτὰ τὰ ξυλώδη, τῶν καρπῶν ἡ γένεσις τοῦ φοίνικος, οὐ τῷ φέρειν ἀπό τινων ἔνων ἢ νέων, ἀλλὰ τῷ κυούμενον [9] πρότερον, ὥσπερ τὰ σταχυηρὰ τῶν 5 σπερμάτων, ἐκφαίνειν· [10] ἐκ γὰρ τῶν ὁμογενῶν, ἃ

[1] μὲν ⟨γὰρ⟩ Schneider.
[2] ⟨νέων⟩ Itali after Gaza.
[3] ἀθρόος ego (ἀθρόα Wimmer): ἀθρόως U.
[4] γὰρ Gaza (*enim*), Itali: τε U.
[5] ἀλλὰ u: ἄλλα U.
[6] ⟨τὰ⟩ ἐκ Schneider.
[7] ἀποβλάστημά τι U^{c}: ἀποβλαστήματι U^{ac}.
[8] παράλογον u: παραλόγον U.
[9] τῷ κυούμενον Schneider: τῶν κυουμενων U.
[10] ἐκφαίνειν Schneider: εκφαίνει U.

[a] *Cf. HP* 1 14. 1: "Trees differ in the bearing of fruit in the following points as well: some bear from the new shoots, some from last year's, some from both. Fig and vine bear from the new, olive pomegranate apple almond pear myrtle and just about all such trees bear from last year's . . .; some twice-

Fruit from This Year's and from Last Year's Shoots

Some trees bear on last year's shoots, some on this year's,[a] the difference in their natures beginning directly with the difference in their special tempering of qualities. All dry, close-textured and woody trees bear from last year's, since the influx of food to the shoot is small and gradual, and then too the new shoot is too weak to transmit or retain[b] the food, as in the olive; on the other hand all trees that are full of fluid, open-textured and (in a word) well-fed bear from the new, for the impulse of growth moves in a mass and is abundant, and this is why the tree can create both products at the same time, the fruit together with the shoot. (Still, even when they bear from the old shoots, the trees produce a small offshoot[c] first, and the fruit does not come directly from the woody part; indeed, barring anomalies, such growth of fruit from a woody part is not in the natural course of things.) 20. 1

Fruit Without a Shoot: the Date-Palm

Most peculiar, both in comparison to the rest and to its own group of woody trees, is the fruiting of the date-palm. The peculiarity does not lie in bearing on a shoot, whether last year's or this year's, but in the tree's first being pregnant with the fruit and then disclosing it, as grains do that have an ear. For 20. 2

bearing apples and other fruit-trees bear from both the new shoots and last year's . . ."

[b] Similarly the olive will not propagate from its twigs, which are too thin and dry (*CP* 1 3. 2).

[c] The pedicel.

καλοῦσί τινες πλοῖα, περιρρηγνυμένων[1] ἐκφαίνεται καθάπερ στάχυς ἡ ῥάβδος ἔχουσα πρὸς ἑαυτῇ τὸν καρπόν. αἰτιάσαιτο δ' ἄν τις τὴν ξηρότητα τοῦ δένδρου καὶ τὴν ὅλην μορφήν· ἐπεὶ γὰρ φυλακῆς 10 οἱ καρποὶ δέονται καὶ ἔξω συνιστάμενοι καὶ ἐξ αὐτῶν προφαινόμενοι, τοῖς μὲν ἄλλοις τὰ φύλλα ταῦτα ποιεῖ (καθάπερ εἴρηται)· τούτου δὲ ἐπείπερ οὐ παρὰ τὸ φύλλον ὁ καρπός, ἀναγκαῖον ἐν τούτῳ[2] πως τὴν σύστασιν γενέσθαι, ὅπως ἰσχύων[3] 15 ἤδη καὶ μεμορφωμένος[4] ἀποδοθῇ τῷ ἀέρι.[5] διόπερ ἐν τῷ συγγενεῖ καὶ οἰκείῳ τὴν κύησιν ἐξέτεκεν.

ἴσως δὲ εἴπερ ὅμοιον τῷ[6] σταχυοβολεῖν, τοῦτο δὲ πλειόνων ἐστίν, καὶ τὴν αἰτίαν κοινήν τινα 20 λεκτέον. ἀλλ' εἰ ἄρα μόνῳ[7] τῶν δένδρων τῷ φοίνικι τοῦτο συμβαίνει; τὸ γὰρ ἴδιον ἐν τοῖς ὁμογενέσιν θαυμάζεται.

περὶ μὲν οὖν τούτου σκεπτέον.

20. 3 παρενιαυτοφόρα δὲ καὶ οὐκ ἐπετειοφόρα[8] τῶν δένδρων (ὡς τύπῳ λαβεῖν) τὰ ξηρὰ καὶ ξυλώδη

[1] ἃ (ἃ u)—περιρρυ(η u)γνυμένων U: N HP omit.
[2] τούτωι U: αὐτῷ Wimmer.
[3] ἰσχύων Schneider: ἴσχύον U.
[4] μεμορφωμένος Schneider: μεμορφομένον U.
[5] ἀέρι u: ἔαρι U.
[6] τῷ HP: τὸ U N.
[7] μόνῳ Scaliger: μόνον U.
[8] παρενιαυτοφόρα—ἐπετειοφόρα Schneider: παρὲνιαυτόφορα—ἐπετειόφορα U.

---

[a] The spathes bearing the fruit. They are "of uniform substance" with the wood of the tree, unlike shoots or leaves.

when the parts of uniform substance, which some call "boats,"[a] break open, the branch[b] bearing the fruit is disclosed like an ear of grain. One would take as cause of this the dryness of the tree and its general conformation: for since fruit, whether formed outside or disclosed from the tree itself, requires protection, this protection in other trees is afforded (as was said)[c] by the leaves, but in the date-palm, since the fruit is not found next to the leaf, it was necessary that its formation should take place within the tree somehow, so that it should already be strong and already have its shape when exposed to the air. And this is why the tree gives birth to its embryo in a part that is of uniform substance with the tree and intimately associated with it.

But perhaps if this is similar to putting forth ears, and such putting forth of ears occurs in a number of plants, we should also give a cause that applies to all. But may it not be that among *trees* this occurs only in the date-palm? For it is what is isolated in plants *of the same kind* that excites wonder.

This, then, is a case to be investigated.

Bearing in Alternate Years

Those trees bear in alternate years and not annually which (roughly speaking) are dry and woody 20. 3

[b] That is, the axis of the panicle.

[c] *HP* 1 2. 1: " Other parts are as it were annual parts serving the production of fruit, as leaf . . ." Aristotle is more explicit: *cf. On the Soul*, ii. 1 (412 b 1–3): "Among organs too are the parts of plants, although quite simple; thus the leaf is a protection for the pericarpion, and the pericarpion a protection for the fruit; " *Physics*, ii. 8 (199 a 25–26): ". . . the leaves (*sc.* are produced) for the sake of sheltering the fruit."

καὶ ὅσα μὴ ἐκ τῶν νέων, ἀλλ' ἐκ τῶν ἕνων φέρει τοὺς καρπούς. οἷον γὰρ προσυλλέξαι [1] δεῖ καὶ
5 προγεννῆσαι τὸ γεννῶν, οὐ δύναται δ' ἅμα ταῦτα διὰ τὴν πρότερον λεχθεῖσαν αἰτίαν, ὥσπερ τὰ εὐαξῆ [2] καὶ εὔτροφα.

μάλιστα δ' ἐπιδήλως [3] ἡ ἐλαία τοιοῦτον, καὶ γὰρ ἀσθενέστατον καὶ ἐπικηρότατον, καὶ ἅμα ῥαβδι-
10 ζομένη πονεῖ καὶ κατακοπτομένη [4] τὰς θαλλείας· [5] ἐπεὶ ὅσοι γε μὴ οὕτω συλλέγουσιν, ἀλλὰ αὐτομάτως ἀεὶ τὴν ἀπορρέουσαν, καὶ ἀποσείοντες, ἐπετειο-

20. 4 φορεῖν φασι μᾶλλον. μέγα δὲ καὶ αἱ χῶραι διαφέρουσιν. ἐν γοῦν τῇ Ὀλυνθίᾳ φασὶν ὡς ἀεί τι καρποφορεῖ, παραλλαγὴν δὲ ποιοῦνται κατὰ τριετίαν· ὅσα ⟨γὰρ⟩ [6] ἰσχυρότερα τῶν ἐκ τῶν ἕνων [7] φερό-
5 ντων, οἵας ἂν ὁ θεὸς ἄγῃ τὰς ὥρας, οὕτως καὶ τὰ τῶν καρπῶν ἀποδιδόασιν, ὥσπερ καὶ τὰ ἐκ τῶν νέων καρποφοροῦντα· καὶ γὰρ ταῦτα ξυνακολουθεῖν [8] ταῖς ὥραις καὶ τῇ κράσει τοῦ ἀέρος.

9–13. Varro, *R.R.* i. 55. 3: qui quatiet (*sc.* oleam), ne adversam caedat, saepe enim ita percussa olea secum defert de ramulo plantam . . . nec haec non minima causa quod oliveta dicant alternis annis non ferre fructus aut non aeque magnos.

2–3. Varro, *R.R.* i. 44. 3: . . . in Olynthia quotannis restibilia esse dicunt, sed ita ut tertio quoque anno uberiores ferant fructos.

1 προσυλλέξαι U^r N HP: προσσυ- U^{ar}.

2 εὐαυξῆ Liddell-Scott-Jones (*s.v.* εὐξυλῆ; εὔχυλα Schneider): εὐξυλῆ U.

3 ἐπιδήλως U^c: ἐπιδηλῶσει U^{ac}.

4 κατακοπτομένη u: κατακοπτομένης U; κατακομένη N; κατακεκομμένη HP.

5 θαλλείας u N (θαλλειας U): θαλείας HP.

and which bear not on new shoots but on those of last year. For to bear annually the tree must (as it were) first form a collection [a] and first generate the generator of the fruit; [b] and these trees, unlike the ones that grow rapidly and feed well, do not have the power to do both [c] in the same year for the reason mentioned above.[d]

The olive is most noticeably such a tree that is unable to do both, for it is very weak and delicate and then too suffers from the cudgelling and breaking off of its branches. Indeed growers who do not use this way of harvesting by cudgelling but gather the fruit as it drops of its own accord or shake the tree, say that it does more annual bearing. Countries too differ greatly in this. Thus at Olynthus it is reported that every year there is some olive crop, but that there is a fluctuation in yield between the odd years, since in the odd years the stronger trees among those bearing from old wood will also bear from the new if the season is good, just as the trees bearing on the new wood will then have a better yield, for these too respond to the seasons and temperateness of the air.

20. 4

[a] The collection of warmth and fluid from which the shoot arises: *cf*. *CP* 1 11. 4.

[b] The generator of the fruit is this year's shoot.

[c] That is, to generate the shoot and the fruit that comes from it.

[d] *CP* 1 20. 1.

---

6 ⟨γὰρ⟩ ego (⟨δὲ⟩ Schneider).

7 ἕνων Gaza (*ramis annotinis*), Itali (ἕνων): νέων U.

8 ξυνακολουθεῖν U[ar]: -εῖ U[r] N P (συνακολουθεῖ H).

20. 5 συμβαίνει δέ, ὅταν μὲν εὐβλαστῶσιν ἄγαν, ἀκαρπεῖν μᾶλλον, ὅταν δ' εὐκαρπῶσιν, ἀβλαστεῖν, ὡς οὐ δυναμένης εἰς ἄμφω διήκειν τῆς φύσεως, ἀλλὰ καὶ καταναλισκούσης θάτερον πρὸς θάτερον.

5 ἴδιον δὲ τὸ ἐπὶ τῆς συκῆς καὶ τῶν λευκῶν ἀμπέλων συμβαῖνον, ὥς τινές φασιν· ταῦτα γὰρ ὅταν εὐβλαστῶσιν, τότε μάλιστα εὐκαρπεῖν. εἰ δὲ τοῦτό ἐστιν, καὶ τὸ πρότερον εἰρημένον ἀληθές (ἡ γὰρ εὐβλάστεια [1] ἀφαιρεῖται [2] τοὺς καρπούς), 10 γίνεται [3] δὲ τοῦτο μάλιστα χώρας ἀρετῇ καὶ ἰσχύϊ τῶν δένδρων (ἐν τῇ [4] μὲν γὰρ ἀφθόνῳ [5] τάδε [6] ἑλκύσαι δεινὰ διὰ τὴν ἰσχύν, ὥσπερ ἡ ἀμυγδαλῆ καὶ ἡ ῥόα), δῆλον [7] ὡς ὅσα τὴν φύσιν ἀσθενῆ, ταῦθ'

20. 6 ἥκιστα ὑπερβλαστάνει,[8] ἀλλὰ σύμμετρος αὐτῶν ἡ εὐβλαστία γίνεται πρὸς τὴν καρπογονίαν (δεῖ γὰρ μήθ' ὑπερβλαστές, μήτε κακοβλαστὲς [9] εἶναι τὸ 5 καρποτοκῆσον), ἡ δὲ συκῆ καὶ ἡ ἄμπελος ἡ τοιαύτη μάλιστα λαμβάνει τὸ ξύμμετρον διὰ τὴν ἀσθένειαν

[1] εὐβλάστεια ego: εὐβλαστία U.
[2] ἀφαιρεῖται (-ρεῦται u) N HP: ἀφαιρῖῆιται U.
[3] γίνεται Schneider: γίνεσθαι U.
[4] τῆι U: γῇ Wimmer.
[5] ἀφθόνῳ Wimmer: ἄφθονος (ἄ- u) U; ἄφθονα N HP.
[6] τάδε Wimmer: τὰ δὲ U.
[7] δῆλον U N: δῆλον δὴ HP; δῆλον δὲ a.
[8] ὑπερβλαστάνει (υ- u) N HP: υπερβλαστάνη U^c (υπερ in an illegible erasure).
[9] ὑπερβλαστὲς—κακοβλαστὲς Wimmer: ὑπέρβλαστες—κακόβλαστες U.

[a] Fig and vine fruit from new wood: *cf*. *HP* 1 14. 1, cited in note *a* on *CP* 1 20. 1.

The Relation Between Exceptionally Good Sprouting and Bearing

20. 5 It so happens that when there is an exceptionally good sprouting of shoots trees tend to produce little fruit, whereas abundant fruiting is attended by a poor production of shoots. This implies that the nature of the tree lacks the power to achieve both objects, and proceeds to expend the provision for the one on the other.

An Apparent Exception

What occurs in the fig and white vine [a] is limited to them, as some assert; these trees fruit best when they sprout well. If this is so; and if what was said before [b] is true (that good sprouting diminishes the crop); and if this is mainly due to excellence of the country and strength of the trees (for when the soil is generous such trees—as the almond and pomegranate [c]—are given to drawing food in profusion by reason of their strength); it is clear that 20. 6 trees naturally weak are the least given to over-sprouting, and that their " good sprouting " turns out instead to be sprouting in the right amount for producing fruit (for what is to produce fruit must neither over-sprout nor sprout poorly), and that the fig and this sort of vine [d] acquire this proper adjustment in their sprouting mainly beause of their weak-

[b] In the preceding paragraph.

[c] Almond and pomegranate fruit from old wood: *cf. HP* 1 14. 1, cited in note *a* on *CP* 1 20. 1.

[d] The white vine. In plants as in animals the light are weaker than the dark: *cf. CP* 3 22. 2.

(δεῖ δὲ ἴσως καὶ χώραν εἶναι μὴ ἀγαθήν, ἵνα μηδὲ ἐκ ταύτης ᾖ ὑπερβολή,[1] διόπερ εἰς [2] τὰς νήσους τὰ τοιαῦτα μᾶλλον ξυμβαίνει). κοινὸς δ' ὁ λόγος 10 περὶ πάντων τῶν ἀσθενῶν.

τούτου μὲν οὖν ἐνταῦθα τοῦτο [3] αἴτιον ὑποληπτέον.

21. 1 ἐν ἅπασι δὲ τοῖς καρποῖς [4] τὸ περικάρπιον πρότερον μᾶλλον ἢ αὐτός γε [5] ὁ καρπὸς καὶ τὸ σπέρμα γίνεται· τοῦτο δ' οὐ μόνον [6] ὅτι ξυλώδη καὶ πυρηνώδη γίνεται τὰ πολλὰ τῶν σπερμάτων, 5 τὰ δὲ τοιαῦτα συνίσταται βραδύτερον, ἀλλὰ καὶ ὅτι τέλος (ὥσπερ ἐλέχθη) τὸ σπέρμα, δεῖ δὲ τὸ ἕνεκα ἄλλου πρότερον ἢ ἐκεῖνό γε εἶναι (ἐν προϋπάρχοντι γὰρ ἡ τῶν τοιούτων γένεσις). διὸ οὐδ' [7] ἅπαντα ἐνίων τὰ περικάρπια σπερμοφόρα, καθάπερ τῶν 10 βοτρύων ⟨αἱ⟩ [8] μικραὶ ῥᾶγες, ὡς οὐκέτι δυναμένης τελειῶσαι τῆς φύσεως ταύτας· γλυκεῖαι [9] δ'

§ 1. 8–12. *Cf.* [Aristotle], *Problems*, xx. 24 (925 b 23–29): διὰ τί τῶν τε μύρτων τὰ ἐλάττω ἀπυρηνότερά ἐστι, καὶ ἐν τοῖς φοίνιξι καὶ ἐπὶ τῶν βοτρύων ἔνθα [δ'] αἱ μικραὶ ῥᾶγες οὐκ ἔχουσιν ἢ ἐλάττους πυρῆνας; ἢ διὰ τὸ ἀτελέστερα εἶναι οὐκ ἔχει ἀποκεκριμένον; τέλος γὰρ ὁ πυρὴν ἔχει τὸ σπέρμα. διὰ τοῦτο δὲ καὶ ἐλάττους εἰσίν, ὡς ὄντα παραφυάδες καὶ ἀτελῆ. καὶ ἧττον δὲ γλυκέα τῶν ἐχόντων πυρῆνας· ἀπεπτότερα γάρ εἰσιν, ἡ δὲ πέψις τελείωσίς ἐστιν.

[1] ᾖ ὑπερβολή HP: ἡ περβολή U^ac; ἡ ὑπερβολή U^c (ἡ ὑ- u N).
[2] εἰς U (*cf.* εἰς Λέσβον *HP* 3 9. 5 "off in Lesbos").
[3] ἐνταῦθα τοῦτο U: ἐντεῦθεν τὸ Wimmer.
[4] καρποῖς U: καρπίμοις Schneider.
[5] αὐτός γε Wimmer (αὐτός τε Schneider): αὐτοῦ γε U.
[6] μόνον u HP: μένον U N.
[7] οὐδ' U: οὐχ' u.
[8] ⟨αἱ⟩ Schneider.
[9] γλυκεῖαι Schneider: γλυκεῖα U.

ness. But perhaps the country too must not be good, to prevent any excess from this source as well, which is why this sort of good sprouting occurs mainly in the islands. But the explanation holds for all weak trees.

So we must take this to be the cause of good sprouting and good bearing here.[a]

Fruiting: The Pericarpion Produced Before the Fruit

21. 1 In all fruits the pericarpion is produced before the fruit proper and the seed. This is not only because most seeds are woody and pip-like formations, and such take longer to form, but also because the seed (as we said)[b] is the end, and what serves an end must exist before the end that it serves (since the production of ends is of a kind where the one relative exists before the other).[c] This again is why in some trees not all the pericarpia contain seeds, as for example small grapes of a cluster, the nature of the tree having no longer been able to bring them to completion. But these grapes are no less sweet than the

[a] The cause is weakness. The solution rests on interpreting " good sprouting " not as " sprouting to excess," but as sprouting to the extent suitable for good fruiting.

[b] *CP* 1 16. 3.

[c] *Cf.* Aristotle, *Categories*, vii. (7 b 15–8 a 12 and especially 7 b 22–25): " But simultaneity in their nature is not held to apply to all relatives; for the knowable would be held to be prior to knowledge, since for the most part we acquire knowledge of realities that exist before we know them . . ."

21. 2 οὐδὲν ἧττον τῶν μεγάλων. ᾗ[1] καὶ δῆλον ὡς ἄρα ῥᾷον[2] ἐκπέψαι[3] τὸ περικάρπιον· ἔοικε γὰρ ὥσπερ[4] ὑπὸ τοῦ ἡλίου καὶ τοῦ ἀέρος καὶ τῆς ὥρας συνέψεσθαι, τὸ δὲ σπέρμα τῆς φύσεως ἰδιώτερον εἶναι.[5]
5 πρὸς ἡμᾶς δὲ τὰ μὲν οὐδέν,[6] τὰ δὲ ἐν ἐλάττονι λόγῳ,[7] ἐλάττων γὰρ[8] χρεία. διὸ καὶ τἀπύρηνα[9] καὶ τὰ μαλακοπύρηνα μάλιστα ζητοῦμεν, καὶ ἐφ' ὅσον δυνάμεθα τοῦτο σπεύδομεν (ὥσπερ οἱ τοὺς βότρυς τοὺς ἀγιγάρτους ποιοῦντες)· ἡ δὲ φύσις
10 δῆλον ὡς ἀμφοῖν ἀποδιδόναι βούλεται τὸ σύμμετρον.

21. 3 ὁπότερον δ' ἂν πλεονάζῃ, θάτερον ἔλαττον· τοῦτο δὲ τῶν μὲν δι' ὑγρότητα, καὶ ἁπλῶς εὐτροφίαν, γίνεται (διὸ καὶ τὴν γεωργίαν ἔφαμεν ξυμπονεῖν), τῶν δὲ διὰ ξηρότητα καὶ πυκνότητα,
5 καὶ τὸ ὅλον ἀτροφίαν.[10] καὶ πρὸς μὲν τὴν ἑκατέρου τελείωσιν καὶ πέψιν, τάχα δὲ καὶ τῶν φυτῶν γένεσιν ὅλως καὶ πρωϊβλαστίαν, ἅπαντα τὰ τοιαῦτα, τήν τε τοῦ ἀέρος καὶ τοῦ ἡλίου δύναμιν αἰτιατέον[11] καὶ τὰς ἰδίας ἑκάστων φύσεις, εἴτ'
10 οὖν ὑγρότητι καὶ ξηρότητι <καὶ πυκνότητι>[12] καὶ

[1] ᾗ Schneider: ἡ U.
[2] ἄρα ῥᾷον Schneider: ἀραιὸν U.
[3] ἐκπέψαι U[r] N HP: ἐκπέμψαι U[ar].
[4] ἔοικε γὰρ ὥσπερ ego (εἴπερ Schneider after Gaza; ἔοικε γὰρ Wimmer): ἔοικεν. ὥσπερ U.
[5] συνέψεσθαι—εἶναι U: συνέψεται—εἶναι <ἔοικε> Schneider.
[6] οὐδὲν U: ἐν οὐδενὶ Link.
[7] λόγῳ a: χρόνωι U (-ω N HP).
[8] γὰρ ego (γὰρ ἡ a): ἡ U (ἡ u N HP).
[9] τὰ ἀπύρηνα Gaza, Itali: τὰ πύρηνα U.

large ones.[a] This also shows that it is easier to carry out the concoction of the pericarpion; for it appears that the sun, air and season lend a hand in the boiling (as it were), whereas the seed is more the private work of the nature of the tree. But for man some seeds count for nothing, others for less than the pericarpion, being of less use; and this is why we look above all for trees with no stones or with soft stones and do our best to breed them, like the growers who produce grape clusters that have no pips. The nature of the tree, on the other hand, evidently aims at giving each of the two the right amount of development. 21. 2

If the one exceeds this amount, the other falls short of it. The pericarpion gets more because of plenty of fluid in the tree, in short because of good feeding (and this is why we said[b] that husbandry lends a hand); the seed gets more because of dryness and close texture, in a word because of poor feeding; and as causes of the perfecting and concocting of each, and perhaps in general of the generation and early sprouting of plants, we must take all such things as we have been mentioning, both the power of the air and of the sun and also the distinctive natures of the various plants, whether the natures differ in 21. 3

[a] And so are complete as pericarpia, since they serve man's use.
[b] *CP* 1 16. 2.

---

10 ἀτροφίαν Gaza (*alimentorum inopiae*), Itali: εὐτροφίαν U.
11 αἰτιατέον U: ἀνακτέον Schneider.
12 ⟨καὶ πυκνότητι⟩ Wimmer.

μανότητι καὶ τοῖς τοιούτοις διαφερούσας, εἴτε θερμότητι καὶ ψυχρότητι, καὶ γὰρ ταῦτα τῆς φύσεως.

21. 4 τούτων δὲ τὰ μὲν ἄλλα σχεδὸν τῇ αἰσθήσει φανερόν·[1] τὸ δὲ θερμὸν καὶ ψυχρόν, ἐπείπερ οὐκ εἰς αἴσθησιν ἀλλ' εἰς λόγον ἀνήκει, διαμφισβητεῖται καὶ ἀντιλέγεται, καθάπερ τὰ ἄλλα τὰ τῷ λόγῳ 5 κρινόμενα· περὶ ὧν καλῶς ἔχει διωρίσθαι πως, ἄλλως ⟨τ'⟩ ἐπεὶ καὶ[2] πολλὰ πρὸς ταύτας ἀνάγεται τὰς ἀρχάς. ἀνάγκη δὲ ἐκ τῶν συμβεβηκότων ἅπαντα τὰ τοιαῦτα σκοπεῖν, ἐκ τούτων γὰρ κρίνομεν καὶ θεωροῦμεν τὰς δυνάμεις.

21. 5 μία μὲν οὖν αἰτία λέγεται τῶν θερμῶν καὶ ψυχρῶν ἡ εἰς τὸ κάρπιμον ἀνάγουσα καὶ ἄκαρπον, ὡς τῶν μὲν θερμῶν καρπίμων ὄντων, καθάπερ καὶ

§ 5. 1. U[m] ᾱ.

[1] φανερόν U N: φανερά HP.
[2] ⟨τε⟩ ἐπεὶ καὶ Schneider in his text (*cf.* Aristotle, *History of Animals*, iii. 2 [511 b 12] ἄλλως τε ἐπειδὴ καὶ); in his commentary Schneider proposes ⟨τε⟩ καὶ ἐπεὶ: ἐπεὶ καὶ U.

---

[a] *Cf. CP* 1 1. 3 (for generation), *CP* 1 10. 3 (for early sprouting), and the beginning of the present paragraph (for development of the pericarpion).
[b] *Cf. CP* 1 2. 4; 1 3. 2–3 (for generation), *CP* 1 10. 3 (for early sprouting), and the beginning of the present paragraph (for development of the pericarpion).
[c] Such as strength or weakness, large or small size.
[d] *Cf.* Aristotle, *On the Parts of Animals*, ii. 2 (648 a 19–25): "We must study the causes which establish the necessity of

fluidity [a] and dryness or in closeness and openness of texture [b] or the like,[c] or else in heat and coldness (these too belonging to a plant's nature).

Heat and Cold

21. 4 Of these differences the rest are (one might say) a matter evident to sense; whereas the difference between the hot and the cold, since it does not fall to the province of sense but of reason, is subject to dispute and denial, like everything else that is decided by reason, and it is well that the question should somehow be settled, especially as many matters are referred to these two principles.[d] But we are compelled to study all such differences in the light of the effects, for it is from the effects that we decide and understand the potencies.[e]

Menestor's Inferences Establishing Heat

21. 5 (1) One causation that is spoken of as working in hot and cold plants is the one inferred from their character of fruitfulness and of failure to bear, the hot plants being fruitful, as in animals with the fertile

possessing blood . . ., and show what the nature of blood is, but only after we have first dealt with the distinction between hot and cold. For the nature of many things is referred to these principles, and many dispute about what are hot and what are cold among animals or their parts."

[e] Aristotle (*On the Parts of Animals*, ii. 2 [648 b 11–12]) begins his investigation of heat with formulating the ἔργον (effect) of the hotter thing.

ἐπὶ τῶν ζῴων τῶν γονίμων καὶ ἀγόνων, καὶ τῶν
5 ζῳοτόκων καὶ ᾠοτόκων.

ἑτέρα δ' ἡ[1] κατὰ τὰς χώρας (οἷον[2] ψυχράς)· τὰ γὰρ ἐναντία ἐν ταῖς ἐναντίαις δύνασθαι διαμένειν, τὰ μὲν [θερμὰ][3] ἐν ταῖς θερμαῖς,[4] τὰ δὲ [ψυχρὰ][5] ἐν ψυχραῖς.[6] οὕτως γὰρ εὐθὺς καὶ τὴν
10 φύσιν γεννᾶν, ὡς ὑπὸ μὲν τοῦ ὁμοίου φθειρομένων διὰ τὴν ὑπερβολήν, ὑπὸ δὲ τοῦ ἐναντίου σῳζομένων, οἷον εὐκρασίας τινὸς γινομένης, ὥσπερ καὶ Ἐμπεδοκλῆς λέγει περὶ τῶν ζῴων· τὰ γὰρ

21. 6 ὑπέρπυρα τὴν φύσιν ἄγειν εἰς τὸ ὑγρόν. συνηκολούθηκεν δὲ ταύτῃ τῇ δόξῃ καὶ Μενέστωρ οὐ μόνον ἐπὶ τῶν ζῴων, ἀλλὰ καὶ ἐπὶ τῶν φυτῶν· θερμότατα γὰρ εἶναί φησιν τὰ μάλιστα ἔνυγρα,
5 οἷον σχοῖνον κάλαμον κύπειρον, διὸ καὶ ὑπὸ τῶν χειμώνων οὐκ ἐκπήγνυσθαι, καὶ τῶν ἄλλων ὅσα μάλιστα ἐν τοῖς ψυχροῖς δύνασθαι διαμένειν, οἷον

§ 5. 6. U^m β̄.
13–14. Empedocles Frag. A 73, Diels-Kranz, *Die Fragmente der Vorsokratiker* (vol. i^10, p. 298. 7–8).
§ 6. 1–§ 7. 12. Menestor Frag. 5, Diels-Kranz, *Die Fragmente der Vorsokratiker*, vol. i^10, pp. 375–376; *cf.* Plutarch, *Quaest. Conv.*, iii. 2. 1 (648 C).

[1] δ' ἡ Wimmer: δὴ U N; δὲ HP.
[2] οἷον U: θερμὰς καὶ (ἢ Wimmer) Schneider.
[3] [θερμὰ] ego.
[4] θερμαῖς U: ψυχραῖς Gaza (*frigidis*), Itali.
[5] [ψυχρὰ] ego.
[6] ψυχραῖς U: ταῖς ψυχραῖς N HP; ταῖς θερμαῖς Gaza (*calidis*), Itali.

as opposed to the infertile, and the viviparous as opposed to the oviparous.[a]

(2) A second cause is shown by the character of the country (as when it is cold): for plants (it is urged) are able to survive in the countries of a character opposite to their own, one set surviving in hot countries, the other in cold ones. For this (it is argued) is how their nature generated them at the outset, since they are killed by the like character in their habitat, owing to the resultant excess, but are preserved by the character opposite to their own, since the result is a kind of tempering. So Empedocles says of animals that the ones with an excess of fire are brought by their nature to water.[b] Menestor too follows this view not only for animals but for plants as well, saying that the hottest plants are those that live most in water, as rush, reed and galingale (which is why they do not freeze out in winter), the hottest of the rest being those best able to survive in cold localities,

21. 6

[a] *Cf.* Aristotle, *On the Generation of Animals*, ii. 1 (733 a 33–b 12): "For the more complete and hotter of animals produce an offspring complete in character . . ., and these animals generate within themselves offspring that are animals from the start. The second group do not generate offspring in themselves that are complete from the start (for they produce an egg before they produce an animal), but do bring the offspring to birth as an animal. Others do not generate a completed animal, but an egg, and the egg is a complete one. Animals with a still colder nature generate an egg, but the egg is not complete, but completed outside the animal . . . The fifth and coldest kind of animal does not even bring forth an egg from itself, but even the egg is formed outside . . ."

[b] *Cf.* Aristotle, *On the Parts of Animals*, ii. 2 (648 a 25–27) [of animals]: "For some assert that the aquatic are hotter than the terrestrial, saying that the heat of their nature makes up for the coldness of their habitat . . ."

ἐλάτην πεύκην κέδρον ἄρκευθον κιττόν, ἐπὶ τούτου γὰρ οὐδὲ τὴν χιόνα τῇ [1] θερμότητι ἐπιμένειν, ἔτι
10 δὲ σκολιὸν εἶναι διὰ ⟨τὸ⟩ [2] τὴν ἐντεριώνην, θερμὴν οὖσαν, καὶ διαστρέφειν.[3]

21. 7 τρίτην δ' αἰτίαν λέγει [4] τοῦ πρωϊβλαστῆ καὶ πρωΐκαρπα εἶναι· φύσει γὰρ καὶ ὁ ὀπὸς αὐτὸς ὢν θερμὸς καὶ βλαστάνειν πρωῒ ποιεῖ [5] ⟨καὶ⟩ [6] πέττειν τοὺς καρπούς· σημεῖον δὲ ποιεῖται [7] καὶ
5 τούτου τόν τε κιττὸν καὶ ἕτερ' [8] ἄττα.

τετάρτη δ' ἡ τῶν ἀειφύλλων· διὰ γὰρ θερμότητα καὶ ταῦτα οἴεται διατηρεῖν, τὰ δὲ ἐνδείᾳ τούτου φυλλοβολεῖν.[9]

προσεπιλέγει δὲ τοῖς εἰρημένοις καὶ τὰ τοῖα [10]
10 σημειούμενος, ὅτι τὰ πυρεῖα [11] ἄριστα καὶ κάλλιστα ἐκπυροῦται, καὶ τῶν δένδρων [12] ὡς τὰ μάλιστα τοῦ πυρὸς ὄντα [13] τάχιστα ἐκπυρούμενα.

καὶ τὰ μὲν ὑπὲρ τῆς θερμότητος λεγόμενα σχεδὸν ταῦτ' ἐστίν.

22. 1 ἔχει δ' ἀπορίαν εὐθὺς ἐπὶ τοῦ πρώτου λεχθέντος, ὡς οὐκ ἔστιν τὰ καρπιμώτερα θερμότερα. τὰ γὰρ

1. Um γ̄.
6. Um δ.

[1] τῇ u: τὴν U N HP.
[2] ⟨τὸ⟩ Schneider.
[3] διαστρέφειν u: διατρέφειν U.
[4] λέγει Ur: λέγειν Uar N HP.
[5] πρωῒ ποιεῖ HP: πρωϊποιεῖ U N
[6] ⟨καὶ⟩ u.
[7] ποιεῖ⟨ται⟩ ego.
[8] ἕτερ' Wimmer: ἕτερα U.
[9] φυλλοβολεῖν u: φύλλον βολὴν U.
[10] τοῖα u N (τοία U): τοιαῦτα HP.
[11] πυρεῖα Schneider: πύρεια U.

as silver fir, pine, prickly cedar, Phoenician cedar and ivy (this last being so hot that snow does not even remain on it, and again, it is crooked because, he says, its pith is hot and warps it).

(3) A third causation that he mentions is that of early sprouting and fruiting; for when the sap is itself naturally hot it makes the plant both sprout and concoct its fruit early, and here too he cites the ivy in confirmation, together with certain others. 21. 7

(4) A fourth causation results in the character evergreen; for he thinks that with these plants too it is heat that makes them keep their leaves, the rest shedding theirs from lack of it.[a]

To the remarks that we have mentioned he adds such proofs as this: that firesticks ignite best and give the finest flame, trees too with the most fire in them catching fire quickest.[b]

So the arguments used to establish that a plant is hot are (one may say) these.

Reply to Menestor
(1) *Fruitfulness*

But in the very point that is first mentioned [c] they involve a difficulty, since it can be argued that the 22. 1

[a] *Cf.* Plutarch, *Quaestiones Convivales*, i. 3. 1 (648 D).
[b] *Cf.* Aristotle, *On the Parts of Animals*, ii. 2 (649 a 27–29): " In another way pine (πεύκη) and fat things are hot, because they quickly change to the actuality of fire."
[c] *CP* 1 21. 5 (first paragraph).

---

12 καὶ τῶν δένδρων U: τὰ ἐκ τῶν ἐνύγρων Itali (*quae ex plantis aquatilibus fuerint* Gaza).
13 ὡς—ὄντα Wimmer (*tanquam . . . quorum natura plus in se caloris contineat* Gaza; ὡς τὰ τοῦ πυρὸς πλεῖον ἔχοντα Schneider): ὥστ' ἀεί τίς τα τοῦ πυρὸς ὄντα U.

θήλεα τῶν δένδρων πολὺ καρπιμώτερα[1] μέν, ἧττον δὲ θερμὰ τῶν ἀρρένων (ὥσπερ[2] ἐκ τῆς τῶν ζῴων
5 ὁμοιότητος ληπτέον κἂν ᾖ[3] ὁμώνυμον).[4] ἔτι δὲ οὐδὲ τὰ ζῷα τὰ πολυγονώτερα θερμότερα· [οὐδε][5] ἀνάπαλιν θερμότερα τὰ ὀλιγογονώτερα,[6] καθάπερ τὰ σαρκοφάγα καὶ λαίμαργα. μόνα γὰρ τῶν θερμῶν δοκεῖ κύων καὶ ὗς πολυτοκεῖν. αὐτῶν
10 δὲ τῶν ὁμογενῶν τὰ ὁμοιοειδέστερα πολυγονώτερα, καθάπερ ἐπὶ τῶν ὀρνίθων· ἡ γὰρ θερμότης ἐξαύξειν φαίνεται καὶ διαρθροῦν τὰ μέλη καὶ σκληρύνειν.

22. 2 ἀλλὰ καὶ εἰς τὴν ζῳογονίαν καὶ εἰς τὴν καρποτοκίαν καὶ πέπανσιν συμμετρίας τινὸς δεῖ τοῦ θερμοῦ καὶ οὐχ ὑπερβολῆς, εἴπερ αὕτη μὲν ξηραίνει καὶ πυκνοῖ μᾶλλον.

[1] πολὺ καρπιμώτερα B, Gaza: πολυκαρπιμώτερα U N HP.
[2] ὥσπερ U: ὅπερ Gaza (*quod*), Schneider.
[3] κἂν ᾖ ego: καὶ μὴ U.
[4] ὁμώνυμον U: ὁμωνύμως Schneider.
[5] [οὐδε] ego.
[6] ὀλιγογονώτερα u: ὀλιγουγονώτερα with an ὑφέν connecting *υγ* (no doubt the ὑφέν of an ancestor was taken as a subscribed *υ*) U.

---

[a] *Cf. HP* 3 8. 1 (of wild trees): "in all trees, taking them by kinds, there are a number of differences; one is common to all, that whereby people distinguish female and male, the first bearing, the other (in some kinds) not. Where both bear, the female bears the finer and more plentiful fruit (except for those who call these trees male, as some do)."

[b] Heat is nowhere else ascribed to male trees, though the following passage is consistent with such an ascription: *HP* 5 4. 1 (of timber): "All wild trees compared with cultivated, and male with female, are closer in texture, harder, heavier and to put it generally stronger . . ."

ıore fruitful plants are not hotter. For females in ·ees are a good deal more fruitful than the males,[a] ut not so hot (as one can assume from their similarity ɔ the females in animals,[b] even if here the word[c] as a different sense).

Again, in animals too it is not true that the more rolific are hotter; on the contrary the less prolific re hotter, as the carnivorous and voracious.[d] For the nly hot animals considered to bear many young are he dog and the swine;[e] and within the same class, s in birds, it is those of the more uniform aspect hat produce the more young, since heat is observed o make the members grow out and to differentiate hem, and to bring about hardness.[f] Rather, what is 22. 2 equired not only for the generation of animals but for he production and ripening of fruit is a right amount f heat, and no excess of it, since the excess leads to oo much dryness and too close a texture.

[c] Aristotle holds the male in animals to be hotter: *cf. On the Generation of Animals*, iv. 1 (765 b 15–17); *On Length and Brevity of Life*, chap. v (466 b 15–16).

[d] That is, birds with crooked talons: *cf.* Aristotle, *On the Generation of Animals*, iii. 1 (749 b 1–6) [of wind-eggs]: "These are produced by those birds that are not good fliers and that do not have crooked talons, but that are prolific, because they have a great amount of residue, whereas in birds with crooked talons this secretion is diverted to producing wings and wing-feathers, their bodies being small, dry and hot . . ."

[e] For the swine as hot *cf.* [Aristotle], *Problems*, x. 21 (893 a 6).

[f] *Cf.* Aristotle, *On the Generation of Animals*, iii. 1 (749 b –33), where the great development of wings and feathers in the crooked-taloned birds, which are not prolific, is contrasted with that of more prolific birds, and the sparer and drier bodies of the better and less prolific breed of hens with the bulkier and more fluid bodies of the more prolific and inferior breed.

5 ὥστε ταῦτα μὲν ἐν ἀμφιδόξῳ, προσδεόμενα τινος διορισμοῦ.

περὶ δὲ τῶν ἐνύδρων ῥᾴων ἡ ἀμφισβήτησις· γὰρ οὔτε γεννᾶν οὔτε εὖ τρέφειν οὔτε σῴζε πέφυκεν τὸ ἐναντίον, ἀλλὰ τὸ ὅμοιον. ἐπεὶ κ
10 Ἐμπεδοκλεῖ πρὸς τοῖς ἄλλοις καὶ τοῦτ' ἄτοπ (ὅπερ καὶ ἐν ἑτέροις εἴρηται), τὸ γεννήσασαν τῷ ξηρῷ τὴν φύσιν μεταίρειν εἰς τὸ ὑγρόν· πο γὰρ ἂν διέμενεν, ἢ πῶς οἷόν τε καὶ διαμένε ὁντιναοῦν χρόνον, εἴπερ ἦν ὁμοία [1] τοῖς νῦν;

22. 3 ἔτι δ' αὐτὸ τὸ συμβαῖνον κατὰ τὴν νῦν γέννησ ἀποσημαίνειν· [2] ἅπαντα γὰρ φαίνεται τὰ ζῷα κ τὰ φυτὰ καὶ διαμένοντα καὶ γεννώμενα ἐν τ οἰκείοις τόποις, ὁμοίως ἔνυδρα καὶ χερσαῖα κ
5 εἴ τις ἄλλη τοιαύτη διαφορά· διὸ καὶ ἀπαθῆ μ ὑπὸ τούτων, παθητικὰ δ' ὑπὸ τῶν ἐναντίων, ἅ μεγάλης τῆς μεταβολῆς γινομένης.

22. 4 ἀσύμφωνοι δὲ καὶ αἱ δόξαι πρὸς αὐτάς,[3] ὅτ ἅμα τε τὰ ἔνυγρα θερμότερα ᾖ, καὶ τὰ καρπιμο

[1] ὁμοία U: ὅμοια N HP.
[2] ἀποσημαίνειν ego: -ει U.
[3] αὑτὰς u P: αὐ- U N H.

---

[a] *CP* 1 21. 5.

[b] For the great distinction between terrestrial and aqua in both plants and animals *cf.* *HP* 1 4. 2 and especially *HI* 14. 3 and *CP* 2 3. 5. For the distinction in animals *cf.* Aristot *History of Animals*, i. 1 (487 a 14–b 32). The distinction com from Plato, *Sophist*, 220 A 7–B 2, where the art of hunti animals is divided into the hunting of the pedestrian kind a of the swimming kind, the latter in turn divided into flyi and aquatic.

There are then two ways of thinking about the first point, and some further distinction is required.

(2) *Aquatic Plants*

It is easier to dispute the point about aquatic plants. For in the natural course of things it is not the quality opposite to that of the plant that generates it or rears it well or preserves it, but the quality that is similar. Indeed in the view of Empedocles (mentioned in another connexion)[a] there is this absurdity among the rest, that the nature of animals, after generating them on dry land, transfers them to water. For how could they have succeeded in surviving? Or how, supposing their nature like that of animals today, was it possible for them to survive for any time whatever?

22. 3 Again what in fact takes place in generation as it is today can be urged against the theory. For all animals and plants, alike whether they are of water or of land[b] or whatever other such division there may be,[c] are observed both to continue to live and to be generated in the places to which they belong. This is why they are not adversely affected by these places, but are adversely affected by the opposite ones, since the change from the one kind of place to the other turns out to be a great one.

Two Inconsistencies

22. 4 The views are also inconsistent with one another, when both aquatic[d] and more fruitful plants[e] are

[c] Such as the animals of the air and conceivably those of fire.
[d] *CP* 1 21. 6 (the second point).
[e] *CP* 1 21. 5 (the first point).

τερα, πολλὰ γὰρ ἄκαρπα τῶν ἐνύδρων· καὶ πάλιν, ὅταν τά τε[1] πρωϊβλαστῆ καὶ πρωΐκαρπα,
5 καὶ τὰ ἀείκαρπα καὶ ἀείφυλλα θερμὰ λέγωσιν· ὀψικαρπότατα γὰρ (ὡς εἰπεῖν) τὰ ἀείφυλλα. καὶ τὸ ὅλον (ὥσπερ πρότερον εἴρηται) πρωϊβλαστῆ καὶ πρωΐκαρπα δι' ἀσθένειαν· ἔνια δὲ καὶ συμπαρακολουθεῖ βλαστάνοντα καὶ ἀνθοῦντα πλείω χρόνον,
10 ὥσπερ καὶ ἐν τοῖς ἐπετείοις ὁ ἠριγέρων, ἔχων δῆλον ὅτι συμμετρίαν τινὰ κατὰ τὴν ἐπιρροὴν τῆς τροφῆς.

22. 5 ὃ δὲ προσεπιλέγει περὶ τῶν πυρείων οὐκ ἄν τις ἴσως φαίη θερμότητος, ἀλλὰ μανότητος[2] εἶναι σημεῖον· ἡ γὰρ τρῖψίς ἐστιν ἡ ποιοῦσα τὸ πῦρ,

[1] τά τε U^css: U^t omits.
[2] ἀλλὰ μανότητος ego (ἀλλὰ ξηρότητος Moldenhawer): ἀλλ' ἀνθηρότητος U (a conflation with θερμότητος).

---

[a] *Énhydra* ("in water"); in *CP* 1 21. 6 they were called *énhygra* ("in fluid"). In Plato the distinction is plain: *énhygros* (occurring only in the compound *enhygrothērikós*) means in fluid, whether air or water; *énhydros* means in water (*Sophist*, 220 A 10–B 5; 221 B 5, E 6). Aristotle takes over this usage of *énhydros* for water animals (*énhygros* appears in the spurious *De Spiritu*, chap. ii [482 a 21, 25]). A certain fondness for *énhygros* in Theophrastus may come from the opposition of *hýdōr* (fresh water) to *thálatta* (sea-water); the traditional *énhydros* might have appeared to exclude the latter.

[b] So the fresh-water goat-willow (*HP* 4 10. 2), female *phleōs* and sedge (*HP* 4 10. 4), a kind of rush (*HP* 4 12. 1, 2).

made out to be hotter, for many aquatic [a] plants do not bear.[b]

Again there is an inconsistency when both groups, the early sprouters and fruiters,[c] and the ever-fruiters [d] and evergreens,[e] are made out to be hot, for evergreens are (so to say) the latest fruiters of all; and (for that matter) plants sprout and fruit early from weakness (as we said before),[f] and some, once they have begun, also continue sprouting and flowering with the rest for some time, as groundsel among annuals,[g] evidently because the influx of food is somehow [h] well adjusted to this continuance.[i]

(3) *Firesticks*

As to his added point about firesticks [j] one would perhaps say that easy ignition is no indication of heat,[k] but of open texture, since it is the attrition that makes the fire, and this is more pronounced 22. 5

[c] *CP* 1 21. 7 (the third point).

[d] Not mentioned in *CP* 1 21. 7, but easily added: *cf. CP* 1 10. 6–7.

[e] *CP* 1 21. 7 (the fourth point).

[f] *CP* 1 10. 2, 4.

[g] *Cf. HP* 7 7. 4: ". . . groundsel also flowers for a long time." In *HP* 7 10. 2 it is called a winter plant.

[h] Whether circumstances are responsible or the plant's nature.

[i] For a similar adjustment of influx in evergreens *cf. CP* 1 10. 7.

[j] *CP* 1 21. 7 (third paragraph).

[k] But *cf. HP* 5 3. 4: " Hot too are ivy, bay and in general the wood used for firesticks; Menestor adds mulberry. Coldest are the wood of aquatic plants and wood that is watery."

σφοδροτέρα δὲ ἐν τούτοις, ᾗ [1] καὶ μᾶλλον ἐξαεροῦν
5 δυναμένη τὸ ὑγρόν.

ἀλλὰ δὴ μάλιστα ἐκεῖνα φαίνεται καὶ κατὰ τὴν αἴσθησιν θερμὰ καὶ κατὰ λόγον· τὰ λιπαρά τε καὶ τὰ δριμέα. καὶ ⟨τὰ⟩ [2] εὔοσμα οὕτως ἔχει· πάντα γὰρ ταῦτα δοκεῖ ἐν θερμότητι [3] εἶναι, διὸ
10 καὶ ξηρά τε ὄντα [4] καὶ ὡς ἐπίπαν πυκνὰ καὶ ἀσαπῆ, καὶ τοὺς χυλοὺς ἔχοντα λιπαροὺς καὶ δριμεῖς, ὅθεν γέ [5] ἐστί ⟨τι⟩ [6] καὶ ἄφορον.[7] καὶ ἡ πεύκη θερμὴ καὶ εὐπύρωτός ἐστιν. οὐ μὴν μόνα γε ταῦτα, ἀλλὰ καὶ ἄλλα δοκεῖ θερμὰ εἶναι, καθάπερ
15 καὶ ἡ φίλυρα καὶ ὅλως ὅσα τὴν τοῦ σιδήρου βαφὴν ἀνίησιν.

22. 6 χρὴ δὲ καὶ ταῖς τοιαύταις δυνάμεσιν ἀθρεῖν τὰ θερμὰ καὶ ἐπικρίνειν, οἷον ὅσα κατὰ τὰς προσφορὰς

[1] ᾗ Wimmer: ἡ U; ῇ u.
[2] ⟨τὰ⟩ ego.
[3] ἐν θερμότητι U: θερμότητος (?) ego.
[4] ξηρά τε ὄντα N HP (ξηρά τέ [-ον erased] ὄντα u): ξηρατέον ὄντα U.
[5] γέ ego: τε U.
[6] ⟨τι⟩ ego.
[7] ἄφορον ego: ἄφρων U.

[a] *Cf.* Theophrastus, *On Fire*, chap. iii. 28–29: ". . . but charcoal and wood cannot burn unless one blows on them because of their earthy and solid character, for in all such bodies the passages are crowded, and the blowing opens them and prepares a path for the fire, and does this the more, the more numerous and small the parts into which the bodies that it encounters are divided. For this reason people sometimes break up charcoal and then put the fragments together and blow on them (for a flame is a stream formed by the con-

when sticks of open texture are drilled, and so is better able to vaporize the fluid.[a]

Solution: The Hot Plants

Rather the case is this. It is another set of plants that appear most of all to be hot both to our senses and to reasoning: oily [b] and pungent [c] plants. Again fragrant [d] plants are in this case, for all instances of fragrance are considered to be the work of heat, which is why the plants are dry and for the most part close in texture and resistant to decomposition, and have flavours that are oily and pungent; and from this comes occasionally even a failure to bear. Again the pine is hot and easily ignited. Still, these are not the only hot plants and trees, but others too are considered to be hot, as the lime tree[e] and in general all that dull the temper of iron.

Tests for Heat and Coldness

22. 6 In looking into the question of heat in a plant and adjudicating claims we must also judge by such

fluence of many rills as it were). So firesticks by attrition of the wood produce the same result . . .;" *cf. ibid.*, chap. ix. 64 and *HP* 5 9. 7: ". . . besides being dry and not juicy it (*sc.* the firestick that is bored into by the drill) must also be more open in texture, to make the attrition more vigorous."

[b] *Cf.* Aristotle, *On the Generation of Animals*, v. 3 (783 b 18–20): ". . . among fluids the oily is most of all of this character (*sc.* hot). This is why in plants the oily ones tend more to be evergreen."

[c] *Cf. CP* 6 1. 3 (the pungent savour heats).

[d] *Cf. CP* 6 9. 4 (all fragrant plants are bitter) and *CP* 6 16. 7–8.

[e] *Cf. HP* 5 3. 3; 5 5. 1.

τοῖς σώμασιν θερμότητάς τινας ἐμποιεῖ καὶ πέψεις ἢ συντήξεις,[1] ἢ καὶ τὸ ὅλον κατὰ τὴν ἁφὴν 5 καὶ τὴν γεῦσιν διαδίδωσιν τὴν αἴσθησιν· οὐ γὰρ ἔτι ταῦτα λόγου δεῖται πρὸς τὴν πίστιν, ἀλλὰ καὶ ἡ τῶν ἰατρῶν χρεία μαρτυρεῖ καὶ ἡ αἴσθησις. ὡς δ' ἐπὶ τὸ πᾶν πλείω ταῦτα ἐν τοῖς θερμοῖς ἢ ψυχροῖς, ἢ οὐκ ἐλάττω, γίνεται· διὸ καὶ πρὸς 10 τὴν ἐν [2] τοῖς ἐναντίοις γένεσιν καὶ τοῦτ' ἀμφισβητεῖται.

22. 7 τὸ μέντοι παρέχεσθαί [3] τινα ῥοπὴν [4] εἰς τὸ διαμένειν ἐν τοῖς ψυχροῖς ἔνια τῶν θερμῶν τάχ' ἄν τις συγχωρήσειεν· ἀλλὰ τὰ ποῖα καὶ πῶς, τοῦτο πειρατέον διορίζειν, εἰ μὴ ἄρα ἁπλῶς ἰσχύϊ τινὶ 5 μᾶλλον ἢ ῥιζῶν ἢ τῶν ὅλων σωμάτων ἡ διαμονή, καθάπερ καὶ τῆς ἀπίου καὶ τῆς ἀχράδος καὶ τῆς ἀμυγδαλῆς, ἃ [5] δὴ καὶ ἥκιστα ἐκπήγνυται.[6]

καὶ περὶ μὲν θερμότητος ἐκ τούτων ληπτέος ὁ διορισμός. [περὶ δὲ τὰς βλαστήσεις καὶ καρπο- 10 τοκίας τῶν δένδρων] [7, 8]

[1] συντήξεις Gaza, Itali: συντάξεις U.
[2] πρὸς τὴν ἐν Wimmer: προστο U; πρὸς u.
[3] παρέχεσθαί ego: παρέπεσθαι U.
[4] ῥοπὴν ego: τροφὴν U.
[5] ἃ Schneider (*et hinc* Gaza): ὧ U; ὃ u; τὸ N HP; καὶ a.
[6] ἐκπήγνυται Schneider: ἐκπήγνυσθαι U.
[7] [περὶ—δένδρων] HP (once the catchwords for a second roll).
[8] U has the subscription θεοφράστου περι φυτῶν αἰτίων.

potencies as these: that a plant taken internally produces certain manifestations of heat in the body and cases of concoction or colliquescence, or else that it simply transmits the sensation of heat to touch and also to taste.[a] For these matters, unlike the rest, need no reasoning to be convincing, but are attested by the practice of physicians and by our own senses. On the whole plants of this character occur in greater (or not in fewer) number in hot countries than in cold; and this is why this point is added to the others in disputing the view that plants occur in places of the opposite character to their own.

22. 7 Still one might perhaps concede that this pairing of opposite characters has a certain weight in determining the survival of some hot plants in cold countries. But an attempt should be made to specify the character of these plants and indicate how their survival is brought about, unless after all it is simply due rather to some strength either of the roots or of the whole tree, as it is in the case of the pear, wild pear and almond, which are the trees least apt to be killed by freezing.

The presence of heat, then, is to be determined by these tests.

[a] *Cf.* Aristotle, *On the Parts of Animals*, ii. 2 (648 b 12–15).

BOOK II

B[1]

1\. 1 περὶ δὲ τὰς βλαστήσεις καὶ καρποτοκίας τῶν δένδρων, καὶ ἁπλῶς τῶν φυτῶν, ὅσα μὴ πρότερον εἴρηται πειρατέον ὁμοίως ἀποδοῦναι, διαιροῦντας χωρὶς ἕκαστα, τά τε κατὰ τὰς ἐνιαυσίους ὥρας 5 γινόμενα καὶ ὅσα κατὰ τὰς γεωργικὰς θεραπείας· δύο γὰρ δὴ μέρη ταῦτ' ἐστίν, τὸ μὲν ὥσπερ φυσικὸν καὶ αὐτόματον, τὸ δὲ τέχνης καὶ παρασκευῆς βουλομένης[2] τὸ εἶναι εὖ. λόγος[3] δ' ἀμφοῖν ἐστιν οὐχ ὁ αὐτός, ἀλλ' ὁ μὲν οἷον φυσικός

[1] το β̄ Um.
[2] βουλομένης ego: βουλόμενον εἰς U; δεόμενον εἰς u.
[3] εὖ. λόγος u: εὔλογος U N HP; ὁ λόγος a (after Gaza).

---

[a] Sprouting from the seed or slip was discussed in *CP* 1 1. 1–1 9. 3, annual sprouting in *CP* 1 10. 1–1 15. 2; fruiting was discussed in *CP* 1 15. 3–1 21. 3.

[b] *CP* 2 1. 1–2 19. 6.

[c] *CP* 3 1. 1–4 16. 1.

[d] The starting-points of "nature" are internal (*CP* 1 16. 10). In the "distinctive natures" the starting-points are internal to the plant. Here however the term "nature" is expanded to include the natural environment, where the starting-points are also internal to itself, as opposed to those of art, which are external to the thing affected.

BOOK II

DISTINCTION BETWEEN THE EFFECTS OF THE ENVIRONMENT AND OF AGRICULTURE

We must endeavour to present similarly all that 1. 1
as not previously been said [a] about the sprouting
nd fruiting of trees and of plants in general, dealing
eparately with two sets of phenomena, those that
ollow the annual seasons [b] and those associated with
he care bestowed in husbandry.[c] For these consti-
ute two divisions of the subject, the one natural (so
o speak) [d] and spontaneous, the other belonging
o art and preparation, which aim at excellence.[e]
But the kind of explanation is different in each: the
irst is what one might call an explanation from nature,
he second an explanation from inventiveness, nature

[e] The object of will is the good: *cf.* Aristotle, *Nicomachean*
thics, iii. 6 (1113 a 24). *Cf.* also Aristotle, *On the Parts of*
nimals, i. 5 (645 a 22–25): ". . . in all animals (*sc.* the lower
nimals included) there is something that belongs to nature
nd is beautiful. For the character of not being at random,
ut directed to some end, is found in the works of nature to an
ven superlative degree; and the end for which the animals
ave been put together or generated occupies the place of the
eautiful."

10 ὁ δὲ ἐπινοητικός, οὔτε γὰρ ἡ φύσις οὐθὲν μάτην, ἡ τε διάνοια βοηθεῖν θέλει τῇ φύσει.

ἐπεὶ δὲ πρότερα τὰ τῆς φύσεως, ὑπὲρ τούτων καὶ ῥητέον πρότερον.

1. 2 μέγιστον μὲν οὖν (ὡς ἁπλῶς εἰπεῖν) παντὶ δένδρῳ καὶ ἡμέρῳ καὶ ἀγρίῳ, καὶ ὅλως δὲ φυτῷ παντί, πρὸς εὐβλάστειαν[1] καὶ εὐκαρπίαν, τὸ χειμασθῆναι χειμῶσιν ὡραίοις καὶ καλοῖς, οὕτως 5 γὰρ αἱ βλαστήσεις κάλλισται καὶ αἱ καρποτοκίαι γίνονται.

καλὸς[2] δὲ χειμὼν ἐὰν πολυυδρίαν τε ἔχῃ βόρειον καὶ χιόνος πλῆθος, καὶ τὸ ὅλον ψύχῃ χωρὶς πάγου· δεῖ γὰρ κεκενωμένα τὰ δένδρα μετὰ τοὺς καρποὺς

[1] εὐβλάστειαν u (no accent U): -είαν N; -ίαν HP.
[2] καλὸς u: καλῶς U.

---

[a] The principle that nature does nothing in vain is explicitly appealed to at *CP* 1 1. 1 and implicitly in many passages where purposiveness or hypothetical necessity (x is required for y) is expressed: *cf.* *CP* 1 7. 3 ("for the sake of protection"); *CP* 1 11. 2 ("for what end"); *CP* 1 11. 8 ("for the sake of fruit production"); *CP* 1 12. 5 (there is no reason for the alternative); *CP* 1 16. 1 ("must be produced"); *CP* 1 16. 3 ("end"); *CP* 1 16. 11 ("nature sets out for the best"); *CP* 1 19. 2 ("in order that"); *CP* 1 20. 2 (reason for the "ear" of the date palm). In Book II the many expressions indicating what is "best" or "good" or "required" or "beneficial" may often have been taken from the writers on agriculture but are probably to be understood in a teleological sense. The end is full concoction of the fruit.

[b] *Cf.* βοηθῶ and related forms in *CP* 2 14. 1; 2 14. 4; 3 2

doing nothing idly [a] and human thought proposing to go to its aid.[b]

The Effects of Natural Environment

Since the natural phenomena are prior to the others we must treat them first.

Winters

Now of the greatest importance (broadly speaking) for good sprouting and fruiting in all trees, whether cultivated or wild, and indeed for all plants in general, is exposure to winters that come in their season and are fine, since it is under these circumstances that trees sprout and bear best.[c] 1. 2

Fine Winters

A winter is fine if there is plenty of precipitation from the north and plenty of snow, in short if there is cold without freezing. For the trees, depleted after bearing their fruit, need to be replenished

1; 3 9. 5; 5 9. 8; 5 9. 11 *bis*; ἐπικουρία *CP* 2 14. 1; συνεργῶ *CP* 3 1. 1; 5 1. 1; 5 1. 2; θεραπεία *CP* 2 14. 2. Theophrastus has a special interest in the forms of assistance that are not (or not ordinarily) resorted to by nature herself, such as slitting, cutting back, hammering in pegs, and applying swine manure (*cf.* *CP* 2 14. 1–2). This perhaps is why the word βοήθεια tends to have the sense of " remedy."

[c] *Cf.* the Hippocratic *De Regimine*, iii. 68 (vol. vi, pp. 596. 18–598.4 Littré): " One should also expose oneself confidently to the cold . . . for it is not good for the body not to be exposed to winter in its season, since trees too cannot bear fruit or be themselves vigorous unless they have been exposed to winter in its season."

10 ἀντιπληρωθῆναι πάλιν τῆς τροφῆς, καὶ ταύτην πέψαι καὶ κατασχεῖν, εἴπερ εὐβλαστῆ καὶ εὔκαρπα μέλλει γενήσεσθαι.[1] τροφῆς μὲν οὖν πλῆθος ἐν

1. 3 ὄμβρου πλήθει· τὸ δὲ κατασχεῖν καὶ πέψαι ταύτην, ἐὰν ὁ χειμὼν πιέσῃ καὶ μὴ εὐθὺς ἡ ἐκδρομὴ γένηται. τὰς γὰρ ῥίζας ὀρεγομένας ἀφθόνου τροφῆς

5 διαδιδόναι δεῖ παντὶ τῷ δένδρῳ, καὶ ταύτην, ὥσπερ κυουμένην καὶ πεττομένην, χρόνον λαμβάνειν σύμμετρον. οὐκ ἔσται δὲ τοῦτο ἐὰν μὴ κατάσχῃ τὰ ψύχη, ταχὺ γὰρ ἡ μαλακότης τοῦ ἀέρος ἐκκαλεῖται τὴν βλάστησιν. διὸ τούς τε ὄμβρους συμφέρει

10 βορείους, μὴ νοτίους, εἶναι, καὶ πλῆθος χιόνος, ὅπως τηκομένη κατὰ μικρὸν διαδύηται πρὸς τὸ ἔδαφος, καὶ μὴ ἀθρόον τὸ ὑγρὸν ἀπορρυῇ προσπεσόν, ἅμα τε καὶ τὴν γῆν ἀναζυμοῖ συγκατακλείουσα[2] καὶ ἐναπολαμβάνουσα[3] τὸ θερμόν.

1. 4 ὃ καὶ τοῖς σπέρμασι συμφέρει· ῥιζωθέντα γὰρ καὶ ἐπισχύσαντα[4] τῇ πιλήσει καὶ τῇ[5] καταπιέσει τοῦ ψύχους, ἅμα τῇ ἡμέρᾳ διαγελώσῃ ταχείας ποιεῖται[6] καὶ ἀθρόας τὰς ἀναδόσεις. ἀλλὰ τὰ

5 μὲν σπέρματα προσεπιζητεῖ καὶ τοὺς ἠρινοὺς[7] ὑετοὺς μᾶλλον κατὰ μικρά τε καὶ πλείους γινομένους διὰ τὴν ἀσθένειαν καὶ τὸ ἐπιπόλαιον τῶν ῥιζῶν, ταχὺ γὰρ ἀναξηραίνονται,[8] καὶ ταχὺ πάλιν

[1] γενήσεσθαι HP: -νν- U N.
[2] συγκατακλείουσα U^c (from -ειοῦσα): -αν u.
[3] ἐναπολαμβάνουσα U N: -αν u H^c(-ἀπ- H^ac)P.
[4] ἐπισχύσαντα Gaza (*firmata*), Scaliger: -ή- U.
[5] τῇι u: τατηι U.
[6] ποιεῖται Schneider: ποιει U.
[7] ἠρινοὺς u (ὀρινοὺς N; ὀρεινοὺς HP): εἰρηνοὺς U.
[8] ἀναξηραίνονται u: ἂν ξηραίνονται U (-ωνται N HP).

again with food and to concoct and retain it if they are to sprout and bear well. Now abundance of food comes from abundance of precipitation, and retention and concoction of the food comes about if the tree is held in check by the wintry weather and does not run up as soon as it is fed. For the roots must seek out plenty of food and pass it on to the whole tree, and this food, as a thing with which the tree is as it were pregnant and which is undergoing concoction, must have sufficient time for this. But this process will not occur unless cold weather restrains the tree, since any mildness of the air is quick to induce sprouting. This is why it is best that the rain should come from the north and not from the south, and that there should be a good deal of snow, so that as it melts the water can sink into the soil gradually, and not all run off as soon as it strikes the ground; and then too so that the snow can ferment [a] the soil by shutting up and enclosing the heat.[b] 1. 3

This is also good for grain,[c] since after it has rooted and gained strength from being compressed and held down by the cold, it all springs up in a quick spurt as soon as the days turn mild. But there is this difference: from its weakness and shallow roots, grain also requires greater lightness and frequency in the spring rains, for the roots are quick to dry out, 1. 4

[a] *Cf. CP* 3 23. 4 and Theophrastus, *On Fire*, chap. ii. 18, where the fermenting of the earth by snow is an example of counter-displacement.

[b] By counter-displacement, for which *cf.* the note on *CP* 1 12. 3.

[c] *Spérmata* (literally "seeds," here rendered "grain") are those herbaceous plants whose seeds are used for consumption. These are cereals, legumes and "summer seeds." Here cereals are meant.

δέονται·[1] τὰ δὲ δένδρα καὶ ἰσχυρότερα καὶ
10 βαθυρριζότερα, καὶ ἅμα διάπλεα τροφῆς ἐν ἑαυτοῖς, ὥστε τρόπον τινὰ μᾶλλον τοῦ συνεργήσοντος δεῖσθαι πρὸς τὴν πέψιν καὶ τὴν βλάστησιν· σημεῖον[2] δέ, τὸ μὴ βλαστάνειν πρὸ τοῦ ἦρος.

1. 5 ὅτι δὲ ἡ πολυυδρία συμφέρει τοῖς δένδροις κἀκεῖθεν φανερόν· ἐν γὰρ ταῖς ἐπομβρίαις ἅπανθ' (ὡς εἰπεῖν) εὐθενεῖ[3] μᾶλλον. ἀλλ' ὅταν μὲν ὦσιν νότιαι διυγραίνονται καὶ ἀσθενέστερα[4] γίνονται,
5 βορείων δ' οὐσῶν ἰσχυρά[5] τε καὶ ἐκπέττει μᾶλλον, ἅτε τῆς μὲν γῆς διακοροῦς[6] οὔσης, αὐτά τε ξυνεστῶτα[7] καὶ ἐναπειληφότα ⟨τὸ⟩[8] οἰκεῖον θερμόν. ὅπου γὰρ ἀεὶ μάλιστα μαλακὸς ⟨ὁ⟩[9] ἀήρ, ἐνταῦθ' ἡ εὐβλαστία καὶ εὐκαρπία γίνεται τῶν
10 δένδρων (ὥσπερ ἐν Αἰγύπτῳ) διά τε τὴν εὐτροφίαν καὶ διὰ τὸ μηδὲν ἀντικόπτειν τῶν ἔξωθεν (ἀφθόνου γὰρ τῆς τροφῆς οὔσης καὶ τοῦ ἀέρος εὐτρεφοῦς,[10] εὐλόγως ἡ εὐβλαστία καὶ ἡ εὐκαρπία)·

[1] δέονται U: δεύονται u N (-ωνται HP).
[2] σημεῖον HP: σημεῖα U N.
[3] εὐθενεῖ Schneider: εὐσθενεῖ U.
[4] ἀσθενέστερα Gaza, Itali: ἀσθενέστεραι U.
[5] ἰσχυρά Gaza, Scaliger: ἰσχυραί U.
[6] διακοροῦς ego: διακόρου U.
[7] ξυνεστῶτα ⟨μᾶλλον⟩ Wimmer.
[8] ⟨τὸ⟩ Schneider.
[9] ⟨ὁ⟩ N HP.
[10] εὐτρεφοῦς ego: ευτρόφου U.

---

[a] The helper is heat.
[b] Most kinds of wheat and barley were sown at the setting of

and quick to need water again. But trees are stronger and have deeper roots, and then too are full throughout with food within themselves, so that in a way their need is more for what will help them to concoct it and sprout;[a] this is shown by their not sprouting before spring.[b]

Winters: Winter Should Fall in Winter and Spring in Spring [c]

1.5 That plenty of water is good for trees is further seen from this: that during the rains practically every kind of tree thrives better. Still, when the winter rains come from the south the trees get soaked with water and are weakened; but when from the north, they are strong and concoct the food better, the ground being saturated but the trees themselves having a firm consistency and retaining their native heat shut up within them.[d] For in places where the air in winter is almost unbrokenly mild, one gets good sprouting and fruiting in the trees, as in Egypt,[e] from the joint effect of being well fed and suffering no setbacks from harsh weather (for when food abounds and the air fosters growth, good sprouting and fruiting is a reasonable consequence); for good quality,[f]

the Pleiades (in early November), and came up about a week later (*cf. HP* 8 1. 2–3, 5).

[c] "Winter" means cold and wet weather, whatever the season. This "winter" is seasonable when it falls in winter and not in spring.

[d] It is not dissipated (so exposing the tree to damage by subsequent cold) but shut in by counter-displacement.

[e] *Cf. CP* 1 11. 5; *HP* 3 5. 4.

[f] That is, fragrance: *cf. CP* 6 18. 3.

πρὸς δὲ τὸ ποιὸν[1] αὐτῶν ἢ πάντων ἤ τινων
1. 6 δεῖταί τινος ἴσως ἑτέρας κράσεως. ἐνταῦθα δέ,
ἐὰν μὴ καθ' ὥραν ἔτους αἱ βλαστήσεις ὦσιν,
ἀλλὰ προεκδράμωσιν δι' εὐτροφίαν καὶ ἄνεσιν
τοῦ ἀέρος, ἐπιγινόμενα ψύχη διελυμήνατο καὶ
5 ἀπέκαυσεν·[2] διὸ καὶ οἱ ὀπισθοχειμῶνες[3] χαλ-
εποὶ τοῖς δένδροις, ὅταν γὰρ ἅπαξ ἐκτέκωσιν,
εὐθὺς οἱ καρποὶ μαλακοῦ τινος ἀέρος δέονται καὶ
εὐμενοῦς εἰς τὴν ἐκτροφήν, καὶ μάλιστα ἐν ταῖς
ἀρχαῖς, τότε γὰρ ἀσθενέστατοι, τὸ δὲ ἀσθενὲς οἷον
10 τιθηνήσεως δεῖται· καὶ γὰρ ὅλως πᾶσα μεταβολὴ καὶ γένεσις δεῖται τῆς τοιαύτης εὐκρασίας.
1. 7 μεταβολαὶ δ' ἅμα καὶ ὥσπερ γενέσεις τινὲς ἥ τε
βλάστησις καὶ ἡ ἄνθησις καὶ εἴ τι τοιοῦτον
ἕτερον, ἐν αἷς καὶ πλεῖσται φθοραὶ γίνονται τῶν
καρπῶν, ἐρυσιβουμένων τε καὶ ἀποκαομένων καὶ
5 ἀποπιπτόντων καὶ τὸ ὅλον χειμαζομένων. ἐπεὶ
καὶ τὰ ἄγρια μάλιστα συμβαίνει πονεῖν ὅταν
μάλιστα ἀρτιβλαστῶν[4] ὄντων ἐπιγίνηταί τι πνεῦμα
ψυχρὸν ἄγαν καὶ θερμόν· ἀποκάει[5] γὰρ ἄμφω
καὶ ἀπόλλυσιν.
10 ἀλλὰ ταῦτα μὲν ἐν τοῖς καθ' ἕκαστα δεῖ θεωρεῖν·
ἡ δ' ὅλη διάθεσις καὶ κατάστασις τοῦ ἀέρος εἰς τὴν
τῶν δένδρων εὐσθένειαν ὅτι ταύτῃ ξυμφέρει,
φανερόν ἐστι διὰ τῶν εἰρημένων.

[1] ποιὸν ego: ποιοῦν U.
[2] ἀπέκαυσεν U: ἀπέκαυσαν u; ἐπέκαυσε N HP.
[3] ὀπισθοχειμῶνες Coray: ὄπισθεν χειμῶνες U.
[4] ἀρτιβλαστῶν u: ἄρτη βλαστῶν U.
[5] ἀποκάει Schneider: ἀπορεὶ U (ἀπορεῖ N; ἀπορρεῖ u HP).

however, in the fruit of all or some,[a] the trees perhaps require a different tempering of the air. But in our part of the world, if the sprouting does not wait for the spring season, but comes out too early because of good feeding and a relaxation of the repression of the air, cold spells follow, harming the sprouts and searing them off. This is why recrudescent winters in spring are bad for trees; for once they have brought forth their fruit, it requires from that moment, if it is to be reared to maturity, a certain gentleness and clemency in the air, and the requirement is greatest at the outset, since then the fruit is weakest, and the weak require what one might call "coddling;" indeed all change and birth requires this sort of well-tempered character in the air. Sprouting, flowering and similar events [b] are changes and at the same time (so to say) births. It is in the course of these that the fruit is lost in the greatest number of ways, getting rust or else getting seared off and dropping and suffering in a word the ill effects of winter. Indeed it so happens that wild trees too suffer most when a very cold or hot wind arises just when they have sprouted, both kinds of wind searing the sprouts off and destroying them.[c]

1. 6

1. 7

But we must study all this by considering the particular cases. That the general disposition and settled ordering of the air is good for the vigour of trees in the way indicated is evident from what has been said.

[a] In all but the myrtle: *cf. CP* 6 18. 4.
[b] Perhaps a hint at the fig, which was not known to have a flower (*HP* 3 3. 8).
[c] *Cf. CP* 5 8. 3 (with second note).

2. 1 ἑπόμενον δέ πως τούτοις ἐστὶν περὶ τῶν ὡραίων ὑδάτων εἰπεῖν. ὡραιότατα μὲν γὰρ τὰ χειμερινά, διὰ τὰς λεχθείσας αἰτίας. δεύτερα δὲ τὰ πρὸ τῆς βλαστήσεως, οὕτω γὰρ ἀθροωτέρα τε καὶ καλλίων 5 ἡ βλάστησις, ἐκπληρωθέντων πάντων ταῖς τροφαῖς, εἰ δὲ μή, ἀβλαστεῖς καὶ ἄμπελοι γίνονται καὶ τἆλλα, τῶν μὲν ἐχόντων, τῶν δὲ λειπομένων. τρίτα δὲ μετὰ [1] τὴν ἀπάνθησιν, ἃ πρὸς τὴν ἐκτροφὴν ἤδη καὶ τελείωσίν [2] ἐστιν, ταῦτα δὲ 10 μὴ εὐθύς, ἀλλ᾽ ὅταν ὁ καρπὸς ἰσχύσῃ· εἰ δὲ μή, συμβαίνει τὰ μὲν ἄλλα καὶ ἀπορρεῖν, κἂν μὴ τοῦτο πάθῃ, διυγραινόμενα χείρω καὶ ἀσθενέστερα γίνεσθαι, τὴν δ᾽ ἐλάαν καὶ ἐπιβλαστοῦσαν ἀποβάλλειν, ἅτε τῆς τροφῆς ἰούσης εἰς τὸν βλαστόν.

2. 2 ἀώρια [3] μὲν οὖν ταῦτα· χείριστα δὲ καὶ παρακαιρότατα ⟨τὰ⟩ [4] περὶ τὰς ἀνθήσεις ἑκάστων, ἅπαντα γὰρ ἀσθενῆ,[5] καὶ πάνθ᾽ (ὡς εἰπεῖν), ἢ τά γε πλεῖστα, ἀπόλλυται καὶ ἀποπίπτει, τὰ μὲν 5 ἐρυσιβούμενα, τὰ δὲ ὑγραινόμενα, τὰ δ᾽ ἐπιμένοντα χεῖρον ἀνθεῖ [6] πλὴν εἴ τινων ὀλίγων.[7] (καὶ τοῦτ᾽ οὐχ ἧττόν ἐστιν ἐν τοῖς φρυγανικοῖς καὶ τοῖς ποιώδεσιν, οἷον τὰ στεφανωτικά, καὶ ὅλως τὰ ἄγρια καὶ αὐτόματα τῶν ἀνθῶν, ἔτι δὲ τῶν ποιωδῶν 10 ἔνια, καὶ τῶν ἡμέρων σπερμάτων τὰ χεδροπά·

[1] ⟨τὰ⟩ μετὰ Wimmer.
[2] τελείωσιν u: -σις U.
[3] ἀώρια HP: ἀωρία U N.
[4] ⟨τὰ⟩ Schneider.
[5] ἀσθενῆ U N HP: -εῖ u.
[6] ἀνθεῖ Gaza (*florent*), Scaliger: -εῖν U.
[7] Schneider transposes πλὴν—ὀλίγων after ποιώδεσιν below (line 8).

Timely and Untimely Rain

2. 1 Next in order (so to say) is the discussion of rain falling at the right time. Winter rains come at the best time for the reasons given.[a] Next is rain that comes before sprouting, since then the sprouting is more uniform throughout the tree and so is finer, all parts having had their fill of food. Otherwise there is failure to sprout both in the vine and in the rest, some parts having their sprouts, some falling behind. Third is rain coming after the flower is shed. From this point on the rain serves the rearing and maturing of the fruit. But it should not fall too soon, but only after the fruit has gathered strength; otherwise what happens in the rest is that the fruit drops (or if it does not go so far as that, is soaked to the point of deteriorating and getting weaker), whereas the olive even puts out a second set of shoots and so loses its fruit, the food passing to the new growth.

2. 2 Such rain, then, is untimely. Worst and most untimely of all is rain that falls when the various trees are in flower. For the flowers of all kinds of tree are weak, and in all of them (so to say), in any case in most, they are lost and drop off when rained on, some getting rust, others getting soaked,[b] and where the flowers remain they bloom worse with a few exceptions. (The exceptions are more numerous in shrubs and herbaceous plants, such as the coronaries and all wild and self-propagating flowering plants, further certain herbaceous plants and among cultivated seed-crops the legumes.[c] Here the flowers

[a] *CP* 2 1. 2, 5.
[b] *Cf. CP* 4 10. 2.
[c] *Cf. CP* 3 24. 3.

ταῦτα δὲ διαμένει δυοῖν θάτερον· ἢ δι' ἰσχὺν ἑαυτῶν τε καὶ τῶν προσφύσεων, καθάπερ καὶ τὸ ῥόδον καὶ τὸ κρίνον καὶ τὰ ἄλλα ὅσα τούτοις ὅμοια, ἢ διὰ ξηρότητα τῆς ὅλης φύσεως ἀναλαμβα-15 νόντων τὸ ὑγρόν, ἐπικρατεῖ γὰρ οὕτω, τὸ δ' ἐπικρατοῦν ἀπαθές.)

2. 3 ὧν δὲ οἱ καρποὶ χρονιώτεροι καὶ πλείονος δεόμενοι τροφῆς καὶ πέψεως, τούτοις ὡραῖα καὶ τὰ ὀψιαίτερα, καθάπερ ἀμπέλῳ ῥόᾳ ἐλάᾳ τοῖς ἄλλοις, ἁπλῶς δὲ ἑκάστοις πρὸς τὴν αὑτοῦ[1] 5 τελείωσιν· διὸ καὶ οὐχ ὁ αὐτὸς ἅπασι καιρός, ὥσπερ ὁ τοῦ χειμῶνος πρὸς τὴν βλάστησιν, ἀλλ' ἕτερος ὁ τοῖς ὀψικάρποις καὶ πρωϊκάρποις, ὥσπερ καὶ τῶν σπερμάτων τοῖς τριμήνοις, καὶ ἁπλῶς τοῖς ὀψίοις καὶ πρωΐοις. ἁπλῶς δ' αἰεὶ τὰ 10 βόρεια βελτίω τῶν νοτίων· καὶ γὰρ ψυχρότερα καὶ τὴν ἀπόλαυσιν ποιεῖ πλείω, ξυνεστηκότων καὶ ἰσχυόντων καὶ τῶν καρπῶν καὶ τῶν δένδρων.

2. 4 ἔτι δὲ ἀφαιρεῖ τὸ περιττὸν καὶ ἀποξηραίνει[2] καὶ οὐκ ἐᾷ προσκαθήμενον διυγραίνειν, οὐδ' ὑπὸ τοῦ ἡλίου συνεψόμενον λυμαίνεσθαι. (διὸ καὶ τὰ ἐπιγινόμενα πνεύματα ὠφελεῖ καὶ μάλιστ' ἐὰν ᾖ 5 βόρεια, περιαιρεῖ γὰρ ταῦτα καὶ ἀπόλαυσιν ποιεῖ πλείω.)[3] διὰ ταῦτα γὰρ καὶ τὰ παραπλήσια καὶ

[1] αὑτοῦ u: αὐτοῦ U; ἑαυτοῦ N HP.
[2] ἀποξηραίνει ("Voluerunt ἀποξηραίνει τε") Schneider: ἀποξηραίνεται U.
[3] Schneider transposes διὸ—πλείω before ἔτι (line 1).

---

[a] *Cf. CP* 4 10. 1, where open texture is mentioned.
[b] An early variety of wheat and barley: *cf. HP* 8 2. 7:

survive for one of two reasons: either because of their own strength and that of the pedicel—as rose, lily and the like—or because, owing to the dryness [a] of the whole nature of the plant, the flower takes up the water. For in so doing the flower masters the water, and what masters a thing escapes harm from it.)

2. 3 For trees with fruit that takes longer to mature and that requires more food and concoction later rains than these are also timely, as for the vine, pomegranate, olive and the rest of them; in short, rain is timely for a given kind of tree when it promotes the tree's own type of maturing. This is why the same time of year is here not the best for every tree (as winter rain is best for their sprouting), but differs for late and early fruiters (just as it differs for three-months grain [b] and for late and early grains in general). Broadly speaking, northerly rain is always better than southerly, since northerly rain is colder and makes for greater consumption, both fruit and tree being then firm and strong. 2. 4 Again it [c] removes the superfluous water on the fruit and dries it, now allowing the water to remain and soak the fruit or burn it by getting overheated by the sun. (This is why wind that comes up after rain is beneficial, especially when the wind comes from the north, since wind at this time removes the moisture from the surface of the tree and brings about more consumption.) For it is these and similar reasons that also make rain at night better than rain in the daytime,

'. . . but in Greece barley matures in the seventh month and in most districts in the eighth, whereas wheat takes still longer."

[c] The wind is thought of as continuing after it has brought the rain.

τὰ νυκτερινὰ βελτίω τῶν ἡμερινῶν, ἀπόλαυσίς τε γὰρ γίνεται πλείων μὴ εὐθὺς ἀφαιρουμένου τοῦ ἡλίου, καὶ τῶν ἄλλων ἀκινδυνότερα.

3. 1 τὸν αὐτὸν δὲ τρόπον καὶ τῶν πνευμάτων τὰ βόρεια τῶν νοτίων βελτίω, καὶ τὰ πόντια τῶν ἀπογείων (ὅτι ψυχρότερα), καὶ τὰ ἀπὸ δύσεως τῶν ἀφ' ἑῴων· καθόλου γὰρ (ὡς εἰπεῖν) τὰ ψυχρὰ τῶν
5 θερμῶν (ἐὰν μὴ ἀρτιβλαστῇ [ᾖι] ᾗ[1] καὶ ἐν ἀνθήσει λαμβάνῃ·[2] τότε γὰρ ἀποκάει τὰ ψυχρά, καθάπερ εἴρηται).
βελτίω δὲ καὶ τὰ ζεφύρια καὶ αἱ τροπαὶ (καὶ ὅλως αἱ αὖραι) τῶν σκληρῶν καὶ διατόνων· τὰ
10 μὲν γὰρ τρέφει, θάτερον δὲ πιλοῖ καὶ κωλύει τὰς αὐξήσεις. ἰσχύει δ' ἕκαστον κατὰ τὴν θέσιν τῆς χώρας· ἄλλα γὰρ ἄλλοις τοιαῦτα (καθάπερ ἐλέχθη καὶ πρότερον).
διὸ καὶ ὡς μὲν ἐπίπαν εἰπεῖν βελτίω τὰ βόρεια τῶν
3. 2 νοτίων· οὐ μὴν ἀλλὰ ἐπείπερ αἱ παραλλαγαὶ καὶ τῆς χώρας ποιοῦσιν τὰς[3] δυνάμεις, καὶ δεῖ τοῦ μὲν χειμῶνος εἶναι θερμά, τοῦ δὲ θέρους ψυχρά

[1] [ᾖι] ᾗ ego: ἦι ἡ U (Hindenlang deletes).
[2] λαμβάνῃ HP^c: -ει U N P^ac(?).
[3] ποιοῦσι τας (-ὰς u) U^r: ποιοῦσιν | ἔτας U^ar.

---

[a] That is, there will be no sun-burn.
[b] *CP* 2 1. 7.
[c] "Gentle westerlies at sunset" renders *zephýria*, the diminutive of *zéphyros*. *Cf.* Theophrastus, *On Winds*, chap. vii. 38: "The zephyros is the smoothest of the winds and blows in the evening and downward toward the earth and is cold , , .;" chap. viii. 47: "The air, when the sun in

since night rains not only bring about more consumption, since the water is not removed at once by the sun, but are moreover attended with less danger.[a]

Winds

3\. 1 As with rain, so with wind: northerly wind is better than southerly, sea wind (being colder) than land wind and wind from the setting sun than wind from the sunrise. For in general (one may say) cold winds are better than hot, unless the wind catches the trees when they have just sprouted or else when they are in bloom, since then (as we said)[b] cold winds kill the parts that have just come out.

Again gentle westerlies at sunset,[c] "returners,"[d] and indeed all breezes are better than harsh and powerful winds, since the breezes nurture the tree, but the other kind compresses it and prevents growth. But this strength in a wind depends on the lie of the land, for in different countries different winds are gentle and harsh (as we said before).[e]

This is why, although it is true that northerlies are as a rule better than southerlies, nevertheless, since 3. 2 variations from one country to the other also have their part in giving the wind a different power, and since the wind should be warm in winter and cool in

his pull no longer holds it, is released and flows. Hence the setting sun leaves clouds behind from which come the *zephýria*."

[d] *Cf.* Theophrastus, *On Winds*, chap. iv. 26: "From the land breeze and the like are produced the 'returners,' when the moist air has been accumulated; for the 'returner' is a 'reflux' as it were of wind, as in narrow seas there is a reflux that follows the flux of water; . . ."

[e] *Cf. HP* 4 14. 11; 8 6. 6; 8 7. 6–7.

(βοηθεῖ γὰρ οὕτως ἑκάτερα[1] πρὸς τὰς ὥρας,
5 ὥσπερ εἴπομεν, ἐὰν δὲ ὅμοια, βλάπτει, ποιεῖ γὰρ ὑπερβολήν), εὔλογον ἤδη[2] μὴ τὰ αὐτὰ πᾶσιν εὔτροφα καὶ ὠφέλιμα καὶ βλαβερὰ γίνεσθαι· διὸ τοῖς μὲν ὁ νότος ἐπισινής, τοῖς δ' ὠφέλιμος, ὡσαύτως δὲ καὶ ὁ ζέφυρος καὶ τῶν ἄλλων ἕκαστος.
10 ἅπασι δὲ χαλεπά (καθάπερ εἴρηται) τὰ κατὰ τὴν βλάστησιν εὐθὺς ἢ θερμὰ λίαν ἢ ψυχρὰ πνέοντα· διαφθείρει γὰρ ἄμφω διὰ τὴν ἀσθένειαν.

3. 3 ὡς δὲ τὸ σύνολον εἰπεῖν, εὔπνουν εἶναι χρὴ τὸν τόπον· ἕτερος δ' ὁ ἄπνους,[3] καὶ ὅλως ὁ προσήνεμος ἀναυξής.

σχεδὸν δ' ὁμολογουμένη τις καὶ ἡ τοῦ ἀέρος
5 διάθεσίς ἐστιν τούτοις· ⟨ὁ⟩[4] γὰρ [ὁ][5] εὐκραὴς [ὁ ἀὴρ][6] (ὡς ἁπλῶς εἰπεῖν) ἄριστος τοῖς δένδροις, εὐβλαστὴς[7] ὢν καὶ εὔκαρπος· οἱ δὲ περισκελεῖς ἐφ' ἑκάτερα διαφθείρουσιν οἱ μὲν τοὺς καρπούς, οἱ δ' ὅλως καὶ τὰ δένδρα, πλὴν ὅσα πέφυκεν οἰκεῖα
10 τούτοις. ἔνια γὰρ δὴ ταῖς ὑπερβολαῖς χαίρει, καὶ τὰ μέν ἐστι φιλόθερμα (καθάπερ φοῖνιξ), τὰ δὲ φιλόψυχρα[8] μᾶλλον, ὥσπερ ὁ κιττὸς καὶ ἡ ἐλάτη.

[1] ἑκάτερα (ἑ- U) N HP: ἑκατέρα u.
[2] ἤδη Uc: δη Uac.
[3] ἄπνους Gaza, Basle ed. of 1541 (εὔπνους ⟨καὶ ὁ προσήνεμος·⟩ Schneider): εὔπνους U.
[4] ⟨ὁ⟩ Schneider.
[5] [ὁ] Schneider: ὁ Uc from ἐστὶ (?) ὁ.
[6] [ὁ ἀὴρ] ego: ὁ ἀὴρ u N; ἀὴρ HP.
[7] εὐβλαστὴς Schneider: συμβλαστὴς U.
[8] φιλόψυχρα Ucc from ψυχό.

[a] Perhaps a reference to *CP* 1 10. 3 or 1 21. 5–6. *Cf.* Theophrastus, *On Winds*, chap. vii. 43: "As to the west wind's

summer (for then, as we said,[a] each is a corrective to the season, whereas the same character in wind and season is harmful, producing excess), it is by now seen to be no problem that the same winds should not everywhere turn out to be the ones that foster and are beneficial or that on the other hand are harmful. This is why the south wind in some countries is destructive, in others beneficial, and similarly with the west wind and each of the rest. But in all countries winds that blow very hot or cold right at the time of sprouting are bad (as we said),[b] since both destroy the sprouts which are then too weak to resist.

3. 3 The region in a word must have good winds, this being not the same as to have no winds, and a windy region is definitely stunting to growth.

Different Temperings of the Air

So too the settled condition of the air should (one may say) agree with what we said about the winds: to put it broadly, air that is well-tempered is best for trees, since it promotes sprouting and fruiting, whereas either extreme, of heat or of cold, in some cases destroys the fruit and in others the whole tree, except for the trees naturally belonging to such a climate. For some trees delight in the one or the other excess, some favouring heat, like the date-palm, others cold, like the ivy and silver-fir. In fact these

destroying some crops and fostering others, it is true to make the general statement that also applies to the rest, that it fosters wherever it blows cold in summer, and destroys where it blows hot. So again in winter, and so too in spring: where it blows cold it destroys, where hot, it fosters and preserves, blowing with a character contrary to that of the season."

[b] *CP* 2 1. 7.

ταῦτα γὰρ ὅλως ἐν τοῖς ἐμπύροις οὐ[1] φύεται, χαλεπῶς[2] δὲ καὶ πύξος καὶ φίλυρα[3] (καθάπερ ἐν ταῖς ἱστορίαις εἴπαμεν). αἴτιον δὲ ἡ θερμότης καὶ ἡ ξηρότης, οἷον γὰρ πῦρ ἐπὶ πῦρ γίνεται, συμμετρίας γάρ τινος δεῖται καὶ τὸ ὅμοιον. ὡσαύτως δὲ οὐδ' ἐν τοῖς ψυχροῖς ἔνια φύεται τῶν ψυχρῶν διὰ τὴν αὐτὴν αἰτίαν. ἔστι δὲ καὶ τῶν ἐναντίων δῆλον ὅτι συμμετρία τις πρὸς ἄλληλα, ὥστε τὰ μὲν δύνασθαι βλαστάνειν·[4] ἀεὶ γὰρ δεῖ λόγον[5] τινὰ ἔχειν τὴν κρᾶσιν τῆς φύσεως πρὸς τὸ περιέχον. ἔοικε δὲ κοινὸν εἶναι τοῦτο καὶ ἐπὶ τῶν ζῴων· καὶ γὰρ τὰ ζῷα καθ' ἑκάτερον τῶν τόπων ἴδια τυγχάνει, τὰ μὲν δεχόμενα, τὰ δ' οὐ δεχόμενα τὴν τοῦ ἀέρος διάθεσιν, ὁτὲ δὲ καὶ τροφὰς οὐκ ἔχοντα τὰς οἰκείας. ἐνδέχεται γὰρ καὶ τοῦτο κωλύειν ὁμοίως καὶ τὰ ζῷα καὶ τὰ φυτά, τάχα δὲ καὶ ἑτέρας πλείους αἰτίας αἳ πρὸς τὰς ἰδίας φύσεις εἰσὶν ἐναντίαι.

3.4

μεγίστη δ'[6] οὖν διαφορὰ κατὰ τὰ ἔνυγρα καὶ χερσαῖα ζῷα καὶ φυτά, περὶ ὧν οὐδὲ ζητοῦμεν

3.5

[1] οὐ U^cm: U^t omits.
[2] χαλεπῶς Gaza, Itali: -ὸς U.
[3] φίλυρα N: φιλύρα U HP.
[4] βλαστάνειν ⟨τὰ δ' οὔ⟩ Basle ed. of 1541 (*germinare . . alia nequeant* Gaza); βλαστάνειν ⟨τὰ δὲ μή⟩ Wimmer.
[5] δεῖ λόγον u (δὴ λόγον U): δῆλον N HP.
[6] δ' U: γ' Schneider.

[a] *HP* 4 4. 1: ". . . thus it is said that ivy and olive [U has ἐλάτην, "silver-fir"] are not found in Asia inland from Syria more than five days' journey from the sea . . . (Harpalus

last do not grow at all in torrid countries, and the box and lime do so only with difficulty (as we said in the History).[a] The cause is the heat and dryness of the trees, since for them to grow in torrid country would be (as it were) to add fire to fire, for even likeness to the climate must not go too far. Similarly some cold trees for the same reason do not grow in cold countries. But even in trees of the opposite character to the climate there is evidently a point beyond which the opposition does not go, allowing some at least to sprout, since the tempering of the tree's nature must always bear some proportion[b] to the blend in the surrounding air. This appears to hold for animals as well, since in both hot and cold climates animals are found peculiar to each,[c] some animals tolerating the climate, and some not doing so[d] or in some cases not getting in that climate their proper food.[e] For this last circumstance too can very well prevent the occurrence of plants as well as of animals; and there could also be a number of other circumstances responsible that are unfavourable to the distinctive natures. 3. 4

At all events the greatest local distinction is that between animals and plants of the water and of the 3. 5

attempted unsuccessfully to grow ivy in the gardens of Babylon.) So the country refuses to admit the plant because of the character of the weather; and it admits box and lime only under compulsion; . . ."

[b] *Cf.* Aristotle, *On the Generation of Animals*, iv. 2 (767 a 16–17): ". . . for everything produced in accordance with art or nature is in a certain proportion (λόγῳ τινί ἐστιν)."

[c] *Cf.* Aristotle, *History of Animals*, viii. 28 (606 a 6–25).

[d] *Cf.* Aristotle, *History of Animals*, viii. 28 (606 b 2–14).

[e] Aristotle, *History of Animals*, viii. 28 (606 a 25–b 2).

λόγον (ὡς εἰπεῖν), πλὴν ὑπὲρ τοῦ πότερα θερμότερα καὶ ψυχρότερα (τοῦτο γὰρ ἀμφισβητεῖται). [καὶ][1]
5 τὰ δ' ἄλλα, ⟨καὶ⟩ οὐ[2] συγκεχωρημένα, τῇ φύσει τίθεται. καίτοι τὰ καθόλου καὶ κοινὰ πρῶτον[3] ἔδει ζητεῖν· εὑρεθέντων γὰρ τούτων, καὶ τὰ κατὰ μέρος φανερά.

τοῦτο μὲν οὖν[4] ἴσως κωλύοιτ' ἂν διὰ τὸ
10 χαλεπόν, εἴτε πλείους αἰτίαι τυγχάνουσιν[5] εἴτε μία· περὶ δὲ τῶν ἐν τοῖς[6] καθ' ἕκαστα μᾶλλον εὐποροῦμεν, ἡ γὰρ αἴσθησις δίδωσιν ἀρχὰς ἐπ' ἄμφω, καὶ ἔτι μᾶλλον καὶ πλείους ἐπὶ τῶν φυτῶν (ἐμφανέστατα γὰρ τὰ συμβαίνοντα περὶ αὐτά).

3. 6 μεγίστη δὲ διαίρεσις κατά γε τὸν αὐτὸν τόπον τοῖς ἡμέροις καὶ ἀγρίοις, ἀεὶ γὰρ ταῦτα[7] μαλακωτέρους καὶ ὑγροτέρους ζητεῖ τοὺς ἀέρας. οὐ μὴν ἀλλὰ[8] ἔνιά γε καὶ τῶν ἡμέρων ἀδυνατεῖ

[1] [καὶ] Gaza, Schneider.
[2] ⟨καὶ⟩ οὐ ego (ὡς Schneider from Gaza): οὐ U.
[3] πρῶτον ego (*prius* Gaza): πρῶτα U.
[4] οὖν U^{cm} (with indices): U^{t} omits.
[5] τυγχάνουσιν u HP: -ωσιν U N.
[6] [ἐν τοῖς] Schneider.
[7] ταῦτα ego (*sativa* Gaza, τὰ ἥμερα Schneider): τε U.
[8] ἀλλὰ U: ἀλλ u (dot over -ὰ).

[a] For the distinction *cf*. *HP* 1 4. 2: "In all plants we must also set down the divisions resting on locality; perhaps it is not even possible to avoid this. Such divisions as these would appear to produce a generic kind of separation (*sc*. applying to all four kinds of plants: tree, shrub, undershrub and herbaceous), for instance the division between aquatic and terrestrial, as they do in animals;" *HP* 1 14. 3: "Such differences as the following seem to be evidently of the whole essence (*sc*. and not of the parts): cultivated and wild; fruitful and fruit-

land;[a] but about these two groups we (so to say) do not even seek any explanation, except for the question which of the two is the hotter or colder (for here a dispute is raised).[b] But all other points about them, although not conceded, are simply ascribed to the aquatic or terrestrial nature. Yet one ought to have sought explanations first for matters of high generality and common to many different kinds, since with the discovery of these the particulars become evident.

Now the general enquiry into this highest of the distinctions might perhaps be prevented by its difficulty, whether the causes here are many or only one; but we are better off in dealing with the particular kinds, since our senses provide us with approaches to these both in animals and in plants, and provide us even better and with even more approaches in the case of plants, since what occurs in particular kinds of plants is most readily observed.

3. 6 The greatest division between plants, that is, of the same place,[c] that affects climatic preference, is that between cultivated and wild, since the cultivated always seek a milder and wetter climate. Still even some cultivated trees are unable to sprout

less; . . . And in a way such matters are in the parts or not without the parts. But the separation that is most peculiarly of the whole plant and in a way the greatest is the one found also in animals, that some are aquatic, some terrestrial . . .;" *HP* 4 6. 1: " We must take the greatest difference of the very nature of trees and woody plants in general to be the one we mentioned before, that in plants, as in animals, some are terrestrial, some aquatic; . . ."

[b] *Cf. CP* 1 21. 5–6; 1 22. 2–4.

[c] Of land, as opposed to water.

5 βλαστάνειν ἐν τοῖς θερμοῖς καὶ ψυχροῖς, οὐ μόνον διὰ τὴν ἀσθένειαν ἢ τὴν κρᾶσιν, ἀλλὰ δι' ἕτερ' ἄττα, καθάπερ ἡ ἐλάα, καὶ [1] θερμὸν καὶ πυκνόν, ἐν τοῖς ψυχροῖς [2] διὰ τὸ μετέωρον τῶν ῥιζῶν, ἐκπήγνυται [3] γάρ, ἡ δὲ ἀχρὰς ἐν τοῖς
10 σφόδρα θερμοῖς, ὥσπερ περὶ Αἴγυπτον· μοχθηραὶ δὲ καὶ αἱ ἄπιοι καὶ μηλέαι καὶ σπάνιαι. τὴν δ' αἰτίαν σκεπτέον, ἐπεὶ οὐκ ἂν θερμὰ δόξειεν εἶναι.

3. 7 τὰ μὲν οὖν ὅλως οὐδὲ βλαστάνειν ἐνιαχοῦ δύναται, τὰ δὲ βλαστάνει μέν, ἄκαρπα ⟨δὲ⟩ [4] γίνεται, καθάπερ ἡ περσέα ἡ Αἰγυπτία περὶ Ῥόδον, προϊόντι δέ, οὕτω [5] φέρει μέν, ὀλίγον δέ,
5 καὶ καλλικαρπεῖ καὶ γλυκυκαρπεῖ ⟨δ'⟩ [6] ἐκεῖ μόνον. ὁμοίως δὲ καὶ ὁ φοῖνιξ, καὶ ἔτι μᾶλλον ἐν τοῖς περὶ Βαβυλῶνα καὶ Συρίαν καλλίκαρπος. ὁ γὰρ ἀὴρ διὰ ψυχρότητα τὰ μὲν ὅλως οὐ δέχεται, τὰ δ' ὅσον εἰς βλάστησιν, ἔνια δὲ εἰς καρπόν·
10 ὁ δὲ οἰκεῖος ἤδη διατελειοῖ τὰ [7] τῆς φύσεως.

3. 8 διὰ τὴν αὐτὴν δὲ αἰτίαν οὐδὲ αἱ συκαῖ περὶ

[1] καὶ U: καίπερ Scaliger.
[2] ψυχροῖς Schneider: φυτοῖς U.
[3] ἐκπήγνυται Schneider: ἐμπήγνυται U.
[4] ⟨δὲ⟩ u HP.
[5] οὕτω U: χρόνῳ Schneider.
[6] ⟨δ'⟩ ego.
[7] διατελειοῖ τὰ Wimmer: διατελειότητατη Uar (-ότητα Ur=c): διὰ τελειότητα u (-οτήτα N) HP.

---

[a] The wild pear and cultivated pear best resist freezing (*CP* 1 22. 7). So of wild fruits, pear and apple are among those most resistant to cold (*CP* 2 8. 2).

[b] If the trees were hot the explanation would be easy: that

in hot or cold regions not merely out of weakness or owing to their tempering of qualities, but for certain other reasons; for instance the olive, though hot and close in texture, cannot grow in cold regions because of the shallowness of its roots, since it freezes out. The wild pear on the other hand cannot sprout in very hot regions, such as Egypt, where the cultivated pear and apple too are poor and rare.[a] (We must look for the cause, for it would not appear that these trees are hot.)[b]

Now whereas some trees cannot even sprout at all in certain countries, others sprout but do not bear, as with the *persea*[c] at Rhodes; but as you proceed southward it begins to bear, but the amount is small, and only in Egypt is the fruit both plentiful and sweet.[d] Similarly with the date-palm, and here the abundance of fruit is still greater in Babylon and Syria.[e] For our climate is too cold to allow some trees to grow at all, and allows others only to reach the stage of sprouting, and a few to reach that of bearing. It is only the climate proper to a tree that brings to completion all that is in the tree's nature. The same cause further makes the fig poor in Egypt 3. 7 3. 8

the heat of the tree coupled with the heat of the climate produces excess. But the actual cause is yet to be found.

[c] *Mimusops Schimperi.*

[d] *Cf. HP* 3 3. 5: "A great difference in bearing or failure to bear is also made by the nature of the regions, as with the *persea* and date-palm. For the *persea* bears in Egypt and in some neighbouring districts, but in Rhodes only reaches the flowering stage."

[e] *Cf. HP* 3 3. 5 (continued): "And the date-palm bears marvellously in Babylon, but does not even ripen its fruit in Greece, and in certain countries does not even promise any."

Αἴγυπτον, οὐδ' ὅλως ⟨ἐν⟩[1] ἐκείνοις τοῖς τόποις, χρησταί· θερμὸς γὰρ ὢν ἄγαν ὁ ἀὴρ περικάει, καὶ οὐ ποιεῖ πέψιν, ἀλλ' ἡ ἐκ τῆς γῆς εὐτροφία 5 διυγραίνει μόνον, ἀπέπαντος οὖσα, διὸ καὶ τοῖς μεγέθεσιν γίνεται μικρά. τοὐναντίον δὲ ὁ ψυχρός· ἐξαιρεῖ[2] γὰρ τὴν ὑγρότητα, τὴν δὲ[3] οὐ δύναται πρὸς τὴν οἰκείαν πέψιν ἀγαγεῖν, ὥσπερ ἐν ταῖς ἐλαίαις, διὸ καὶ οἱ[4] μὲν ἄσαρκοι πάμπαν μεγαλο- 10 πύρηνοι[5] δὲ τῶν ἐν τοῖς ψυχροῖς, οἱ[6] δὲ σαρκώδεις μέν, ἀνέλαιοι δέ· πλείονος γὰρ τοῦτο θερμότητος δεῖται πρὸς τὴν πέψιν.

αἱ μὲν οὖν τοῦ ἀέρος κράσεις καὶ διαθέσεις τοιαύτας τινὰς παρέχονται δυνάμεις.

4. 1 ἐπεὶ δὲ καὶ τὰ ἐδάφη μεγάλας ἔχει διαφοράς, λεκτέον καὶ περὶ τούτων· καὶ γὰρ αὐτὰ τῆς φύσεως. ἔτι δὲ ἀβλαστῆ[7] δι'[8] ἄμφω γίνεται, καὶ διὰ τὸν ἀέρα, καὶ διὰ τὴν γῆν πολλάκις· 5 ὁτὲ[9] μὲν γὰρ τὰ κάτω χρηστότατα, τὰ δ' ὑπὲρ γῆς φαῦλα, ⟨ὁτὲ δὲ χρηστά, τὰ δὲ φαυλότατα⟩,[10] καθάπερ ὅταν ἀμμώδης ᾖ κεραμὶς ἢ κατακεκαυμένη τις τυγχάνῃ, ῥίζωσιν[11] γὰρ καὶ τροφὴν οὐδεμία[12] τῶν τοιούτων ἔχει. καὶ σχεδὸν αἱ μὲν

1 ⟨ἐν⟩ Schneider.
2 ἐξαιρεῖ Schneider: ἐξερει U; ἐξαίρει u N; ἐξάρει HP.
3 δὲ U: δ' οὖσαν Schneider.
4 οἱ U (*sc.* καρποί): αἱ u.
5 μεγαλοπύρηνοι ego (*fructum . . . oleo refertum* Gaza; ἐλαιηρότεροι Wimmer): μεγαλόριζοι U.
6 οἱ ego: αἱ U.
7 ἀβλαστῆ Wimmer: εὐβλαστῆ U.
8 δι' Schneider: δὲ U.
9 ὁτὲ u: ὅτε U N HP.

and in that part of the world in general. For the air, which is extremely hot, burns the outside of the fruit all around and fails to bring about concoction; instead the abundant food from the ground fails to get concocted and merely soaks the fruit.[a] This also accounts for its small size. Cold air does the opposite: it removes the fluid and cannot bring the remainder to its proper concoction. So with the olive: this is why some olive fruits in cold regions have very little flesh, but large stones, whereas others have flesh but no oil,[b] oil requiring greater heat for its concoction.[c]

Such then are the effects of different temperings and dispositions of the air.

Soils

4. 1 Since soils too differ widely we must also speak of them, for these too come under the head of nature.[d] Furthermore trees fail to sprout for both reasons, not only owing to the air, but often owing to the soil. Thus sometimes when this happens conditions in the ground are very good but conditions above ground poor, and at other times conditions above are good but in the soil very poor, as when it is sandy, clayey or burnt out, for no soil such as these allows a tree to root or feed. And these (one may say) are

[a] *Cf. CP* 6 17. 5.
[b] *Cf. CP* 6 8. 6.
[c] *Cf. CP* 6 8. 4, 6 8. 6.
[d] *Cf. CP* 2 1. 1.

---

10 ⟨*ὁτὲ—φαυλότατα*⟩ ego: ⟨*ὁτὲ δὲ ὁ μὲν ἀὴρ εὔτροφος τὰ δὲ τῆς γῆς φαῦλα*⟩ Wimmer.
11 *ῥίζωσιν* Wimmer: *ῥιζῶσιν* U^ar N (-*σι* U^r HP).
12 *οὐδεμία* Heinsius: *οὐ δὲ μίαν* U.

10 ἀβλαστεῖς εἰσιν αὗται, καὶ εἴ τις ἄρα δίυγρος ὅλως ἢ πηλώδης.

4. 2 [δη][1] βλαστητικῶν ⟨δὴ⟩[2] καὶ ἐγκάρπων οὐ κακῶς ἡ διαίρεσις ἡ πρὸς τὰ σπέρματα[3] καὶ τὰ δένδρα λέγεται, τῷ τὴν μὲν πίειραν ἀμείνω σιτοφόρον, τὴν δὲ λεπτοτέραν, δενδροφόρον εἶναι. 5 λαμβάνει γὰρ (ὥσπερ καὶ πρότερον εἴπομεν) ὁ σῖτος (καὶ ἁπλῶς τὰ ἐπέτεια) τὴν ἐπιπολῆς τροφήν, ἣν δεῖ μὴ ὀλίγην μηδ' εὐξήραντον εἶναι (καθάπερ ἐν ταῖς λεπταῖς), τὰ ⟨δὲ⟩[4] δένδρα, διὰ τὸ μεγάλας καὶ ἰσχυρὰς ἔχειν τὰς ῥίζας, καὶ τὴν ἐκ

4. 3 βάθους. αὕτη δέ, ἐν μὲν τῇ πιείρᾳ[5] πολλὴ[6] λίαν οὖσα, βλάστην μὲν ποιεῖ καλὴν καὶ μέγεθος τοῖς δένδροις, καρπὸν δ' οὐ ποιεῖ, διὰ τὸ μὴ ἐκπέττειν· ἐν δὲ τῇ λεπτοτέρᾳ ξύμμετρος γίνεται πρὸς ἄμφω, 5 καὶ κρατοῦντα τὰ δένδρα δύναται καρποτοκεῖν. ἡ δὲ πίειρα πάμπαν οὐδενὶ ξυμφέρει φυτῷ, ξηραίνει γὰρ μᾶλλον τοῦ δέοντος, ὥσπερ καὶ Μενέστωρ φησίν· τοιαύτην δ' εἶναι τὴν πλυντρίδα,[7] χρῶμα δ' ὑπόλευκον. ἀρίστη δὲ δῆλον ὡς ἡ ἄριστα κεκραμένη,[8] 10 καὶ ὅλως μανή τις οὖσα καὶ μὴ ψυχρὰ καὶ ἔνικμος· καὶ εὐδίοδος γὰρ οὕτω ταῖς ῥίζαις ἐστὶ καὶ εὔτροφος, ὅπερ βούλονται καὶ αἱ ἡμερώσεις καὶ αἱ κατεργασίαι καὶ κοπρίσεις ποιεῖν.

4. 4 ἄλλη δὲ πρὸς ἄλλα τῶν δένδρων ἁρμόττει μᾶλλον, ὥσπερ καὶ διαιροῦσιν· οἷον ἡ σπιλάς,

§ 3. 7. Menestor Frag. 6, Diels-Kranz, *Die Fragmente der Vorsokratiker*, vol. i[10], p. 376.

[1] [δη] N HP: τῶν δὴ Schneider; τῶν δὲ Wimmer.
[2] ⟨δὴ⟩ ego.
[3] τὰ σπέρματα U[c]: τά////ματα U[ac].
[4] ⟨δὲ⟩ Wimmer here (after δένδρα Schneider).

the soils that prevent sprouting, together with such as are quite water-logged or muddy.

4.2 Turning to soils that allow sprouting and bear crops a good distinction is made that refers to their production of grain or trees: fat soil is a better producer of grain, leaner soil of trees. For (as we said before)[a] grain (and indeed all annuals) get the food that is near the surface. But this food must not be scanty or dry out easily, as in lean soils. Trees on the other hand, because of their great and powerful roots, can also draw food from far below. 4.3 But in fat soil this food is far too abundant, and whereas it produces a fine foliage and good height in trees, it produces no fruit, since the tree does not fully concoct the food. But in leaner soil the food turns out to be of just the right amount for producing both, and the trees master it and so are able to bear fruit. Extremely fat soil is good for no plant, drying it up more than is wanted, as Menestor says, fuller's earth, which is whitish in colour, being (he says) of this kind. The best-tempered soil is evidently the best, being in general of open texture, not cold, and containing moisture. In this way it is not only easily penetrated by the roots, but also feeds the plant well, these being the aims of land reclamation, tillage and manuring.

4.4 Among soils good for trees some are better suited to one tree, some to another, and such are the distinc-

[a] *CP* 1 18. 1–2; *cf. CP* 2 4. 1.

---

5 πιείρα U[r] N HP: πιείραν U[ar].
6 πολλὴ Schneider (δαψιλὴς Wimmer): λευκὴ U.
7 πλυντρίδα Gaza, Schneider: πληντρίδα U.
8 κεκραμένη u: κρεμαμένη U N HP[c](ρεμ in an erasure).

καὶ ἔτι μᾶλλον ἡ λευκόγειος,[1] ἐλαιοφόρος, ἰκμάδα τε γὰρ[2] ἔχει καὶ πνεῦμα πολύ, δεῖται δὲ καὶ
5 ἀμφοῖν· ἡ δὲ λειμωνία καὶ ἔφαμμος[3] ἀμπελοφόρος ἀγαθή, καὶ ὅλως ἥτις ἂν ᾖ μανὴ καὶ κούφη καὶ λεπτὴ καὶ ὕφυδρος[4] οὕτως ὥστε τὸ οὐράνιον ὕδωρ συνικνεῖσθαι[5] πρὸς τὸ ἐν αὐτῇ, τροφῆς γὰρ πολλῆς ἡ ἄμπελος δεῖται διὰ τὸ θερμὴ καὶ μανὴ
10 καὶ ὑγρὰ καὶ πολύκαρπος εἶναι, τάχα δὲ καὶ δι' αὐτὰ ταῦτα καὶ πολύκαρπος· ἔτι δὲ οὐκ εὔσηπτοι τῶν ὑγρῶν αἱ ῥίζαι, καθάπερ αἱ τῶν ξηρῶν, ὥστε δύνασθαι καὶ ἐπισπᾶσθαι καὶ ἀντέχειν καὶ
4. 5 ⟨δια⟩διδόναι.[6] τὸν αὐτὸν δὲ τρόπον καὶ τῶν ἄλλων ἑκάστοις ἐστίν τις οἰκεία πρὸς τὴν φύσιν, καὶ ἡ αὐτὴ τοῖς μὲν μᾶλλον, τοῖς δ' ἧττον.

ὡς δ' ἁπλῶς εἰπεῖν τοῖς μὲν πίοσιν[7] οὐ συμφέρει
5 ἡ πίειρα, δοκεῖ γὰρ καταξηραίνειν μᾶλλον τοῦ μετρίου, διὸ καὶ πημαίνεσθαι καὶ νοσεῖν· ὅσα δὲ λυπρά, τούτοις ξυμφέρει· σημεῖον δέ, ὅτι τὰ λάχανα καὶ ὁ Δημήτριος καρπὸς ἐν ταῖς τοιαύταις εὐθενεῖ,[8] πάντα δὲ ταῦτα λυπρὰ τῇ φύσει, τὰ
10 γὰρ φύσει λυπρὰ πιοτέρας[9] τροφῆς δέονται.
4. 6 δηλοῦν δὲ οἴονται καὶ τὴν τῶν ἀνθρώπων τοιαύτην φύσιν· τοὺς γὰρ ἐκλίμους[10] καὶ χαίρειν μάλιστα

[1] λευκόγειος u: -καργ-(?) U.
[2] γὰρ U^cm (no index): U^t omits.
[3] ἔφαμμος HP (ὕφαμμος Schneider): ἐφᾶμμιν U (ἐφ' ἅμμιν N).
[4] ὑφ' ὗδρος U: ἔφυδρος u (ο from ω N) HP.
[5] συνικνεῖσθαι U: μὴ διϊκνεῖσθαι Schneider (*cf. CP* 3 11. 3).
[6] ⟨δια⟩διδόναι Schneider.
[7] πίοσιν ego: πλείοσιν U.
[8] εὐθενεῖ Schneider: εὐθηνεῖ U.
[9] πιοτέρας Scaliger: προτέρας U.
[10] ἐκλίμους Gaza, Scaliger: ἐκλίμνους U.

tions that the agriculturists make. For instance stony soil, and white[a] soil still more, is a good producer of the olive, since it has moisture and a good deal of *pneuma*,[b] and the olive requires both. Meadow and and sandy soil are good producers of the vine,[c] nd so in general is any soil that is open-textured, ight, lean, and with a water table within easy reach f the rain,[d] since the vine requires plenty of food ecause it is hot, open-textured, fluid and an abundnt bearer (in fact it is perhaps these very features hat make it an abundant bearer). Furthermore, the oots of fluid trees do not easily decompose, like those f dry trees, and so are able to attract, retain and ransmit the food. 4. 5 So too with the rest: for every tree here is a soil appropriate to its nature, and the same oil is more appropriate to some, less so to others.

Broadly speaking, fat soil is not good for fat plants, ince it dries them out overmuch (it is held),[e] and hey suffer and become diseased. But it is good for ll spare plants. This is proved by the following: egetables and cereals thrive in fat soils, and all these lants are by their nature spare; for the naturally pare requires fatter food. 4. 6 And experts[f] suppose hat this beneficial character of rich food is shown by he similar spare constitution in man: starvelings

[a] That is, calcareous.

[b] That is, "warm air." The soil is doubtless loose and not ompact. The olive needs water (the raw material) and heat the agent) to make its oil.

[c] *Cf. CP* 3 6. 8.

[d] A high water table keeps the rain water from sinking oo deep and so from becoming unavailable (*cf. CP* 3 11. 3–4).

[e] By Menestor: *cf. CP* 2 4. 3. The rest of this section and he whole of the next are no doubt derived from him.

[f] Perhaps Menestor.

ταύτῃ, καὶ ἐπιδιδόναι πρὸς εὔχροιαν καὶ ἰσχύν·
λυπρὰ [1] γὰρ ὄντα τὰ σώματα δεῖσθαι τροφῆς
5 πολλῆς καὶ πιείρας, ὑπὸ δὲ τῶν ξηρῶν καὶ
λυπρῶν οὐθὲν ὠφελεῖσθαι, διὰ τὸ μὴ ἀπολαύειν,
ἀλλὰ καὶ ἐπικίνδυνα εἶναι πρὸς νόσους ἄλλας τε καὶ
μάλιστα δὴ τὰς τῆς κοιλίας. ὁμοίως δὲ καὶ ἐπὶ
τῶν δένδρων ἔχειν [2] τοῦτο, πλὴν ταῦτα μὲν ὅμοια
10 διατελεῖν,[3] τὸ δὲ σῶμα, ὅταν ἀνακομισθῇ, μεταβαίνειν [4] εἰς τὴν τετρυμένην [5] καὶ ἄκνισον.

4. 7 οὗτος μὲν οὖν καθόλου τις διορισμός· διαφορα
δὲ πολλαὶ καὶ τῆς γῆς καὶ τῶν δένδρων, ὥσπερ καὶ
τῶν ἀμπέλων ταῖς μὲν ἡ πεδιεινή, ταῖς δ' ἡ
ὀρεινὴ μᾶλλον ἁρμόττει, καὶ ἐν αὐταῖς ταύταις αἱ
5 τοιαίδε ταῖς τοιαῖσδε, μικραὶ δὲ παραλλαγαὶ
φαινόμεναι μεγάλας ποιοῦσιν ῥοπὰς εἰς τὴν φύσιν·
τὸ δ' ἁπλοῦν ῥᾴδιον εἰπεῖν, ὥσπερ καὶ κελεύουσιν
τὰ μὲν στερεὰ καὶ πυκνὰ φυτεύειν ⟨ἐν τοῖς
ξηροῖς καὶ πυκνοῖς⟩,[6] τὰ δ' ἀραιὰ καὶ ὑγρὰ ἐν
10 τοῖς μαλακωτέροις καὶ ἐφυγροτέροις. ἑκατέροις
γὰρ οὕτως αἱ τροφαὶ δῆλον ὅτι σύμμετροι, τοῖς
μὲν πολλῆς [7] δεομένοις, τοῖς δ' ὀλίγης.[8]

4. 8 καὶ τὸ [9] καθόλου λεχθέν, ὑπὲρ πάντων ἴσως τῶν
δένδρων ἀληθές, ὅτι καθ' ἑκατέρας τὰς χώρας

[1] λυπρὰ ego (ξηρὰ Gaza, Itali; ἰσχνὰ Wimmer): ὑγρὰ U.
[2] ἔχειν U^ar: -ει U^r N HP.
[3] διατελεῖν U: -εῖ u.
[4] μεταβαίνειν U^ar: -ει U^r N HP.
[5] τετρυμένην ego: τετρυμμενην U.
[6] ⟨ἐν τοῖς ξηροῖς καὶ πυκνοῖς⟩ added by Wimmer before φυτεύειν, placed here by me. Schneider (after Gaza), add ⟨ἐν τοῖς ξηροῖς⟩ before φυτεύειν.
[7] πολλῆς u: -οῖς U N; -ῶν HP.

not only take the greatest delight in this fat food but gain in colour and strength; for their spare bodies require plenty of fat food, but get no benefit from dry and spare food because they do not assimilate it, being apt instead to contract various diseases, especially those of the digestive tract. The case is the same (they say) for trees as well, except that the tree continues as before, whereas when the human body has recuperated it passes to the austere and fatless diet.

4. 7 Now this distinction [a] is a somewhat general one, and there are many differences both in the soil and in the trees. So with the vine: the soil of a plain suits some better, mountain soil suits others, and among these very types of soil different sorts are better for different sorts of vine, and what appear to be slight differences greatly influence the nature of the vine. The broad rule is easily given, and is what the authorities recommend: to plant the solid and close-textured trees in dry and compact soils, the open-textured and laxer ones in the softer and more humid soils.[b] Thus the trees of either group will evidently get the right amount of food, the latter requiring much, the former little.

4. 8 Again the general precept [c] is doubtless true of trees in the aggregate, that in each of the two types

[a] *CP* 2 4. 2–6: that fat soil is best for grain, leaner soil for trees.
[b] *Cf. CP* 3 11. 1–4.
[c] *CP* 2 4. 7.

---

8 ὀλίγης u: ὀλίγεις U; ὀλίγοις N; ὀλίγων HP.
9 τὸ ⟨μὲν⟩ Schneider.

ἑκάτερα δεῖ καὶ[1] τῶν δένδρων φυτεύειν· ἀλλ' ἐν τοῖς καθ' ἕκαστα τὸ ἀκριβὲς μᾶλλον ἴσως
5 αἰσθητικῆς δεῖται συνέσεως, λόγῳ δὲ οὐκ εὐμαρὲς ἀφορίσαι. ἐπεὶ καὶ ταῖς πρὸς τὸν ἥλιον διαφοραῖς, οἷον ἀνιόντα ἢ δυόμενον ἢ μεσοῦντα ἤ πως ἄλλως ἔχοντα, δεῖ μὴ ἀγνοεῖν ποῖα[2] τῶν φυτῶν τὰ οἰκεῖα καὶ ὅλως καὶ τῶν ὁμογενῶν, ὅπερ[3] οἵ γ'
10 ἀμπελουργοὶ πειρῶνται διαιρεῖν, ὅταν συνάγκειάν τινα λαβόντες φυτεύσωσιν, οὐ γὰρ ταὐτὰ[4] τιθέασιν εἰς ἑκάτερον τὸ μέρος, ἀλλὰ διαιροῦσι, καὶ ποιεῖ μεγάλην διαφορὰν οὕτω τε φυτευθέντα καὶ ἀνάπαλιν. ὁμοίως δὲ τοῦτ' ἔχει καὶ ἐπὶ τῶν λοιπῶν.
4. 9 ἀλλ' (ὥσπερ ἐλέχθη πρότερον)[5] αἰσθητικῆς δεῖται ταῦτα συνέσεως.

ὡς δ' ἁπλῶς εἰπεῖν ἡ μέσην ἔχουσα τῶν ἐναντίων κρᾶσιν, ⟨ἀραιοῦ καὶ⟩ πυκνοῦ[6] καὶ ξηροῦ καὶ
5 ὑγροῦ καὶ κούφου καὶ βαρέος, ἔτι δὲ τὰ ἄνω πρὸς τὰ κάτω σύμμετρα τούτοις, πασῶν ἀρίστη πρὸς ἅπαντα (ὡς εἰπεῖν) δένδρα τε καὶ σπέρματα, φαίνεται γὰρ ὁμοίαν τινὰ[7] ἔχειν τῷ ἔαρι[8] πρὸς τὰς ἄλλας ὥρας.[9] οὐ μὴν ἀλλ' αὐτῶν γε τούτων ἡ
10 πρὸς θάτερον μέρος ἀποκλίνουσα τῆς ἐναντιώσεως

[1] δεῖ καὶ Wimmer (δεῖ Gaza, Schneider): δείται U.
[2] ποῖα Schneider: ποῖος U.
[3] ὅπερ U: ὥσπερ Gaza (*quemadmodum*), Schneider.
[4] ταυτὰ HP: ταῦτα U N.
[5] πρότερον u: πότερον U.
[6] ⟨ἀραιοῦ καὶ⟩ πυκνοῦ Gaza (*soluti et spissi*): πυκνοῦ ⟨καὶ ἀραιοῦ⟩ HP.
[7] ὁμοῖαν τινα U: ὁμοιότητα Heinsius.
[8] ἔαρι ego: ἀέρι U.
[9] ὥρας ego: χώρας U.

of country we must plant each of the two types of tree. But when we come to particulars the precise suiting of the tree to the land is perhaps more a matter for acuteness of sense perception and not easy to determine by theory. Take for instance the differences in the lie of land with regard to the sun—such as whether it faces the rising, setting or midday sun or the sun in some other position—: one must not fail to recognize which of the slips—whether of different kinds of trees or of different types of the same tree—are appropriate to the one position or the other, the last a distinction that vine-growers endeavour to make when they plant the two slopes of a ravine, for they do not put the same type of vine on both, but distinguish the types, and it makes a great difference whether the one type is planted on this slope, the other on that, or the other way round. And similarly with the rest. But (as we said before) [a] these are matters for acuteness of perception.

4. 9

Put broadly, land so tempered as to be intermediate between the opposites of open and close texture, of dry and wet and of light and heavy, and further with conditions above ground so related to those below as to be adjusted to them in these matters, is the best of all land for practically all kinds of trees and grains, since in contrast to the land resulting from other combinations of these extremes it is seen to possess a tempering like that of spring compared to the other seasons. Nevertheless among these intermediate kinds of land the kind that inclines more to the one side of the opposition is better than

[a] *CP* 2 4. 8.

κρείττων,[1] ἣν δὴ καὶ ἁπλῶς τίθενταί τινες ἀρίστην, οἷον τὴν κούφην καὶ μανὴν καὶ ἔνικμον, ἔχει τε γὰρ τροφὴν ἐν ἑαυτῇ καὶ εὐδίοδός ἐστι ταῖς ῥίζαις· ἡ δὲ πυκνὴ καὶ βαρεῖα καὶ ξηρὰ δῆλον ὡς ἐναντία.

4. 10 κατὰ δὲ τὰς παραλλαγὰς τῶν ἀντιθέσεων, καὶ τὰς πρὸς ἕκαστον ἕξουσιν ἤδη διαφοράς· οἷον ἐὰν ᾖ μανὴ μὲν καὶ λεπτή, βαθύγεως δὲ καὶ ξηρὰ καὶ ἄϋδρος, δενδροφόρος μὲν ἀγαθή, σιτοφόρος δὲ 5 κακή· διὰ γὰρ τὴν μανότητα διίησιν [2] εἰς βάθος τὸ χειμερινὸν ὕδωρ, ὥστε τὸν μὲν σῖτον μὴ ἐφικνεῖσθαι, διὰ τὸ ἐπιπολῆς εἶναι, τὰ δένδρα δ', εἰς βάθος καθιέντα τὰς ῥίζας, ἐφικνεῖσθαι καὶ ἕλκειν. ὡσαύτως δὲ καὶ εἴ τις ἄλλη τοιαύτη διαφορά· τὸ 10 γὰρ πρόσφορον ἀποδώσει τινὶ γένει διὰ τὴν παραλλαγήν.

4. 11 οὐκ ἀηδῶς [3] δὲ οὐδ' ὅσοι ταύτην ἀρίστην ὑπολαμβάνουσιν ἥτις ἂν ᾖ θερμή τε καὶ ἔνικμος· ἄμφω γὰρ ἔοικεν ἔχειν ἃ δεῖ, τροφήν τε καὶ τὸ κατεργαζόμενον·[4] εὐλόγως δὲ καὶ μετὰ τὸ πρῶτον 5 ὕδωρ ἀτμίζειν. ἀλλὰ τὴν μὲν ἔνικμον αὐτῆς εἶναι, τὴν δὲ ξηράν· τὴν μὲν οὖν ἔνικμον σιτοφόρον ἀγαθήν, ἱκανὸν γὰρ ἔχειν πρὸς τὴν ἐκτροφὴν ὑγρὸν τῷ σίτῳ, τοῖς δὲ δένδροις ἔλαττον· τὴν δὲ ξηρὰν σιτοφόρον φαύλην διὰ τὸ μὴ ἔχειν μηδὲ 10 τούτῳ τροφὴν ἀρκοῦσαν.

[1] κρείττων u: κρεῖττον U.
[2] διίησιν ego: δίεισιν U.
[3] ἀηδῶς U: *imperite diiudicant* Gaza; ἀπείρως Schneider; ἀλόγως Wimmer.
[4] κατεργαζόμενον ⟨. . .⟩ Wimmer.

---

[a] So the mean of moral virtue is sometimes closer to the one

that inclining to the other,[a] and this is the land that some set down without qualification as best, namely land that is light, of open texture, and moist, since it contains food in itself and is easily penetrable to the roots.[b] Land of close texture, heavy and dry is evidently of the opposite character.

4. 10 Permutations in the opposites involved go on to yield the soils good for this or that particular plant. If the soil for instance is loose-textured and lean, but deep and dry and without rain, it is a good producer of trees but a bad one of grain, since because of its open texture it lets the winter rain sink deep; thus grain with its shallow roots is unable to reach the water, whereas trees, with their deep roots, reach and absorb it. So with any other such variation as this: take one side of the opposition instead of the other, and the permutation will make the soil suitable for some kind of plant.

4. 11 Those too who suppose that warm and moist soil is the best make no unattractive supposition, since such soil appears to possess both the requirements: food and the agent that prepares it. It is also reasonable to suppose that this soil sends up vapour after the first rain.[c] But some of the soil that sends up vapour (they say) is moist, some dry. The moist (they say) is a good producer of grain, possessing enough fluid to rear it to maturity, but too little for trees; whereas the dry is a poor producer of grain, not possessing enough food for even that.

extreme than to the other: *cf.* Aristotle, *Nicomachean Ethics*, ii. 9 (1109 a 30–b 1, b 23–26).

[b] *Cf. CP* 2 4. 3.

[c] The first rains of autumn (*cf.* CP 1 10. 5), which come after the dry summer. The heat vaporizes the rain.

4. 12 ἐπὶ ταὐτὸ δέ πως φέρονται καὶ ὅσοι φασὶ δεῖν πίειράν τε εἶναι καὶ μὴ παγώδη, μηδὲ πυκνήν, μηδ' ἁλμυράν, ἀλλὰ πότιμον[1] καὶ ψαθυράν· τροφήν τε γὰρ οἴονται[2] δεῖν ἔχειν[3] καὶ θερμότητα, 5 καὶ ἔτι[4] ταῖς ῥίζαις εὐδίοδον εἶναι, πάντα δὲ ταῦτα οἰκεῖα πρὸς αὔξησιν καὶ καρποτοκίαν. ὁμοίως δὲ καὶ οἱ τὴν μελάγγεων ἐπαινοῦντες, ὥσπερ Λεωφάνης·[5] εὐθὺς γὰρ ἀποδιδόναι πειρᾶται τὰς αἰτίας, ὅτι δύναται καὶ ὄμβρον καὶ αὐχμὸν 10 φέρειν, δοχὸς οὖσα καὶ τοῦ θερμοῦ καὶ τοῦ ὑγροῦ.

τὴν μὲν οὖν ἀρετὴν τῆς χώρας σχεδὸν ὥσπερ εἴρηται διὰ τῶν αὐτῶν πως καὶ ἐν τοῖς αὐτοῖς ἐστιν λαβεῖν ἐν οἷς καὶ ἀποδιδόασι πάντες· αἱ διαφοραὶ δέ,[6] ἐπεὶ πλείους εἰσὶν καὶ ταύτης καὶ 15 τῶν φυτῶν, πειρατέον πρὸς ἕκαστον λαμβάνειν καὶ θεωρεῖν.

5. 1 τὰς δὲ τῶν ὑδάτων διαφορὰς τῶν ἐπιγείων, καὶ γὰρ ταῦτα οὐ μικρὰν ἔχει μερίδα πρὸς αὔξησιν καὶ τροφήν, ὁμοίως τούτοις ληπτέον.

7–10. *Geoponica*, ii. 9. 1: ἀρίστη γῆ ἡ μελάγγειος, ὑπερεπαινουμένη παρὰ πᾶσιν, ὅτι καὶ ὄμβρον φέρει καὶ αὐχμόν.

[1] πότιμον ego (ποτίμην Schneider): ποτὶ μὲν U.
[2] οἴονται N HP (οἴ- u): οἷόν τε U.
[3] ἔχειν u HP: ἔχει U N.
[4] ἔτι u: αἵ τι (?) U; εἴ τι N HP.
[5] Λεωφάνης a: λεωφανής U; λεοφάνης u HP; λεοφανὴς N.
[6] δε Uc: δ Uac.

[a] *Cf.* Virgil, *Georgics*, ii. 238.
[b] Literally "potable," not containing any unpalatable taste.

Much the same result is also reached by those who say that the land should be fat and not given to freezing or yet close in texture or salty,[a] but sweet [b] and crumbly, since they think that it should contain both food [c] and warmth [d] and moreover be readily permeable to the roots,[e] and that all this is conducive to growth and production. So too with those who commend black soil, like Leophanes,[f] since he at once endeavours to give the reasons: that it can withstand both rain and drought, since it absorbs both heat and water. 4. 12

To conclude: one can formulate excellence of land much as we have done [g] by resorting to the same reasoning (so to say) and the same qualities as all the authorities do in giving their answer. But as for the differences, since there are a good number of them both in this kind of land and in the plants, we must endeavour to find and account for them separately for each plant.

Surface Waters

We must now formulate the differences in surface waters, since surface water too has no small share in growth and nurture, just as we did [h] with the differences of soils. 5. 1

[c] Implied by " fat " and perhaps by " sweet."
[d] It is not liable to frost.
[e] It is crumbly and not close in texture.
[f] *Cf.* Menestor's rejection of the whitish fuller's earth (*CP* 2 4. 3).
[g] *CP* 2 4. 9–10.
[h] *CP* 2 4. 1–12.

ὅσα μὲν οὖν θερμά, καὶ νιτρώδη, καὶ στυπτη-
5 ριώδη, καὶ εἴ[1] τις ἄλλος τοιοῦτος χυλός, ἄτροφα
καὶ ἄγονα φυτῶν ὡς ἁπλῶς ἔστιν εἰπεῖν, πλὴν εἴ τι
συγγενὲς[2] αὐτοῖς ἐκτρέφειν δύναται, καθάπερ καὶ
ἡ θάλαττα· βραχὺ δέ τι τοῦτο ἢ οὐδὲν (ὡς εἰπεῖν)
ἐστιν· ἐνιαχοῦ δὲ ποτίμων πλείστων ὄντων ἡ
10 ἐκτροφή, καθάπερ τοῖς ἐν Θρᾴκῃ θερμοῖς, καθ'
αὑτὰ δ', ὥσπερ καὶ ζῴων, καὶ φυτῶν ἄγονα.[3]

5. 2 ἡ δὲ θάλαττα πολλὰ καὶ παντοῖα φύει καὶ
ἔστιν, ὥσπερ ζῴων τι γένος ἐν αὐτῇ, καὶ φυτῶν.

ὅσα δ' ἐν τῇ ἀμπώτει δένδρα μέγεθος ἔχοντα καὶ
καρπὸν τυγχάνει, περὶ τούτων οὐκ ἄν τις ἴσως
5 ἀποδοίη τῇ θαλάττῃ τὴν τροφήν, ἀλλ' ἐνδέχεσθαι
πότιμον ἕλκειν ἐν τῇ γῇ τὰς ῥίζας, τὴν δὲ θάλατταν

[1] εἴ Uc: ἤ Uac.
[2] συγγενὲς N HP: σύνεγγενες Uar; σύνγενες Ur.
[3] ἄγονα u N HP: ἀγῶνα U.

---

[a] "Hot" no doubt in the medicinal sense: *cf. CP* 1 22. 5–6.

[b] *Nítron* is apparently carbonate of soda. Such water is also mentioned in Theophrastus, *On Odours*, chap. xiv. 65.

[c] For marine plants *cf. HP* 1 4. 2, 4 6. 1—4 7. 8.

[d] *Cf.* (in a different sense) Aristotle, *On the Generation of Animals*, iii. 11 (761 a 24–26): "But plants as a class grow in the sea and in such waters in small numbers and are practically non-existent there; instead its members all grow on land."

[e] Spontaneous generation involves rain water or sweet water: *cf.* Aristotle, *On the Generation of Animals*, iii. 11 (762 a 11–12) [ὕδωρ "fresh water" is mentioned at 762 a 19, b 12].

[f] Mangroves.

[g] *Cf. HP* 4 7. 4–7: "They say that in the islands that are covered at high tide tall trees grow . . .; and that the trees have . . . fruit . . . 5. . . . In Persia on the coast of Carmania, where there is a tide, there are trees of good

All surface water, then, that is hot [a] or soda-like [b] or astringent or possesses any such flavour neither (to put it broadly) fosters plants nor generates them, except for the cases where such water (like sea-water) [c] is able to rear some plant with a similar flavour to itself; but such plants are few or (so to say) non-existent.[d] In a few places the plant is reared when potable water is the largest component, as with the hot waters in Thrace. But when unmixed such water generates plants no more than it generates animals.[e]

Sea-water grows many living things of all sorts; and as it has its own kind of animal, so it has its own kind of plant. 5. 2

As for the tall fruit-bearing trees [f] found in tidal waters, one would perhaps not assign their feeding to the sea-water, but say that it is possible that the roots draw potable water from the ground,[g] and that the sea-water surrounding the tree does it no more

height . . .; and they bear abundant fruit . . . All these trees have been eaten away in their central part by sea-water, and they stand supported by their roots, like an octopus on its tentacles. 6. The district has no rain whatever, but the tide leaves certain channels by which the natives travel by boat from one part to another, and these are filled with sea-water. And some people think that this shows that the trees are fed by sea-water and not by fresh water (except for what they may get from the ground by drawing it up by their roots; and that it is reasonable to suppose that even this water is brackish, for the roots do not go at all deep). They sum the matter up by saying that plants growing in the sea and those growing on land covered by the tide are of one and the same kind . . . 7. They say that on the eastern side of the island of Tylos the trees are so numerous that they constitute a barricade. All these trees have the height of a fig-tree . . . and fruit that is inedible."

ἀβλαβῆ περιέχουσαν εἶναι, καθάπερ καὶ τὸ ὕδωρ τοῖς ἐνύγροις·[1] ἀλλὰ τούτων μὲν πέρι λόγος ἕτερος.

10 καὶ ὅσα δὴ πρὸς αὐτῇ τῇ θαλάττῃ φύεται χρῆταί πως τῇ ἁλμυρίδι πρὸς εὐσθένειαν καὶ τροφήν, ἔοικεν γὰρ καὶ τοῦτο ἴδιόν τι γένος εἶναι (καθάπερ[2] τὰ εἰρημένα), καθάπερ ἀνὰ μέσον ὄντα.

5. 3 τὰ δ' ἁλυκὰ τῶν ὑδάτων τρέφει μὲν καὶ τὰ ἔγγαια,[3] χεῖρον δὲ τῶν γλυκέων, ἀναυξῆ γὰρ ποιεῖ καὶ ἐπικάει. τὸ δ' ἀλυπότερα τοῖς δένδροις ἢ τοῖς λαχανώδεσιν (ἢ ὅλως ἐπετείοις) εἶναι, 5 καθάπερ τινές φασιν, οὐκ ἄλογον, ὅσῳπερ ἰσχυρότερα[4] (τάχα γὰρ κἂν εἴη τισὶ πρόσφορα[5] τὸ ὅλον, ὥσπερ τοῖς φοίνιξιν, εἴπερ καὶ οἱ ἅλες παραβαλλόμενοι· παραπλήσιος γὰρ ⟨ἢ⟩[6] ὁ αὐτὸς χυλός). οὐ μὴν ἀλλὰ καὶ τῶν λαχανωδῶν ἐστί τισι τὰ 10 ἁλυκὰ πρόσφορα, καθάπερ ῥαφάνῳ, τευτλίῳ,[7] πηγάνῳ, εὐζώμῳ· βελτίω γὰρ γίνεται τοῖς ἁλυκοῖς ἀρδόμενα ταῦτα (διὸ πρός γε τὴν ῥάφανον νίτρον τινὲς παραμιγνύουσιν ἐν τῷ βρέχειν, ὥσπερ οἱ ἐν Αἰγύπτῳ, καὶ γίνεται πολλῷ γλυκυτέρα καὶ ἁπαλω- 15 τέρα, καθάπερ καὶ ἡ ἑψομένη).

5. 4 τοῦτο δὲ συμβαίνει (καὶ ὅλως ἡ ἁλυκότης

[1] ἐνύγροις U: ἐνύδροις Schneider.

[2] καθάπερ U: παρὰ Schneider after Gaza.

[3] ἔγγαια u: ἔγγεα U; ἔγγεια N HP.

[4] ὅσωπερ ἰσχυρότερα u HP: ὅσωπερι ἴσχυρότερα U; ὅσω περὶσχυρότερα N.

[5] κἂν εἴη τισὶ πρόσφορα ego (ἄν τισι πρόσφορα Itali from Gaza): κἂν εἴ τις πρόσφορα U; κἂν ἤ τίς προσφορὰ u; κἄν τις προσφορὰ N; ἄν τις προσφορὰ HP.

[6] ⟨ἢ⟩ Schneider.

injury than the water surrounding freshwater plants. But the discussion of these matters belongs elsewhere.[a]

Again, the plants that grow at the seashore[b] make a certain use of the brine for strength and food; for it seems that here too we have a special class of plant (as we did in the case of the tidewater trees just mentioned), intermediate (as it were) between plants of the land and those of the sea.

5. 3 Saline waters feed land plants too, but do so worse than fresh water, stunting and scorching them. That they should do less harm to trees than to vegetables (or annuals in general), as some assert, is not unreasonable, to the extent that the trees are stronger. In fact for some trees they are perhaps even beneficial, as for the date-palm (inasmuch as the application of lumps of salt is good for it,[c] and the flavour of the salt water and of the lumps is similar or the same). Still, saline water is beneficial even for some vegetables, as cabbage, beet, rue and rocket, since these improve when watered with it (which is why, at least for cabbage, some people add soda when they water it,[d] as in Egypt, and the cabbage gets much sweeter and tenderer, just as it does when boiled in such water).

5. 4 This improvement occurs, and in a word salinity is

[a] *Cf. CP* 6 10. 2 with the first note.
[b] *Cf. HP* 1 4. 2: "of the lesser plants some like the seashore."
[c] Discussed in *CP* 3 17. 1–8.
[d] *Cf. CP* 3 17. 8; 6 10. 9.

---

[7] τευτλίωι U[c]: τεύτλωι U[ac].

πρόσφορος) ὅτι πικρότητά τινα ἔχουσιν[1] ἐν τῇ φύσει, ταύτην δὲ διαδυόμενον καὶ ὥσπερ ἀναστομοῦν τὸ ἁλυκὸν ἐξάγει (διὸ καὶ ἐν ταῖς ἁλμυρίσιν ἡ
5 ῥάφανος ἀρίστη), τὸ δὲ πικρὸν ἄτροφόν τε καὶ δύσχυλον, ὥστε, ἐξαγομένου τούτου, καὶ γλυκύτερα καὶ ἁπαλώτερα καὶ εὐαξέστερα γίνεται, τὰ δ' ἄλλα χείρω, διὰ τὸ μὴ ἐξάγειν τὸ ἀλλότριον, ἀλλὰ τὸ οἰκεῖον, ἐπικάουσαν.[2] δῆλον δ' ὅτι καὶ ἐπὶ τῶν
10 λοιπῶν πικρῶν ὁμοίως ἂν ἔχοι (καθάπερ καὶ ὁ κορωνόπους παρὰ[3] τὰ κιχόρια). τοῖς ⟨δὲ⟩[4] δριμέσιν (οἷον κρομμύῳ, σκορόδῳ, τοῖς ἄλλοις) οὐκέτι πρόσφορον· καὶ[5] γὰρ ἀφαιρεῖ τὴν δριμύτητα διὰ τὴν ὁμοιότητα τῶν χυλῶν· τὸ γὰρ ὅμοιον ἐπὶ
15 τὸ ὅμοιον φέρεται καὶ εἰς τοὺς πόρους ἐνδύεται, καθάπερ ἐπὶ τῶν καθαιρόντων καὶ ἐξαγόντων τὰς ἐπιχροΐσεις.

5. 5 καὶ τὰ μὲν ἁλυκὰ τοῖς τοιούτοις πρόσφορα διὰ τὴν εἰρημένην αἰτίαν.

εἰ δὲ ἀληθὲς ὃ ἔλεγεν Ἀνδροσθένης ὑπὲρ τῶν ἐν Τύλῳ τῇ νήσῳ τῇ περὶ τὴν Ἐρυθρὰν Θάλατταν,
5 ὅτι τὰ ναματιαῖα μᾶλλον συμφέρει τῶν οὐρανίων, ἁλυκὰ ὄντα, καὶ τοῖς δένδροις καὶ πᾶσι τοῖς

§ 5. 3. Frag. 3 Müller.

[1] ἔχουσιν Wimmer: -σα U.
[2] ἐπικάουσαν Schneider: ἐπικάουσα U.
[3] παρὰ U: καὶ Gaza, Wimmer.
[4] ⟨δὲ⟩ Wimmer (after δριμέσι Basle ed. of 1541).
[5] καὶ ego: οὐ U.

---

[a] Hartshorn and the chicories were doubtless not cultivated, and Theophrastus cannot appeal to experience. Presumably hartshorn, a medicine, would have been rendered less useful,

good for these vegetables, because they have a certain bitterness in their natures, and the salt water, by penetrating the plants and as it were opening outlets, extracts it (which is why cabbage is best in briny soil), and what is bitter is non-nutritious and has an evil flavour, so that on the removal of the bitter ingredient the plants get sweeter and more tender and grow larger, whereas the rest deteriorate because the salt water does not extract an alien ingredient but removes what is proper to the plant and burns it to do so. (The same distinction would also hold of the rest of the bitter plants, as with hartshorn in comparison with the chicories.)[a] In pungent plants, on the other hand, as onion, garlic and the rest, salt water ceases to be beneficial, since it robs them of their pungency, owing to the similarity between the two flavours, the one of the two similars moving toward the other[b] and entering into the passages, as with the cleansing agents that remove stains.

5. 5 Saline water, then, is good for bitter plants for the reason given.

If Androsthenes'[c] report about the island of Tylos[d] in the Indian Ocean is true—that the spring water, although saline, is better than rain not only for the trees but for all the other crops as well, and this is why,

whereas the chicories, which were foods, would have been improved.

[b] The saline moves toward the pungent.

[c] *Cf. HP* 4 7. 8 (of Tylos): "They report that although there is rain, the natives do not use it for their crops, but there are many springs in the island and all crops are watered from these, and it suits the grain and trees better. This is why after a rain they open the ditches for the spring water, as if rinsing the rain water off."

[d] Bahrain, the Dilmun (Ṭilmun) of Akkadian.

ἄλλοις, διὸ καὶ ὅταν ὕσῃ, τούτοις ἀποβρέχειν,[1] αἰτιάσαιτ' ἄν τις τὴν συνήθειαν, τὸ γὰρ ἔθος ὥσπερ φύσις γέγονεν. συμβαίνει δὲ τὰ μὲν
10 οὐράνια σπάνια γίνεσθαι, τούτοις δ' ἐκτρέφεσθαι καὶ τὰ δένδρα καὶ τὸν σῖτον καὶ τἆλλα, διὸ καὶ πᾶσαν ὥραν σπείρουσιν. ταῦτα μὲν οὖν ὡς ἐξ ὑποθέσεως εἰρήσθω.

6. 1 τῶν δὲ ποτίμων τὰ ψυχρὰ βέλτιστα, καὶ γὰρ πέψιν ποιεῖ μάλιστα, διὰ τὴν ἀντιπερίστασιν [2] τοῦ θερμοῦ, καὶ κατάψυξιν ταῖς ῥίζαις, δεῖται δὲ τὰ δένδρα ταύτης. ὅτι δὲ βελτίω, σημεῖον τὸ γλυκύτερα
5 γίνεσθαι καὶ εὐχυλότερα τοῖς ψυχροῖς ἀρδευόμενα καὶ λάχανα καὶ ῥίζας καὶ καρποὺς καὶ τἆλλα ὁμοίως. ἔοικεν δὲ κατὰ λόγον ἔχειν, ὥσπερ καὶ ἐπὶ τῶν ἐκ Διὸς ὑδάτων τὰ νυκτερινὰ [3] καὶ τὰ βόρεια, καὶ ἐπ' αὐτῶν δὲ τῶν πνευμάτων ὅταν τὰ [4]
10 βόρεια καὶ μὴ νότια, καὶ ὅλως ὅταν ψυχρὰ καὶ μὴ θερμά· τὴν γὰρ αὐτὴν ὄνησιν πάντα ἔχει, καὶ διὰ

2—*CP* 2 9. 6 l. 12 μάλιστα—ἄλλων. Between these words six pages of H are lost.

[1] ἀποβρέχειν Schneider: εἰ παρέχειν U.
[2] ἀντιπερίστασιν aP: -σπα- U N.
[3] νυκτερινὰ Heinsius: νυκτερίδια U.
[4] τὰ U: ᾗ Wimmer.

---

[a] *Cf. CP* 3 7. 7; 3 8. 4; 4 11. 5.

[b] *Cf.* James H. D. Belgrave, *Welcome to Bahrain, A Complete Illustrated Guide for Tourists and Travellers*, Printed in England by Mark & Moody Ltd. Stourbridge, Worcestershire and Published by James H. D. Belgrave, 1953, p. 120: " The soil is thin and salty, the water is somewhat brackish . . . All gardens are irrigated as the rainfall is very small. Rain, when it does fall, often does more harm than good, it splashes

after a rain, the natives rinse the rain water off with spring water—one would give habituation as the reason, habit having turned into nature;[a] and it so happens that rain is there infrequent, so that not only the trees but also cereals and the rest are reared on spring water (which is why the natives sow them at all seasons).[b] This explanation is to be taken as given on the assumption that the report is true.

6. 1 Of fresh waters the cold are the best,[c] since they not only do most for concoction, owing to their counter-displacement of the heat,[d] but also for cooling the roots, and trees require this.[e] A proof of their superiority is this: the improvement in sweetness and succulence not only of vegetables but of roots and seeds or fruit and the rest. There appears to be a superiority here like that among rains, where night rain [f] and rain from the north [g] and (considering the winds apart from the rains) northerly [h] winds are better than southerly (and in a word, cold than hot), for all bring the same benefit, owing to the same necessary causes;[i] whereas the warm ones relax and

the salty surface soil on to the stems of plants, shrubs and sometimes trees which causes small plants to die and often damages trees and shrubs. It is to prevent this damage that gardeners irrigate the ground immediately after a rain, a labour which appears rather superfluous to people who do not appreciate the reason."

[c] *Cf. HP* 7 5. 2: "Of waters the fresh and the cold are best."

[d] *Cf. CP* 1 12. 3 with note *a*.

[e] *Cf. CP* 1 18. 1.

[f] *Cf. CP* 2 2. 4.

[g] *Cf. CP* 2 2. 3.

[h] *Cf. CP* 2 2. 4, 2 3. 1.

[i] Necessary is opposed to final as the mechanical or instrumental or blind to the purposive: *cf. CP* 1 11. 4, 2 11. 2, 2 11. 7, 2 13. 5.

τὰς αὐτὰς ἀνάγκας· τὰ ⟨δὲ⟩ [1] θερμὰ διαχεῖ καὶ ἀνυγραίνει καὶ ἀσθενὲς ποιεῖ τὸ σύμφυτον θερμόν.

6. 2 ὑπερβολὴ δέ τίς ἐστιν ἴσως καὶ τοῦ ψυχροῦ, καὶ πρὸς τροφὴν καὶ πρὸς γένεσιν (ὥσπερ καὶ ἐπὶ τῶν ζῴων), εἴπερ ἀφαιρεῖται τὴν τοῦ θερμοῦ δύναμιν, ἀφαιρεῖται γὰρ τὴν ζωήν· οὐ μὴν πολλή γε, οὐδὲ 5 ἐν πολλοῖς (ἐὰν μή τινα καὶ ἄλλον ἔχῃ χυμὸν ἀλλότριον). οὐδὲ γὰρ οὐδ' ὅμοιον καὶ [2] ἐπὶ τῶν ζῴων ἐστὶν τῶν [3] ἐν αὐτοῖς· [4] ἐκεῖνα μὲν γὰρ ὅλως ὑπὸ τοῦ ὑγροῦ περιέχεται, καὶ ἐν τούτῳ ξυνίσταται· ταῦτα [5] δ' ἐν τῇ γῇ καὶ ἐν τῷ ἀέρι, 10 μιγνυμένου δὲ τοῦ ὕδατος τούτοις, οὔτε τὴν ἐκ τῆς γῆς οὔτε τὴν οἰκείαν ἀφαιρεῖται [6] θερμότητα (πλήν γ' εἴ που καὶ ὁ ἀὴρ ὅλως [7] τοιοῦτος· οὕτως [8] γὰρ ἀβλαστὴς καὶ ἄγονος ἡ γῆ τὸ ὅλον).

[1] ⟨δὲ⟩ ego (⟨γὰρ⟩ Schneider).
[2] ὅμοιον καὶ U: ὅμοια τὰ Wimmer.
[3] τῶν U^ar N aP: τὸ U^r.
[4] αὐτοῖς U: ὑγροῖς Itali; φυτοῖς Wimmer.
[5] ταῦτα N aP: ταῦτ' U.
[6] ἀφαιρεῖται u N a (P omits): ἀναφερεῖται U.
[7] ὅλως U^ar: -ος U^r N aP.
[8] οὕτως u: οὗτος U N aP.

[a] *Cf.* Aristotle, *On the Generation of Animals*, iv. 2 (767 a 28–35): "Also countries and waters differ among themselves in their effect on these matters (*sc.* fertility or sterility, birth of males or females) for the same reasons. And the food chiefly acquires its character and the body its disposition from the tempering of the surrounding air and from that of what is taken into the body, especially the consumption of water, since water is consumed in the greatest amount, and is a food present in all things consumed, even when they are dry. This is why hard (*atéramna*: resistant to concoction) and cold waters in some cases bring about failure of progeny and in others the birth of females."

soak the firm consistency of a plant and enfeeble its native heat.

6. 2 But cold water too can perhaps be excessive both for nutrition and for generation, just as it is for animals,[a] since it deprives the creature of the working of its heat, for this is to deprive it of life.[b] But the excess is not great nor is it found in many waters unless these have some other prejudicial flavour as well. Indeed the case with animals living in such water is not even similar to that of the plants watered by it: the animals are completely surrounded by water and are formed in it, whereas plants are in the earth and in the air,[c] and when water is mixed in with these two it removes neither the heat coming from the earth nor the native heat of the plant, except where the air as well is unrelievedly cold, since in that case the earth produces no sprouting or generation at all.[d]

[b] *Cf.* Aristotle, *On Life and Death*, chap. iv (469 b 18–20): " Of necessity, then, life and the preservation of this (*sc.* native) heat go together, and what is called death is the destruction of it; " chap. vi (470 a 19–29): " Since every living thing has soul, and this involves the presence of native heat . . ., in plants on the one hand there is sufficient help for the preservation of their native heat, coming from their food and the surrounding air . . . And if the surrounding air is excessive in cold because of the season, there being intense frosts, the plants wither . . ."

[c] *Cf.* Aristotle, *On the Generation of Animals*, iii. 11 (761 a 26–32): " For they (*sc.* testacea) have a nature analogous to that of plants, and the nature of testacea differs from that of plants to the extent that fluid is more conducive to life than the dry, and water than earth, since as plants are to the earth, so testacea aim at being to the fluid, plants being as it were land oysters and oysters aquatic plants."

[d] As in the arctic region.

6.3 τὰ μὲν οὖν ψυχρὰ διὰ ταῦτα βελτίω.
τὰ δὲ πότιμα τῶν ἀπότων, ὅτι τροφιμώτερα. μιγνυμένη γὰρ ἡ ἁλμυρὶς καὶ πᾶς ὁ τοιοῦτος χυλὸς οὐ μόνον ἀτροφίας, ἀλλὰ καὶ ἄλλας βλάβας 5 ἐμποιεῖ, καθάπερ ἐπὶ τῶν σωμάτων. ἡ δὲ μῖξις τῶν ὑδάτων ἡ τοσαύτη[1] χρήσιμος, ὅταν τὸ μὲν ἄγαν σκληρὸν ᾖ,[2] καὶ ὥσπερ ἄπεπτον καὶ ὠμόν, τὸ δὲ θῆλυ καὶ μαλακόν· ᾗ[3] καὶ ἐάν τινα γεώδη συνεπιφέρῃ χυλόν, ὅπερ ἔχει τὰ θολερὰ καὶ κοπ- 10 ριώδη, καὶ γὰρ τῆς τοιαύτης δέονται τροφῆς. διὸ πολλάκις ἂν ἁρμόσειεν[4] λεπτὸν καὶ καθαρὸν μὴ καθαρῷ καὶ παχεῖ, καὶ φρεατιαῖον ναματιαίῳ, καὶ ῥυτὸν καὶ ὄμβριον ἱμητῷ[5] καὶ ἁπλῶς στα- σίμῳ.[6]

6.4 θαυμασιώτερον δ' ἂν δόξειεν εἴ τι γλυκὺ καὶ πότιμον[7] ᾖ ὅλως ἄτροφον, ἢ μὴ τελεσφόρον ἐστίν, ὥσπερ ἐπὶ τῷ[8] περὶ τὴν Πυρραίαν[9] (ὃ καὶ ἐν ταῖς ἱστορίαις εἴρηται). τὴν δ' αἰτίαν ἐν 5 δυσὶν ἄν τις λάβοι τούτων· ᾖ ὅτι πάντως ἀσθενές (ὅπερ οὐκ ἔοικεν, ἐπεὶ καὶ ὁ ἀὴρ δοκεῖ τρέφειν), ᾖ ὅτι χυλόν τινα ἔχει[10] κακοποιόν, ὃ λανθάνει τὴν

[1] τοσαύτη U: τοιαύτη Schneider.
[2] ᾖ u P: ἥ U^ac; ᾗ U^c N a.
[3] ᾗ ego: ἧι U.
[4] ἂν ἁρμόσειε Schneider: ἐναρμόσειεν U^ar (-ε U^r).
[5] ἱμητῷ ego (λιμναίῳ Wimmer): εἰ μὴ τῶι U.
[6] στασίμῳ Wimmer: αἰτῖωι U.
[7] πότιμον Gaza, Scaliger: ἀπότιμον U.
[8] ἐπὶ τῷ ego (τὸ Schneider): ἐπι το U.
[9] Πυῤῥαίαν Schneider: πυραίαν U.
[10] ἔχει aP: ἐχοι U (ἔ- u N).

Cold water, then, is better for these reasons. 6. 3

Potable water is better than unpotable because it is more nutritious. For when salinity or any such flavour is mixed in water, it not only fails to feed the tree but also occasions various injuries, just as it does to our own bodies. Only to the following extent is mixture of waters serviceable: when one water is very harsh and as it were unconcocted and raw, the other mawkish and soft, or when the water that is added carries with it some earthy flavour, which is the case with muddy water and water containing manure, for plants need earthy food too. This is why it would often be proper to add thick and dirty water to water that is thin and pure, ditch water to well water, and water drawn up in buckets, in short standing water, to running water and water from rain.

More surprising[a] would be a case where sweet and palatable water is either quite unnutritious or fails to bring what it feeds to maturity, as with the water in the district of Pyrrha that is mentioned in the History.[b] There are two causes that one might find for this: either the water is quite lacking in strength (which seems unlikely, since even air is considered to be nutritious); or else it has some harmful flavour, a fact not noticed by our taste, and that such is the 6. 4

[a] Than the nutritiousness of dirty water.

[b] *HP* 9 18. 10: " For it is reported that in some places the water furthers child-bearing in women, as at Thespiae, and in others prevents it, as at Pyrrha (for the physicians blamed the water)." *Cf.* also Athenaeus, epitome ii. 15 (41 F) [Theophrastus, Frag. 159 p. 208. 4–6 Wimmer]: ". . . he (*sc.* Theophrastus) says that of sweet waters too some are productive of no offspring or of but few, as the water at †Pheta (ἐν Φέτᾳ καὶ; read Ἀφροδίσιον? [*cf. Pliny, N. H.* 31. 10]) and at Pyrrha."

γεῦσιν, ὅπερ ἐκδηλούμενον φαίνεται καὶ ἐπὶ τούτου τοῦ ὕδατος· καὶ γὰρ ἄνθρωποι λουόμενοι λεπροὶ 10 γίνονται, καὶ τὰ φυτὰ παραπλησίαν τινὰ λαμβάνει διάθεσιν· οὐ γὰρ δὴ τό γε ἄγαν τρόφιμον εἶναι λεκτέον, ὡς διὰ τὸ μὴ δύνασθαι κρατεῖν ἀτροφῆ [1] καὶ χείρω γίνεται.

περὶ μὲν οὖν τούτων ἱκανῶς εἰρήσθω.

7. 1 τοὺς δὲ τόπους ζητεῖ τοὺς οἰκείους οὐ μόνον τὰ περιττὰ καὶ ἴδια τῶν δένδρων (ὥσπερ εἴπομεν), ἀλλὰ καὶ τὰ κοινότερα γινόμενα· τὰ μὲν γὰρ φιλεῖ ξηρούς, τὰ δὲ εὐύδρους, τὰ δὲ χειμερινούς, 5 τὰ δὲ προσείλους, τὰ δὲ παλισκίους, καὶ ὅλως τὰ μὲν ὀρεινούς, τὰ δὲ ἑλώδεις (ὥσπερ καὶ διαιροῦσιν).

οὐκ ἀεὶ δὲ πάντα [2] τοὺς αὐτοὺς ἴσως, οὐδὲ διὰ [3] μίαν αἰτίαν ἀλλὰ διὰ πλείους (ὥσπερ καὶ ἐν τοῖς πρότερον ἐλέχθη)· καὶ γὰρ [4] τὸ συγγενὲς τῆς 10 φύσεως ἕκαστον ἄγει πρὸς τὸ [5] οἰκεῖον, ἐν ᾧπερ καὶ τὰ αὐτόματα φύεται (τὸ δ' αὐτόματον μηνύει τὴν φύσιν, ὥσπερ [6] ἐκ τῶν αὐτῶν αἱ τροφαὶ καὶ αἱ γενέσεις)· καὶ αἱ [7] καθ' ἕκαστον διαιρέσεις, οἷον ἡ θερμότης καὶ ἡ ψυχρότης, καὶ ἡ ξηρότης καὶ ἡ 15 ὑγρότης (ζητεῖ γὰρ πρόσφορα κατὰ τὴν κρᾶσιν),

[1] ἀτροφῆ U: ἀτρεφῆ Scaliger; ἀτραφῆ Schneider.
[2] πάντα Schneider: -ας U.
[3] διὰ U^c: δι U^ac.
[4] γὰρ Schneider: παρὰ U.
[5] τὸ U: τὸν N aP.
[6] ὥσπερ U: εἴπερ Gaza, Schneider.
[7] αἱ u: ἡ U; αἱ ⟨τῶν⟩ Schneider.

---

[a] *CP* 1 9. 2–3 (the pomegranate in Cilicia and Egypt); *CP* 2 3. 3–8 (the date-palm and persea).

explanation is made quite clear in the case of this water of Pyrrha as well, since not only do people who bathe in it become scaly, but plants watered with it also acquire a similar condition. For we are surely not to say that it is far too nutritious, and that the plants are undernourished and deteriorate because they cannot master it.

So much for waters.

Localities

7. 1 It is not only trees of a superior and distinctive type (as in the cases mentioned)[a] that seek their appropriate localities; this is also done by trees that never turn out to be more than ordinary. For some the preferred locality is dry, for others well-watered or wintry or sunny or shady: in a word, some favour the mountain, some the swamp, this being the division that people make.[b]

But perhaps not all trees of a given type always seek the same localities and for this there is no single reason, but several (as we said before).[c] For not only does the kinship of its nature bring the tree of that nature to the appropriate locality, which is where the tree grows spontaneously (and spontaneous production shows where a tree naturally belongs,[d] just as a tree is fed from the sources that produced it);[e] but so too do the particular classifications of the tree as hot or cold, dry or fluid (for trees seek localities with the tempering of qualities that suits them),

[b] *Cf. HP* 3 4. 1.
[c] *CP* 2 3. 6.
[d] *Cf. CP* 1 16. 10.
[e] *Cf. CP* 2 9. 5; 2 9. 6; 3 22. 4.

ἔτι δὲ τὰ ἀσθενῆ καὶ ἰσχυρά, καὶ βαθύρριζα καὶ ἐπιπολαιόρριζα καὶ εἴ τις ἄλλη διαφορὰ κατὰ τὰ μέρη.

7. 2 πολλάκις δὲ καὶ διὰ πλείω τούτων,[1] καὶ ἐνίοτε τὰ μὲν ἔχοντα, τὰ δὲ οὐκ ἔχοντα, πάντα γὰρ ταῦτα.[2]

ἔτι δὲ καὶ τὰ ὅμοια ζητεῖ[3] τὸν[4] ὅμοιον, καὶ τὰ ἀνόμοια μὴ τὸν αὐτόν, ὅταν ᾖ τις παραλλαγὴ τῆς 5 φύσεως. ἐν οἷς καὶ τὰ περὶ τὴν ἐλάτην καὶ πεύκην ἐστίν· ἡ μὲν γὰρ χαίρει παλισκίοις, ἡ δὲ πεύκη τοῖς προσείλοις, ἐν ἐκείνοις δὲ οὐ φύεται πάμπαν[5] ἢ[6] κακῶς. θερμὰ μὲν ἴσως ἄμφω, καθάπερ φασίν, ἀλλ' ἡ μὲν ἐλάτη ξηρόν, ἡ πεύκη 10 δὲ[7] ὑγρόν· σημεῖον[8] δὲ καὶ ἡ τῆς πίττης[9] γένεσις· ἅμα δ' ὑγρότητος πλῆθος[10] καὶ ὁ πρόσειλος τόπος[11] οἰκεῖος, πέψις γὰρ οὕτω μᾶλλον,

7. 3 ἀεὶ δὲ κατὰ τὸ ὑπεραῖρον ἡ ὄρεξις. ᾗ[12] καὶ δῆλον ὡς ἐν τοῖς δοκοῦσι παραλλάττειν τῶν ὁμογενῶν ἀεί τινα ζητητέον τοιαύτην διαφοράν.

ἐπεὶ ὅσα μὴ ὁμογενῆ τὸν αὐτὸν ζητεῖ, καθάπερ ὁ 5 κιττὸς καὶ τὰ πάρυδρα καὶ ἀλσώδη, ῥᾴδιον εἰπεῖν· ὁ μὲν γὰρ θερμὸς καὶ ξηρός, τῶν δὲ ἡ φύσις ὅλως συγγενής· ἔνια δὲ καὶ τούτων ἐν ταῖς καθ'

[1] τούτων U^r N aP: τουτοῦτων U^ar.
[2] ταῦτα U: οὐ ταὐτά Wimmer.
[3] ζητεῖ u: ζητεῖν U.
[4] τὸν Schneider (after Gaza): το U.
[5] ⟨πάμ⟩παν Schneider: παν U.
[6] ἢ u aP: ῆ U; ἧ N.
[7] πεύκη δὲ ego: δε πεύκη U.
[8] σημεῖον u or U^c: σήμερον U or U^ac.
[9] πίττης Gaza (*picis*), Wimmer (πίσσης Schneider): πεύκης U
[10] πλῆθος U: πλήθει Schneider (after Gaza).
[11] τόπος U^r aP: τρόπος U^ar N.
[12] ᾗ aP: ἡ U; ῆ u; ἧ N.

and there are the further classifications of weak or strong and of deep or shallow-rooted [a] (and others that concern its parts). Often moreover different trees seek the same locality because each tree has several of these special differences; and occasionally the special differences hold of some of the trees, but not of others, for all these cases occur. 7. 2

Again, a tree similar to another seeks a similar locality, and a tree dissimilar [b] to another seeks a locality not the same, when there is a certain disparity in the nature of the two trees. Here belongs the case of the silver-fir and the pine: the silver-fir likes shade, the pine the sun, and does not grow at all in shady places or grows there poorly. Now both trees are perhaps hot, as people say; but the silver-fir is dry, the pine fluid, as is confirmed by its production of pitch. But abundance of fluid and the appropriateness of a sunny station go together, since the sunny place means more concoction, and pursuit always follows the higher degree of what is aimed at. This shows moreover that when trees of the same kind are held to vary in their local preferences we must always look for some special difference of the sort. 7. 3

Indeed as for trees of different kinds that seek the same locality, such as the ivy and its host, and trees growing by the water and in groves, the explanation is easy: ivy is hot and dry,[c] whereas the others have a general affinity of nature with one another [d] (but

[a] *Cf. CP* 2 3. 6 for weakness, tempering of qualities, and shallow roots as excluding trees from certain regions.

[b] The two trees look alike but their natures are unlike in a certain respect.

[c] Being dry, it needs the food; being hot, it can concoct it.

[d] They are all weak and fluid (*cf. CP* 2 11. 1).

ἕκαστα διαφοραῖς ἔδειξε τὴν αἰτίαν. ὁμοίως δὲ καὶ ὅσα φιλόσκεπα[1] τυγχάνει, καὶ τῶν ἀγρίων καὶ 10 τῶν ἡμέρων, ὥσπερ ἡ ῥόα καὶ ὁ μύρρινος, καὶ ὁ μὲν πυκνὸς ὢν καὶ ξηρός, ἡ δὲ μανὴ καὶ οὐ ξηρά.

7. 4 τὸν μὲν γὰρ καρπὸν ἀμφότερα πυρηνώδη καὶ οὐχ ὑγρὸν ἔχει· τῷ μὲν οὖν ἡλίῳ παραδιδόμενος ταχὺ καταξηραίνεται, σκιατροφούμενος δὲ καὶ μετρίως εἰσλάμποντος, σῴζει τε τὴν οἰκείαν ὑγρότητα καὶ 5 πεπαίνει, διὸ καὶ τὰς ῥόας πυκνὰς φυτεύουσιν καὶ τοὺς μυρρίνους, ἵνα συσκεπάζωσιν ἄλληλα[2] καὶ προβολὴν[3] ἔχωσι τοῦ ἡλίου· ἅμα δὲ καὶ τῷ μὴ πολύρριζα τυγχάνειν οὐκ ἐνοχλοῦνται κατὰ τὰς τροφάς. εἰ γὰρ αὖ τἀναντία τις οὕτω φυτεύοι, 10 καθάπερ ἄμπελον καὶ συκῆν, οὐκ ἂν ὁμοίως εὐκαρποῖεν· ὑγρὰ[4] γὰρ ὄντα καὶ πέψεως δεῖται πλείονος.

7. 5 αἱ μὲν οὖν τῶν τόπων παραλλαγαὶ διὰ τοιαύτας τινάς εἰσιν αἰτίας, ὅπου μὴ καὶ ἄλλο τι συμβαίνει κώλυμα καὶ σίνος πρὸς εὐκαρπίαν, ὥσπερ περὶ Τάραντα ταῖς ἐλάαις· ἧ[5] γὰρ ἄπνοια[6] κατὰ τὴν

§ 4. 1–7. Plutarch, *Quaest. Conv.*, v. 8. 2 (683 D): . . . διὸ καὶ μόνον τοῦτό φησιν Θεόφραστος τὸ δένδρον (that is, the σίδη or pomegranate) ἐν τῇ σκιᾷ βέλτιον ἐκπέττειν τὸν καρπὸν καὶ τάχιον.

[1] φιλόσκεπα U^c: φυ- U^ac.

[2] ἄλληλα aP: ἀλλήλα U N.

[3] προβολὴν U^r N^2 aP: προσβολὴν U^ar N^1.

[4] εὐκαρποῖεν· ὑγρὰ ego (εὐκαρποῖ· ἔνυγρα Schneider): εὐκαρποι. ἔνυγρα U.

[5] ἧ Wimmer: ἡ U.

[6] ἄπνοια U: ἀπνοίᾳ Wimmer.

here too the cause for the preference in some is seen to lie in special differences). So too with all the trees, not only wild but like the pomegranate and myrtle cultivated,[a] that share a preference for shelter from the sun, the myrtle being close-textured and dry, the pomegranate open-textured and not dry.[b] For in both trees the fruit has a large stone and is not fluid. So the fruit, when exposed to the sun, soon dries out, but when reared in the shade and reached by only a moderate amount of sunshine, retains and ripens its native fluid. This is why both pomegranates and myrtles are each planted in close formation, so that the trees can shelter one another and be screened from the sun.[c] Then too they do not have many roots and so do not interfere with one another's feeding. For if one should plant trees of the opposite character, such as the vine and the fig, in this close formation, the crop would deteriorate, since the trees are full of fluid and require more concoction. 7. 4

To conclude: the localities vary for such reasons as these, except for special circumstances that prevent or spoil the crop, as at Tarentum with the olives.[d] For either there is no wind at all at flowering time (thus many flowers are scorched and drop),[e] or when 7. 5

[a] Ivy and the riparian trees were not cultivated.

[b] That is, they have special differences separating them from others of these descriptions.

[c] *Cf. CP* 3 7. 1–2.

[d] *Cf. CP* 5 10. 3 and *HP* 4 14. 9 (of the olive): "At Tarentum the trees always promise an abundant crop, but most of it perishes at the time of flowering."

[e] This is why Theophrastus prefers windlessness to mist in his explanation: windless heat is more likely than a mist to scorch the flowers.

5 ἄνθησιν,[1] ἀποκάεται γὰρ[2] πολλά, καὶ[3] ὅταν πνέῃ, τοιοῦτόν τι πνεῖ πόντιον ὃ τῇ[4] ἅλμῃ τῇ ἐπιφερομένῃ κατεσθίει καὶ λυμαίνεται τὰ ἄνθη· δοκεῖ δὲ καὶ ὁμίχλη τις ἄνευ πνοῆς ἐκβαίνειν, ἣ ὅταν ἅψηται τῶν ἀνθῶν ἀποκάει, διὸ καὶ μάντεις 10 θύουσιν ὥστε μὴ ἐκβῆναι, καί φασι κωλύειν. τὸν αὐτὸν δὲ τρόπον καὶ εἴ τι πάλιν σωτήριον ἢ [μὴ][5] πρόσφορόν ἐστιν τοῦ ἀέρος· ἐνίοτε γὰρ τῶν ἐδαφῶν[6] ὄντων φαύλων ὁ ἀὴρ ἐκτρέφει τῇ εὐκρασίᾳ (καθάπερ ἐλέχθη) καὶ τοῖς οἰκείοις 15 πνεύμασιν.

8. 1 εἰ δὲ ἥ γε πέψις τῶν καρπῶν τοῖς μὲν ὑπὸ θερμοῦ δοκεῖ, τοῖς δ' ὑπὸ ψυχροῦ γίνεσθαι, κατὰ συμβεβηκὸς ἥ γε ὑπὸ τοῦ ψυχροῦ γινομένη· τὸ γὰρ θερμὸν ἐν ἀμφοῖν[7] πέττει, καὶ μία τις ἡ 5 αἰτία, φανερὸν δ' οὐχ ὁμοίως διὰ τὴν ἀντιπερίστασιν. ὅλως γὰρ πάντων τῶν τοιούτων τὰς

[1] ἄνθησιν Schneider: αἴσθησιν U.
[2] ἀποκάεται γὰρ U: ἀποκάει τὰ Schneider; ἀποκάεται τὰ Wimmer.
[3] καὶ U: ἢ Wimmer.
[4] ὃ τῇ u: ὅτι U; ἢ τῇ N aP.
[5] [μὴ] Schneider; καὶ Wimmer.
[6] ἐδαφῶν Schneider: ἐδάφων U.
[7] ἀμφοῖν u: αμφουν U.

---

[a] *HP* 8 7. 6 (translated in last note on *CP* 3 23. 4); *CP* 1 13. 11–12.

[b] *Cf.* Theophrastus, *On Fire*, chap. ii. 13–14 (when the potency of a thing is collected and works all at once, it is stronger): " One can find many instances of the sort in the case of heat. Thus dressing-rooms and baths are hotter in winter than in summer, and in a north wind than in a south wind, since in the winter the heat has been concentrated and shut in by the surrounding air. Again, our bodies concoct food better and are in general stronger in winter, because the

a wind does blow, it is a sea-wind of a kind to corrode and ruin the flowers with the brine it carries; and it is believed that there is also a mist arising from the sea, unaccompanied by wind, that scorches the flowers when it touches them, making them drop, and for this reason diviners offer sacrifice to keep it from arising and assert that this stops it. Similarly again there are salutary or beneficial effects of the air; for sometimes, in spite of poor soil, the air rears the crop to maturity by its equable tempering of qualities (as we said) [a] and by the appropriate winds.

Concoction and Failure to Concoct

8.1 Views to be sure are divided, some persons thinking that concoction of the fruit is performed by heat, others by cold; but the concoction performed by cold is only incidentally performed by it, since in both cases it is heat that concocts, and the cause is one and the same, but the fact is not so evident in the second case because the heat is counter-displaced.[b] For in all such occurrences [c] we must take

heat has been gathered together and undergone counter-displacement. As a result of this cause moreover cold is in some cases held to do the same as heat, both taken simply and in excess: thus people say that both cold spells and heat concoct fruit and ' burn ' parts off. But they are mistaken; for the cold burns and concocts in this way not in its character of cold but incidentally, because it contracts and collects the heat, which does the work, and when the hot increases in amount it is stronger."

[c] That is, in all occurrences of concoction.

αὐτὰς δυνάμεις ὑποληπτέον[1] αἰτίας εἶναι· ξυμβαίνει δὲ δὴ τοῖς ὀψικάρποις ὑπὸ τοῦ χειμῶνος πεπαίνεσθαι περικαταλαμβανομένοις τῇ ὥρᾳ.
10 ὀψίκαρπα δ' (ὥσπερ ἐλέχθη) διὰ πλείους αἰτίας. ὅσα μὲν οὖν ὑγρὰ τῇ φύσει, συντονωτέρων δεῖται τῶν ψυχρῶν, ὥσπερ ἡ ἄμπελος, οὕτω γὰρ μᾶλλον ἡ πέπανσις· ὅσα δὲ ξηροκαρπότερα, καθάπερ ὁ μύρρινος[2] (καὶ γὰρ τοῦτο τῶν ὀψικάρπων),
15 ἐλαφροτέρων, ἀποξηραίνει γὰρ καὶ ἀποστύφει τὸ ἄγαν, ἡ δ' εὐκρασία καὶ ὁ ὑγρότερος καὶ ὁ ὑπὸ
8. 2 νότων[3] ἀὴρ εὐτροφώτατος. ἐπεὶ οὐδὲ τὰ ὑγρότερα τῇ φύσει πέττουσιν αἱ ὑπερβολαὶ τῶν χειμώνων, ἀλλὰ τὰ μὲν ὅλως ἀναξηραίνουσιν, τῶν δ' ἐξαιροῦνται τὸν οἰκεῖον[4] χυλόν, ὥσπερ τῶν σύκων.[5]
5 ἀντέχει δὲ μάλιστα καὶ δύναται τά τε ἐν ὑγρότητι μᾶλλον πίονι,[6] καθάπερ τῶν ἀγρίων[7] τὰ μιμαίκυλα, καὶ ἔνια γεώδη καὶ στρυφνὰ καὶ ἰσχυρὰ τὴν φύσιν, οἷον βάλανος ἀχρὰς οὖον· ὀψὲ γὰρ[8] ταῦτά γε λαμβάνει τὴν οἰκείαν ὑγρότητα. τοιαῦτα δὲ καὶ
10 τὰ μέσπιλα καὶ τὰ μῆλα τὰ ἄγρια καὶ πάνθ' ὅλως ⟨τὰ μετὰ τὴν ἀφαίρεσιν δοκοῦντα⟩ πεπαίνεσθαι,[9]

§ 2. 8–12. *Cf.* Varro, *R.R.* i. 68. 1: sorbum maturum mite conditum citius promi oportet: acerbum enim suspensum lentius est, quod prius domi maturitatem adsequi vult, quam nequit in arbore, quam mitescat.

[1] ὑποληπτέον u: -λειπ- U.
[2] μύρρινος aP: μυρρίν U; μυρρὴν u; μυρὴν N.
[3] νότων U^ar: νότον U^r N aP.
[4] τὸν οἰκεῖον u: τῶν οἰκείων U.
[5] σύκων u: συκῶν U N aP.
[6] μᾶλλον πίονι ego (λιπαρόν τι ἔχοντα Wimmer): μᾶλλον πλείονι U.
[7] ἀγρίων Wimmer: ὕγρῶν U^ac; ὑγριῶν U^c.

the same powers to be responsible; and so it is incidental to late fruits that they are overtaken by the change of season and thus ripened by winter. Fruit is late (as we said)[a] for a number of reasons. Now the late fruits that are fluid in their nature require the cold to be more intense (like the vine), since then the ripening is better performed; drier fruits, like that of the myrtle (the myrtle being another tree that fruits late) require milder cold, since severe cold dries and puckers them, and what rears them best is an equable tempering of qualities and air that inclines to humidity and is brought by south winds. Indeed excessive cold does not concoct even the fruits that are more fluid in their nature, but dries some of them out completely and removes from others the proper juice of the fruit,[b] as from figs. 8. 2

The fruits that hold out best against the cold and have the best ability to concoct under these circumstances are those whose fluidity is fatter, as among wild fruits[c] the fruit of the strawberry-tree[d] and a few that are earthy, astringent and strong in their nature, as the acorn, wild pear and sorb-apple, for these certainly are all late in acquiring their proper juice. Of this description moreover are medlars and wild apples and in general all fruits that are held

[a] *CP* 1 17. 4–9.

[b] *Cf. CP* 6 17. 5.

[c] Wild trees were held not to ripen their fruit: *cf. CP* 1 15. 4.

[d] *Cf. HP* 3 16. 4: "The fruit (*sc.* of the strawberry-tree) takes a year to ripen, so it turns out that this fruit and the flower of the next are on the tree at the same time."

---

[8] γὰρ U^cc^: γε U^ac^.

[9] ⟨τὰ—δοκοῦντα⟩ πεπαίνεσθαι ego (⟨ἃ⟩ πεπαίνεσθαι ⟨δύναται ἀφαιρεθέντα⟩ Schneider): παίνεσθαι U; πεπαίνεσθαι u.

καθάπερ ἡ ἀχρὰς καὶ τὸ οὖον, οὐ τὴν αὐτὴν μὲν πέπανσιν ἥνπερ καὶ ἐπὶ τῷ δένδρῳ, τὴν φυσικήν, ἔχουσαν δέ τινα γλυκύτητα τὴν ποιοῦσαν ἐδώδιμα, 15 εἴτ'[1] οὖν σῆψιν αὐτὴν χρὴ λέγειν (ὥσπερ ἐπὶ τῶν δρυπεπῶν[2] ἐλαῶν φασιν), εἴτε καὶ ἄλλην τινὰ

8. 3 διάθεσιν ἡντινοῦν.[3] οὐ μὴν ἀλλ' ἴσως οὐθὲν ἂν κωλύοι[4] καὶ τῇ ἔσωθεν[5] θερμότητι πέττεσθαι, καθάπερ τὰ οὖα, πεπαυμένης ἤδη τῆς ἐπιρροῆς ἐκ τῶν δένδρων· τότε μὲν γάρ, αἰεί τινος ἐπι- 5 ούσης, οὐκ ἐκράτει,[6] κωλυόμενον ἅμα[7] διὰ τὰ ψύχη· μὴ προσγινομένης δ' ἑτέρας, ἅμα δὲ καὶ τοῦ θερμοῦ συγκατακλειομένου, δι' ἄμφω πέττεται καὶ λαμβάνει τὴν μεταβολήν· ἐπεὶ καὶ οἱ ἐπὶ τῶν δένδρων καρποὶ πεπαινόμενοι, καθάπερ οἱ βότρυες, 10 ἀφαιρεθέντες γλυκύτεροι γίνονται, τοῦ ὑδατώδους ὑπὸ τοῦ ἡλίου καταξηραινομένου, ἐπ'[8] αὐτῶν δὲ τῶν ἀμπέλων, ὅταν ἐπιστρέφωσιν[9] ἢ καὶ γηράσαντες

8. 4 ἀποσταφιδωθῶσιν. σχεδὸν δὲ καὶ ἐν τοῖς ἄλλοις γίνεταί τις τοιαύτη μεταβολή, τῶν μὲν ἐπ' ἔλαττον, τῶν[10] δ' ἐπὶ πλεῖον.

ἔνια δὲ ἐναντίως· καὶ γὰρ ἐπ' αὐτῶν τῶν δένδρων 5 ἐν τόποις θερμοῖς καὶ οἰκείοις ἀπέπαντα ⟨μὴ⟩[11]

[1] εἴτ' aP: εἰ γ' U (εἴ γ' N).
[2] δρυπεπῶν u N: δρυπέπτων U; δρυπετῶν aP.
[3] ἡντινοῦν ego: ἥνπερ οὖν U.
[4] κωλύοι Schneider: κωλυει U.
[5] ἔσωθεν Link: ἔξωθεν U.
[6] οὐκ ἐκράτει u aP: οὐ κεκράτει U N.
[7] κωλυόμενον ἅμα u (κωλυώμενον ἅμα U): κωλυόμενα Schneider; κωλυόμενα ἅμα Wimmer.
[8] ⟨καὶ⟩ ἐπ' Schneider.
[9] ἐπιστρέφωσιν aP: ἐπιτρέφουσι U; ἐπιστρέφουσι u (-ιν N).
[10] τῶν u: τὸν U.
[11] ⟨μὴ⟩ Wimmer.

o " ripen " after removal from the tree, as the wild
pear and sorb-apple, a ripening which, although it is
not the same as the natural ripening on the tree, produces nevertheless a certain sweetness that renders
the fruits edible, whether we should call such ripening " decomposition,"[a] as people do with tree-
ripening olives, or whether we should rather call it any
condition but that. Nevertheless nothing perhaps 8. 3
would prevent concoction taking place also by the
heat within, as in sorb-apples, when the influx of
food from the tree has already ceased. For before
that, since food was constantly flowing in, the fruit
failed to master it (cold weather also preventing it
from doing so); but when new food is not being
added, and when the heat moreover is being shut
up inside,[b] the fruit gets concocted and undergoes
its change for both reasons. Indeed even those
fruits that are undergoing ripening on the tree, like
grape-clusters, become sweeter when removed from
it, since then the watery part is dried out by the sun;
and the clusters get sweeter even on the vine when
they get twisted or else grow old and turn into
raisins. And some such change as this occurs also 8. 4
(one may say) in the rest, in some to a lesser extent,
in others to a greater.

In a few instances the opposite occurs: the fruit,
though remaining on the tree, fails to ripen in regions

[a] For "decomposition" in black olives *cf.* *CP* 6 8. 4, *HP* 4 14. 10 (where worms in such olives are called σαπροί, or products of decomposition). Theophrastus is reluctant to apply the word to anything that is an improvement: *cf.* his alternatives at *CP* 1 1. 2; 1 5. 2; 2 9. 14; 5 4. 5.

[b] By counter-displacement.

ἐπικνισθέντα καὶ ἐπαλειφθέντα ἐλαίῳ (καθάπερ καὶ πρότερον ἐλέχθη) διὰ τὴν εὐτροφίαν, ὥσπερ καὶ τὰ ἐν Αἰγύπτῳ συκάμινα· τοῦτο δὲ παθόντα, καὶ ἀπέρασίν [1] τινα ἔλαβεν ὑγροῦ καὶ πνεύματος, καὶ 10 τὸ θερμὸν εἰσδέχεται. τῶν δὲ λαχανωδῶν ἔνια καὶ τὸ ὅλον ἀπέπαντα καὶ ἀμετάβλητα, καθάπερ ἡ κολοκύνθη.

πέψεως μὲν οὖν καὶ ἀπεψίας, καὶ ἁπλῶς τῶν γινομένων μεταβολῶν, ἐν τούτοις αἱ αἰτίαι.

9.1 τῶν δὲ δένδρων αὐτῶν ἀεὶ τὰ ἐν τοῖς ἀπνόοις καὶ παλισκίοις ὀρθὰ καὶ ἀστραβῆ καὶ λειότερα καὶ εὐμηκέστερα γίνεται, τὸν αὐτὸν δὲ τρόπον κἂν πυκνὰ τυγχάνῃ πεφυκότα, τὰ δ' ἐν τοῖς εὐπνόοις 5 καὶ προσηνέμοις καὶ εὐειλοις,[2] ἔτι δ' ἐν [3] μανοῖς πεφυκόσιν,[4] ἧττον. ἥ τε γὰρ εἰς βάθος αὔξησις κωλύει τὰ μήκη,[5] καὶ τὰ πνεύματα τραχύνει καὶ

[1] ἀπέρασίν U N aP: ἀπέρυσίν u.
[2] εὐείλοις Schneider: εὐήλοις U.
[3] δ' ἐν ego: δε U.
[4] μανοῖς πεφυκόσιν (-φυϊκ-U) u: καὶ μανὰ πεφυκότα Schneider.
[5] κωλύει τὰ μήκη ego (*altitudinem vetat* Gaza; κωλύεται τὸ μῆκος Schneider; κωλύει τὸ μῆκος Wimmer): κωλύεται τακῆ U.

---

[a] *CP* 1 17. 9.
[b] *Ficus Sycomorus.*
[c] Perhaps the *pneuma* made the fruit swell and become too large to concoct properly.
[d] Presumably when the gourd is not covered with earth: *cf.* *CP* 5 6. 4; *HP* 2 7. 5.
[e] The trees in view here are those whose " crop " is timber; hence the stress on height of the tree and absence of knots.
[f] *Cf. HP* 1 8. 1–2: " For some trees have many nodes, some

that are hot and appropriate, because of the good feeding, unless (as we said before)[a] it is scarified and smeared with oil, as with the Egyptian mulberries.[b] When so treated the fruit is not only relieved of some of its fluid and *pneuma*[c] but also admits the heat. But in a few vegetables there is no ripening or change at all, as in the gourd.[d]

The reasons then for concoction and failure to concoct and in a word for the changes that occur lie in these circumstances.

Habits of Trees[e]

9. 1 To pass from fruit to the trees themselves: in windless and shaded places the trees always grow up erect and undistorted, with fewer knots and taller; so too if they grow close together; whereas in well-ventilated, windward and sunny places, and furthermore when among trees growing far apart, they do not do this to the same extent,[f] since not only does lateral growth prevent height, but also the winds make the trees rough, producing knots, be-

few, and this differs in degree either by the nature of the tree or because of its station . . . Now such trees as elder . . . have few nodes naturally, whereas olive, fir and wild-olive have many. Of these two groups some individuals grow in shady, windless and wet places, others in sunny, cold, windy and lean and dry places, for among trees of the same kind some have fewer nodes, some more. In general mountain trees have more than trees of the plain, and dry trees than those of the marsh. Furthermore the character follows the spacing: crowded trees have few nodes and grow straight, spaced trees have more and are more crooked. For crowding results in shade, spacing in plenty of sun . . ." *Cf.* also *HP* 1 9. 1; 4 1. 4–5.

ὄζους[1] ἐμποιεῖ (καθάπερ ἐλέχθη) διὰ τὴν ἐπίστασιν· ἐκείνοις δέ, ἀφῃρημένων τούτων, ἡ εἰς τὸ
9. 2 μῆκος αὔξησις μόνη γίνεται. διὸ καὶ τὰ μὴ ὁμοίως ὀρθοφυῆ μηδ' εὐμήκη τὴν αὐτὴν διάθεσιν λαμβάνει τοῖς ὀρθοφυέσιν καὶ μακροῖς ὅταν ἐν τόποις γένηται τοιούτοις, ὥσπερ αἱ δρῦς· λεῖαι
5 γὰρ καὶ εὐθεῖαι καὶ σχεδὸν ἰσομήκεις γίνονται ταῖς ἐλάταις, καθάπερ φασὶ καὶ περὶ τὸν Αἷμον. ἅμα δὲ καὶ μανότερα καὶ ὑγρὰ καὶ ἀσθενέστερα τὰ τοιαῦτα γίνεται, διὰ τὸ μήτε ὑπὸ [τε][2] τοῦ ἡλίου μήτε ὑπὸ τῶν πνευμάτων καὶ τοῦ ψύχους
10 λαμβάνειν πύκνωσιν.

9. 3 ἀποβλητικὰ δὲ μάλιστα τῶν καρπῶν πρὶν πεπᾶναι συκῆ καὶ φοῖνιξ καὶ ἀμυγδαλῆ, καὶ[3] διὰ τὴν ἐναπόληψιν ὑγρότητός τέ τινος καὶ πνεύματος, ὥσπερ ⟨αἱ⟩[4] συκαῖ, τὰ δὲ τῷ τὴν προσάρτησιν
5 ἔχειν ἀσθενῆ, τοὺς δ' ὄγκους μείζους, ὥσπερ ἡ ἀμυγδαλῆ καὶ μηλέα καὶ ἄπιος· ἱκανὸν γὰρ καὶ ὁτιοῦν διυγρᾶναι καὶ ἀσθενὲς ποιῆσαι· καὶ ἅμα τούτοις γε[5] καὶ πνευμάτων ἐπιγίνεται μέγεθος.

ἡ δὲ ῥόα τοῖς μὲν κυτίνοις[6] εὐαπόπτωτος,[7]
10 ἀσθενὴς γὰρ αὐτῶν ἡ πρόσφυσις, ὥσθ'[8] ὅταν ψακάδια καὶ δρόσοι πέσωσιν, εἰσδυόμενα κατὰ τὸ

[1] ὄζους (ὄ- u) aP: ὄξους U (ὄ- N).
[2] [τε] aP.
[3] καὶ U: τὰ μὲν Basle ed. of 1541.
[4] ⟨αἱ⟩ Schneider.
[5] γε U: εἴ γε Schneider.
[6] κυτίνοις Gaza (*flores*), Schneider: αὖ τινος U.
[7] εὐαπόπτωτος Schneider (*facile amittit* Gaza): εὐατμωτος U.
[8] ὥστ' Schneider (*itaque* Gaza): ὡς δ' U.

cause the winds (as we said) [a] check the movement of the food. But in the first group, where lateral growth and winds are eliminated, growth in height alone remains. This is why trees not ordinarily so straight or tall acquire the same character as straight and tall trees when they grow in the sort of places described, as oaks; for these come to have fewer knots and to grow straight and almost to the same height as silver firs, as is reported to be the case on Mt. Haemus. But such trees also come to have an opener texture, to be fluid and to be weaker, because they acquire no condensation either from the sun or from the winds and the cold. 9. 2

Fruit Drop

Fig, date-palm and almond are most apt to drop their fruit before it is ripe,[b] both because a certain fluid and *pneuma* [c] is caught in the fruit (as in figs), and in others because they have a weak attachment and fruit too bulky for it (as with almond, apple and pear). Thus the slightest moisture is enough to soak the tree and make it too weak to maintain its hold; then with all this there follows strong wind. 9. 3

In the pomegranate it is the flower that drops easily, since the pedicel is weak, so that when there is a drizzle or fall of dew the water soaks the pedicel by

[a] *CP* 1 8. 3–4.

[b] *Cf. HP* 2 8. 1: " Trees that drop their fruit before ripening it are the almond, apple, pomegranate, pear and above all the fig and the date-palm; and these are the trees for which remedies are sought."

[c] The *pneuma* causes the fruit to swell: *cf. CP* 2 9. 5 (" distension "), 2 9. 6.

9. 4 ἄνθος ἀνυγραίνει καὶ ποιεῖ τὴν βολήν. διὸ καὶ κατάγουσιν τὰ δένδρα, καὶ οὐκ ἀφιᾶσιν εἰς ὕψος, ὅπως οἱ κύτινοι μή, ὀρθοὶ γινόμενοι, δέχωνται τὸ ὑγρόν· οἱ δὲ καὶ ἀνάπαλιν κελεύουσι 5 φυτεύειν τὰς ῥάβδους τούτου χάριν, ὅπως εὐθὺ κατανεύσωσιν. ὅτι δ' ἡ ὑγρότης αἰτία τῆς ἀποβολῆς κἀκεῖθεν δῆλον· αἱ γὰρ ἄπιοι καὶ ἀμυγδαλαῖ, κἂν μὴ βρέχῃ, νότιος δὲ ὁ ἀὴρ ᾖ καὶ ἐπινεφής, ἀποβάλλουσι καὶ τὰ ἄνθη, καὶ τοὺς 10 πρώτους καρποὺς ἐὰν εὐθὺς μετὰ τὴν ἀπάνθησιν ᾖ.

καὶ τούτων μὲν καὶ τῶν τοιούτων ἐν τῇ προσαρτήσει τε καὶ τοῖς ὄγκοις ἡ αἰτία.

9. 5 τῶν δὲ συκῶν καὶ τῶν φοινίκων οὐκ ἐν τούτοις, ᾗ[1] μόνον, ἀλλὰ καὶ ἐν τῇ αὐτῶν διαστάσει· διὸ καὶ ἐρινάζουσιν τὰς συκᾶς. τοῦτο δὲ ποιοῦσιν ὅπως οἱ ψῆνες οἱ ἐκ τῶν ἐρινῶν[2] τῶν ἐπικρεμαν-

§ 4. 1–3. Pliny, *N.H.* 16 109: qua de causa inflectunt ramos eius (*sc.* punicae), ne subrecti (*sc.* flores) umorem infestum excipiant atque contineant.

§ 5. 3–12, *CP* 2 9. 6 lines 1–2. Pliny, *N.H.* 15. 80–81 (*of* caprificus): ergo culices parit; hi fraudati alimento in matre, putri eius tabe, ad cognata evolant morsuque ficorum crebro, hoc est avidiore pastu, aperientes ora earum, ita penetrantes intus solem primo secum inducunt cerialesque auras inmittunt foribus adapertis. mox lacteum umorem, hoc est infantiam pomi, absumunt. quod fit et sponte: ideoque ficetis caprificus praemittitur (per- MSS) ad rationem venti, ut flatus evolantes in ficos ferat. 81. inde repertum ut inlatae quoque aliunde et inter se colligatae inicerentur fico . . .

[1] ᾗ ego: ἧι U.
[2] ἐρινῶν U N aP: -νεῶν u.

---

[a] *Cf. HP* 2 6. 12: "In planting other trees they place the slips upside down, as with the branches of the vine. Now some

entering the flower and causes the drop. This is why growers bend the trees down, not allowing them to grow upwards, to prevent the blossoms from standing upright and thus admitting the water. Others recommend planting the cuttings upside down,[a] so that the blossoms may hang downwards from the very outset. That humidity causes the drop is also shown by the fact that pears and almonds, even when there is no rain, but only a south wind and a cloudy sky, drop not only their blossoms but the young fruit as well when this happens just after the blossom has been shed.[b] 9. 4

The cause of these occurrences and the like lies in the attachment and in the bulk of the fruit.

The Fig: Caprification: The Open Fig Theory

In the fig and the date-palm the cause does not lie in these circumstances, or in these alone, but also in the distension of the fruit itself. This is why people resort to caprification of the fig-trees; [c] this is done so that the gall-insects that are produced from the wild 9. 5

say that this makes no difference, least of all with the vine; but others say that the pomegranate gets leafier then and shades the fruit better, and is moreover not so apt to drop its blossom."

[b] That is, when the fruit still has a weak pedicel.

[c] For a description of caprification *cf. HP* 2 8. 1–3. *Cf.* also Aristotle, *History of Animals*, v. 32 (557 b 25–31): "Wild fig-trees have the insects that are in their fruits, the so-called gall-insects. First to be produced is the grub; the skin then breaks and the gall-insect leaves it behind and flies out and enters the undeveloped fruit of the cultivated tree, and by making openings in it keeps it from being dropped. For this reason growers attach the wild fruits to the cultivated trees and plant the wild trees near the cultivated ones."

5 νυμένων γινόμενοι διοίγωσι[1] τὰ ἐπὶ τῆς συκῆς· ἡ μὲν γὰρ γένεσις ἐξ ἐκείνων, ζητοῦντες δὲ τὴν ὁμοίαν τροφὴν ἐκπέτονται[2] καὶ προσίπτανται τοῖς ἐπὶ τῆς συκῆς ἐρινοῖς[3] (συμβαίνει δὲ τοῦτο καὶ ἐπὶ τῶν[4] ἐριναζομένων ὅταν ἐπ᾽ αὐτῶν τῶν συκῶν
10 ἐπιγένωνται)· διὸ καὶ παραφυτεύουσιν ταῖς συκαῖς ἐρινοὺς[5] ἐπὶ τῶν ἄκρων ὅπως κατ᾽ ἄνεμον ἡ πτῆσις[6] οὐρία γένηται, ταῖς μὲν πρωΐαις πρωΐους, ταῖς δ᾽ ὀψίαις ὀψίους, ταῖς δὲ μέσαις μέσους, ἵνα κατὰ τὴν οἰκείαν ὥραν[7] ἑκάστοις ὁ ἐρινασμὸς ᾖ.

9. 6 διοιχθέντος δὲ τοῦ μέσου, τὴν ὑγρότητα ἐκβόσκονται τὴν πλείω, καὶ τῷ ἔξωθεν ἀέρι δίοδον διδόασιν, καὶ τὸ ὅλον εὐπνούστερα ποιοῦσι· συμβαίνει γὰρ ἅμα τῇ θερμότητι τῇ κατεργαζομένῃ συγκατα-
5 κλείεσθαί τι πνεῦμα (καθάπερ τοῖς ἑψομένοις), οὗ χωρισθέντος ἅμα τῇ ὑγρότητι καὶ διεκπνεύσαντος, ἐπιμένει· τὰ γὰρ αἴτια τοῦ ἀποβάλλειν ταῦτ᾽ ἐστίν.

[1] διοίγωσι Gaza, Heinsius (-γουσι a): δὲ οἴσουσι U; διοίσουσι u N P.
[2] ἐκπέτονται Gaza, Heinsius: -έττ- U.
[3] τοῖς—ἐρινοῖς Wimmer: τοῖς—ἐρρινοῖς (-νεοῖς u) U; τῶν—ἐρινοῦς (-νῶν aP) N.
[4] τῶν ⟨μὴ⟩ Gaza, Scaliger.
[5] ἐρινοὺς N aP: -νοῦς U; -νεοὺς u.
[6] πτῆσις u: πτῶσις U N aP.
[7] ὥραν Gaza, Itali: χώραν U.

---

[a] *Cf. CP* 2 9. 6; 3 22. 4; 5 10. 5; 6 4. 4. They have consumed all the seeds of the caprifig fruit from which they were produced.

[b] Erineón (erinón) is both the name of the wild fruit (the tree being called erineós or erinós) and also of the cultivated fruit that needed caprification and had not yet undergone it

figs hung on the cultivated tree may open the fruit. For the insects are produced from the wild fruit, but in their quest for food like that which produced them [a] they fly out and alight on the undeveloped fruit [b] of the cultivated tree. (This happens also in orchards where wild figs are hung on the tree when the insects from the wild trees alight on the cultivated trees directly.) This is why wild fig-trees are planted on eminences adjoining the orchard, to give the insects an easy flight down-wind. Early, intermediate and late ripening wild fig-trees are planted respectively near early, intermediate and late ripening cultivated trees, so that caprification may occur for each group of the latter in its own season. On opening the centre of the fruit the insects consume the excess fluid and allow a passage for the outside air and in a word make the fruit better ventilated. For it so happens that along with the heat that performs the task of concoction there is shut up in the undeveloped fig a certain amount of *pneuma*,[c] just as in things that are boiled, and when this has been removed through dissipation, together with the excess fluid, the fig remains on the tree, since these were the causes of the drop.

9.6

(*cf.* Aristotle, *History of Animals*, v. 22 (554 a 15); v. 32 (557 b 28, 29) [first instance]). The immature fruit that had undergone caprification was called ólynthos (*cf. HP* 3 7. 3; 4 14. 5; *CP* 5 9. 12); perhaps this term was also used of immature fruit that needed no caprification.

In view of Theophrastus' love of variation the forms of the names of the tree and the fruit that are found in U have been retained. In the *HP* the forms of the name of the fruit in U are as follows: 2 8. 2 *ἐρινων* (*-νεῶν* u); 2 8. 3 *ἔρινα*; 3 3. 8 *ἐρινεοῖς*; *ἐρινα*; 4 2. 3 *ἐρινου* (*-νεοῦ* u); *ἐρινοις* (*-νεοῖς* u); 4 14. 4 *ἔρινα*; 4 14. 5 *ἔρινα*.

[c] That is, gas.

εὐλόγως δὲ καὶ γίνονται καὶ εἰσδύονται πάλιν οἱ ψῆνες εἰς τὰ ἐρινεά·[1] γίνονται μὲν γὰρ διὰ τὸ μὴ
10 δύνασθαι πεπαίνειν μηδὲ τελειουργεῖν τοὺς ἐρινεούς[2] (ὥσπερ γὰρ καὶ τῶν ἄλλων σηπομένων, καὶ ἐν τούτοις ζῳοποιὸς ἡ φύσις)· οὐκ ἔχοντες δὲ[3] τροφήν, ζητοῦντες δὲ τὴν οἰκείαν, φέρονται πρὸς τὸ ὅμοιον· ἡ γὰρ ἐπιθυμία πᾶσι τοῦ συγγενοῦς
15 (ὥσπερ τοῖς φθειρσὶν αἵματος, ἐξ οὗ διαφθαρέντος ⟨ἡ⟩[4] γένεσις).

9. 7 εὐλόγως δὲ καὶ ἐν ταῖς λεπτογείοις καὶ καταβόρροις οὐκ ἐπιζητοῦσι τὸν ἐρινασμόν,[5] ξηρὰ γὰρ γίνονται τῇ φύσει δι' ὀλιγότητα τῆς τροφῆς· οὐδὲ δὴ εἴ τις ἑτέρα χώρα τοιαύτην ἔχει τὴν
5 κρᾶσιν ὥστε σύμμετρον ἐκδιδόναι τὴν τροφήν, ἡ γὰρ ἀποβολὴ δι' ἀπεψίαν καὶ τὸ μὴ κρατεῖν.

10–13. Pliny, *N.H.* 15. 79–80: caprificus vocatur e silvestri genere ficus numquam maturescens, sed quod ipsa non habet alii tribuens, quoniam est naturalis causarum transitus fitque (atque MSS) ut e putrescentibus gignatur aliquid. (continued on *CP* 2 9. 7).

12. ἄλλων] with this word H resumes.

1–8. Pliny, *N.H.* 15. 81 (continued): quod in macro solo et aquilonio non desiderant, quoniam sponte arescunt loci situ rimisque eadem quae (que MSS) culicum opere causa perficit, nec ubi multus pulvis . . . namque et pulveri vis siccandi sucumque lactis absorbendi. quae ratio pulvere et caprificatione hoc quoque praestat ne decidant, absumpto umore tenero et cum quadam fragilitate ponderoso.

[1] ἐρινεά U N aP: -νὰ C.
[2] ἐρινεοὺς u N aP: ἐρινέους U.
[3] ἔχοντες δὲ P (ἔχοντες H): ἔχονται δε U (ἔχονται δὲ u N).
[4] ⟨ἡ⟩ Schneider.
[5] ἐρινασμὸν u (ἐρινισμὸν N HP): ἐρανισμὸν U.

It is also reasonable that the gall-insects should not only be produced by the *erineá* [a] but should also enter the *erineá* [b] once more. They are produced in them because the wild fig-tree is unable to ripen or complete the development of its fruit, since in the wild fruit, just as in other cases of decomposition, the nature of the tree is productive of animals. These animals, having no food in the wild fruit, and seeking their proper food, set out for what is similar to the matter that produced them, since in all animals desire is directed to what they were bred and born from, just as lice desire blood, from the corruption of which they are produced.[c]

9. 7 It is also reasonable that where the soil is lean and sheltered from the north the trees should not require caprification,[d] since they are then made dry by natural means, owing to the small amount of food. Again no caprification is required in any other country with a tempering of qualities such as to provide food in the right amounts,[e] since the drop is due to failure to concoct and inability to master the food.

[a] The fruits of the wild fig.

[b] The undeveloped fruit, not yet subjected to caprification, of the cultivated fig: *cf.* note *b* on *CP* 2 9. 5.

[c] *Cf.* Aristotle, *History of Animals*, v. 31 (556 b 28): ". . . lice (*sc.* are produced) from the flesh."

[d] *Cf. HP* 2 8. 1: " For in southern Italy people say that the fruit is not dropped, and so make no use of caprification; and again, that it does not drop in regions sheltered from the north and with thin soil . . ."

[e] Caprification is not used in southern Italy, at Halycus in the Megarid, and in certain districts of the territory of Corinth (*HP* 2 8. 1).

ὡσαύτως δὲ καὶ ὅπου κονιορτὸς πολύς, ἀναξηραίνει γὰρ καὶ οὗτος.

ἄτοπον δ' ἂν δόξειεν ὅτι βορείοις ἀποβάλλουσι 10 μᾶλλον ἢ νοτίοις, ξηροτέρων ὄντων· αἴτιον δὲ ὅτι πυκνούμενα μᾶλλον τὰ ἐρινεὰ [1] κωλύει διεκπνεῖν· ἅμα δ' ἴσως καὶ πῆξίς τις γίνεται τοῦ ὀποῦ, ταύτην [2] γὰρ καὶ τῆς φυλλοβολίας αἰτίαν φέρουσί τινες (ὥσπερ εἴπομεν).

9. 8 ὅσα δ' ὄψια [3] πάμπαν τῶν γενῶν οὐκ ἀποβάλλει διὰ τὴν ὀψιότητα τῆς βλαστήσεως· οὐ γὰρ ἔτι συμβαίνει κατακλείεσθαι καὶ ἐναπολαμβάνεσθαι τὸ πνεῦμα, διὰ τὴν ὥραν, ἀλλ' ἐπικρατεῖν [4] ἀντι- 5 περιιστάμενον τὸ θερμόν· ἅμα δὲ καὶ φύσει ξηρά πώς ἐστιν, καὶ ὀψὲ διυγραίνεται, διὸ καὶ τὴν ἄρδευσιν αἱ τοιαῦται ζητοῦσιν καὶ δέχονται μᾶλλον.

ἔοικεν δ', εἴπερ ἡ ἄνοιξις ποιεῖ τὴν ἐπιμονήν, 10 εὔπνοιάν τε [5] καὶ ἀπέρασιν [6] ποιοῦσα, παραπλησίως [7] τρόπον τινὰ συμβαῖνον [8] καὶ ἐπὶ τῶν ἐν Αἰγύπτῳ συκαμίνων.

[1] ἐρινεά U: ἔρινα Schneider, ἐρινὰ Wimmer.
[2] ταύτην HP: ταῦτης U (ταύτης N); ταύτηι u.
[3] ὄψια HP: ὄψιω U (ὀψίω u; ὀψίω N).
[4] ἐπικρατεῖν C: -εῖ U N HP.
[5] τε Schneider: γε U.
[6] ἀπερασιν U: ἀπέρυσιν u; ἀπέρισι N; ἀπερισίαν HP.
[7] παραπλησίως U: παραπλήσιον Schneider.
[8] συμβαῖνον u HP (-αίνον U; -αίνων N): ⟨τὸ⟩ συμβαῖνον Schneider.

[a] *Cf. HP* 2 8. 3: ". . . growers append (*sc.* the wild figs) to the (*sc.* cultivated) fruit needing caprification after there has been rain. But where there is most dust, there the *eriná* (that

Again, no caprification is required where there is a good deal of dust, since dust too dries the fruit.[a]

It might appear odd that the trees tend more to drop their figs in a north wind, although the fruit is then drier, than in a south wind; [b] but the reason for this is that the north winds tighten the texture of the undeveloped figs and so prevent them from releasing the gas; then too perhaps a certain congealing of the sap occurs as well, this being (as we said)[c] the reason assigned by some for the shedding of leaves.

9. 8 Very late varieties of fig do not drop their fruit because they sprout too late; [d] since then it no longer happens that the gas is shut up and caught inside the fruit, but instead that the heat, because of the season, is counter-displaced and gains the upper hand. Then too, the fruit of the later trees is in a way dry [e] by its nature and only becomes juicy late, which is why trees of this sort like watering and take to it better.

Seeing that it is the opening of the fruit that makes it remain on the tree by producing ventilation and drainage, it appears that the process in the Egyptian mulberry [f] is in a way similar.

is, cultivated figs not subjected to caprification) are most numerous and strong (that is, least subject to drop)."

[b] *Cf. HP* 2 8. 1 (continued from note *d* on p. 269): "So with the prevailing winds. For the trees drop their fruit more in a north wind than in a south wind . . ."

[c] *CP* 1 21. 7.

[d] *Cf. HP* 2 8. 1 (of the drop of the fig): "Further there is the nature of the trees themselves: the early varieties drop their figs, but the later ones do not . . ."

[e] Dryness is one of the causes of late sprouting (*cf. CP* 1 10. 3) and late fruiting (*cf. CP* 1 17. 7).

[f] *Cf. CP* 1 17. 9; 2 8. 4; *HP* 4 2. 1.

9.9 ἀλλὰ τοῦτο [1] διαμφισβητοῦσί τινες, ὡς ἄρ᾽ [2] οὐκ ἀνοίγουσιν οἱ ψῆνες, ἀλλὰ συμμύειν ποιοῦσιν ὅταν εἰσδύωσιν, ὅθεν καὶ τὴν αἰτίαν ἐστὶν ἐκ τοῦ ἐναντίου φέρειν, ὡς τούτου χάριν ἐριναζομένων· [3]
5 ἐὰν γὰρ συμμύωσιν οὔθ᾽ ἡ δρόσος οὔτε τὰ ψακάδια δύναται παραφέρειν, [4] ὑφ᾽ ὧν ἀποπίπτουσι διαθερμαινομένων, ὥσπερ καὶ οἱ κύτινοι τῶν ῥοῶν. ὅτι [5] δὲ ταῦτα αἴτια μηνύει τὸ συμβαῖνον, ἃ δὴ καὶ λέγουσί τινες· ἀποβάλλουσι γὰρ μᾶλλον
10 ὑδατίων ἐπιγινομένων. σημεῖον δ᾽ ἔτι [6] κἀκεῖνο, ὅτι [7] τοῦ συμμύσαι χάριν· ἐὰν γὰρ μὴ ἔχωσι περιάπτειν, τῇ ἄμμῳ ὑποπάττουσιν [8] ἵνα συμμύσῃ· καὶ ὁ κονιορτὸς δὲ ποιεῖ διὰ τοῦτο ἐπιμένειν, ὅτι συμμύει κονιορτούμενα. τὰ δὲ τῶν ὀψίων οὐ
15 διοίγεται κατ᾽ ἐκεῖνον τὸν καιρόν, ἀλλὰ συμμύει, διὸ καὶ ἐπιμένει, καὶ οὐδὲ ὅλως ἐρινασμοῦ [9] δέονται· μετὰ ταῦτα ἰσχῦον [10] ἤδη, καὶ ἅμα τῆς ὥρας μεταβεβλημκυίας, [11] ἀνοίγεταί τε καὶ οὐκ ἀποπίπτει.
20 τὴν μὲν οὖν αἰτίαν ἀμφοτέρως λαβεῖν ἐνδέχεται.

[1] τοῦτο Ur N HP: -ω Uar.
[2] ἄρ᾽ HP: ἀν U (ἂν N).
[3] ἐριναζομένων U N HP: -νεαζ- u.
[4] παραφερειν U: διαφθείρειν Wimmer.
[5] ὅτι HP: ἔτι U N.
[6] ἔτι ego (Schneider deletes): ὅτι U.
[7] ὅτι ego (ἐστιν, ⟨ὅτι⟩ Schneider): εστιν U.
[8] ὑποπαττουσιν U: ἐπιπάττουσιν Heinsius.
[9] ἐρινασμοῦ Gaza (*caprificationem*), Basle ed. of 1541: θαυμασμοῦ U.

The Closed Fig Theory

9.9 But some dispute this fact of opening and say that when the insects enter the fig they do not make it open but make it shut; and so one can give the opposite cause for retention and assert that caprification aims at closing the fruit. For once the fig is closed neither dew nor drizzle can make it miscarry, and it is dew and drizzle that get warmed and cause the drop, as with the pomegranate blossom. That these are responsible (and they are cited by some persons) is indicated by what happens: there is more dropping of the fruit when light rain follows its first appearance. There is also this proof that the purpose of caprification is to close the fruit: if growers have no wild figs at hand to attach to the tree they sprinkle the cultivated fruit with sand to make it shut; and the reason why dust[a] too prevents drop is that the fruit closes up when dusted. As for the late trees, the fruit does not open at that time but is shut, and this is why it remains on the tree and growers have no need to resort to caprification at all; only later, when the fruit already has a firm hold and the season has changed, does the fruit open, and then it does not drop.

So it is possible to get the explanation on either theory.

[a] *Cf. CP* 2 9. 7.

---

10 ἴσχὺον U (ἰ- u N): ἰσχύοντα Hc(-αι Hac)P; ⟨δ'⟩ ἰσχύοντ' Wimmer.
11 μεταβεβληκυίας N HP: -κυας U; -κύας u.

9.10 τάχα δ' ἄν τις φαίη ταῦτά γε οὐδὲν ὑπεναντιοῦσθαι· καὶ γὰρ ὑδάτων ἐπιγινομένων ἀσθενέστερα τὰ ἐρινεά,[1] καὶ πλείων ὑγρότης γίνεται, δι' ἣν ἐκπνευματουμένην ἡ ἀποβολή· καὶ τῇ ἄμμῳ 5 παττέον [2] ἀποξηραίνειν βουλομένους, διὸ καὶ τὸν κονιορτὸν ὠφελεῖν, ἐξαιρεῖν [3] γὰρ τὸ ὑγρόν, ὃ τῆς ἀποβολῆς αἴτιον.

9.11 εἰ μὲν οὖν μηθὲν ἀντιλέγει,[4] δῆλον ὡς ἐν ἐκείνῳ τὸ αἴτιον· εἰ δ' ἐναντιοῦται, συμβαίνοι ἂν ἐκείνως [5] μὲν ἀπὸ τῶν ἐντὸς εἶναι τὴν ἀρχήν (καὶ εἴ τι [δὰν] [6] προσεπιγίνεται τῶν ἐκτός), οὕτως δ' 5 ⟨ἂν⟩ [7] ἀπὸ τῆς ἔξωθεν ὑγρότητος, ὑφ' ἧς μάλιστ' ἂν ἀποπίπτοι [8] τὰ ὀρθὰ πεφυκότα καὶ μὴ κατακλινῆ, καθάπερ οἱ κύτινοι· πιθανὸν δὲ καὶ ἐξ αὐτῶν τινας αἰτίας εἶναι τῆς ἀποβολῆς, ὥσπερ νοσησάντων, πλὴν ἡ μὲν νόσος ἴσως κοινὴ πάντων 10 τῶν καρπῶν.

9.12 τοῦ δὲ συμμύειν ὅταν εἰσδύωσιν οἱ ψῆνες ἀνάγκη τινὰ λέγειν αἰτίαν· ἐπεὶ [9] τό γ' ἐκβόσκεσθαι [10] τὴν ὑγρότητα τὴν ἐνυπάρχουσαν, οἰκείαν οὖσαν,

[1] ἐρινεὰ u HP (ἔρινα Schneider; ἐρινὰ Wimmer): ἐρίνεα U N.
[2] παττέον U: πάττειν u HP; πάττεν N.
[3] ἐξαιρεῖν u: ἐξαιρειν U *re vera*.
[4] ἀντιλεγει U^r from -ειν.
[5] ἐκείνως u HP: ἐκεῖνος U; ἐκείνοις N.
[6] [δὰν] Schneider.
[7] ⟨ἂν⟩ Schneider.
[8] ἀποπίπτοι Wimmer: -ει U H^ac; -η u N H^cP.
[9] ἐπεὶ u HP: ἐπι U N.
[10] ἐκβόσκεσθαι Schneider (*cf. CP* 2 9. 6): ἐνβ- U.

Examination of the Second Theory

Perhaps one might say that this evidence (at least) does not militate against the case for opening. So when light rain follows the first appearance of the untreated fruit, the fruit has a weaker hold and gets too much fluid, the drop being due to the change of the fluid to gas; and in order to dry the fruit one should sprinkle it with sand (which moreover is why dust is beneficial: it removes the fluid, and it is the fluid that causes the drop). 9. 10

Now if the earlier theory stands unrefuted, the cause of the drop must evidently be looked for in the opening of the fruit. But if a contradiction exists the upshot is that on the earlier theory the source of the drop is internal [a] (together with any external factor that supervenes),[b] whereas on the second theory the source is external fluid, and this would chiefly cause drop of what grows erect and not at an angle, like the erect pomegranate blossom. But it is plausible that some of the causes of the drop should come from the fruit itself, after it had (as it were) become affected with a disease [c] (except disease is perhaps something to which all fruit is liable).[d] 9. 11

But the proponents of the closed-fig theory are under obligation to give some cause for the shutting of the fruit after the insects have entered it. As for the other point, that the insects consume the fluid that they find there, which is the matter from which 9. 12

[a] The gas.

[b] Additional fluid from the air.

[c] Suffering from gas.

[d] The explanation should be restricted to fruit treated by caprification.

ἀληθές, διόπερ πρὸς τὴν ἐπιμονὴν χρήσιμον, τὰ
5 μέντοι σῦκα χείρω ποιεῖ, κενωθέντα γὰρ τὰ ἐρινὰ[1]
μᾶλλον ἐπισπᾶται καὶ πλείω τὸν ὀπόν, ὅθεν
ἔνιοί γ' οὐδ' ἐρινάζουσιν,[2] ἀλλὰ καὶ πωλοῦντες
κηρύττουσιν ὡς ἀνερίναστα,[3] καὶ δοκεῖ πολὺ
διαφέρειν. ἔτι δ' ἐχρῆν τοῖς βορείοις ἧττον
10 ἀποπίπτειν, συμμύει γὰρ μᾶλλον, εἰ μὴ ἄρα τῷ
ξηραίνεσθαι διαχάσκει.

9. 13 φαίνεται δ' οὖν πλείους ἀντιλογίας ἔχειν τῶν
πρότερον (εἰ μὴ ἄρα ἀμφοτέρως συμβαίνει καὶ
ἐνδέχεται).

τὸ δ' ἐνίους τόπους μὴ δεῖσθαι τῶν ἐρινασμῶν[4]
5 οὐκ ἄτοπον (ὥσπερ ἐλέχθη), συμμετρίαν ἔχοντας[5]
ἅμα τῆς ἔκ ⟨τε τῆς γῆς⟩ τροφῆς[6] ⟨καὶ⟩[7] τοῦ
ἀέρος. ἐπεὶ ὅτι γε ἡ ὑγρότης πολλὴ καθ' ὅλον τὸ
γένος καὶ ἐξ αὐτῶν τῶν ἀγρίων δῆλον· καὶ γὰρ ἡ
ἀπεψία διὰ τοῦτο γίνεται, καὶ ἐρινάζουσιν[8]
10 κἀκείνους ὅπως ἐπιμείνωσιν. διὰ τὰς αὐτὰς δ'

§13. 10–12. Athenaeus iii. 12 (77 E): ἐν δὲ τῷ δευτέρῳ περὶ φυτῶν ὁ Θεόφραστος καὶ τὸν ἐρινεὸν εἶναί φησι δίφορον· οἱ δὲ καὶ τρίφορον, ὥσπερ ἐν Κέῳ.

10–14. Pliny, *N.H.* 16. 114: sunt et biferae (*sc.* fici) in isdem (*sc.* Athenis); in Ceo insula caprifici triferae sunt; primo fetu sequens evocatur, sequenti tertius.

[1] ερίνα U: ἐρινεὰ u N HP (ἐριναστὰ Schneider).
[2] εριναζουσιν U: ἐρινεάζουσιν u.
[3] ἀνερίναστα N HP: ἀνεριιστὰ U; ἀνερινέαστα u.
[4] ἐρινασμῶν U: ἐρινεασμῶν u.
[5] ἔχοντας HP: -ος U N.
[6] ἔκ ⟨τε τῆς γῆς⟩ τροφῆς ego (τε τροφῆς Wimmer; τροφῆς Schneider): ἐκτροφῆς U.
[7] ⟨καὶ⟩ Schneider.
[8] ἐρινάζουσιν U (-σι N H[-ι illegible]P): ἐρινεάζουσι u.

they were bred, this point is true,[a] and is why their intervention is useful for preventing drop. Their consumption of the fluid, however, produces inferior figs, since the unripe figs, when empty, do more attracting of fig-sap and attract it in greater amounts.[b] (So some growers practise no caprification at all, but when they offer their figs for sale proclaim that they are free from caprification, and such figs are considered far superior.) Again, there should be less drop when the wind is from the north, since the figs are then more tightly shut (unless the drying splits them open).

9. 13 In any case the second theory appears to be open to more objections than the first, unless after all the thing occurs or can occur in both ways.

It is not strange that in a few districts no caprification is needed (as we said),[c] since the districts have just the right amount of food from both the ground and the air. Indeed it is evident that the entire class of fig-trees has a great deal of fluid, even from considering the wild varieties of fig alone. For the failure of the wild figs to concoct [d] is due to their over-abundant fluid, and growers use caprification even on the wild trees to prevent drop. From the same causes some wild fig-trees bear two crops and a

[a] This is the cause of the opening of the fruit on the other theory (*CP* 2 9. 6).

[b] The opening was due to the insects, an external cause; the shutting is presumably due to the nature of the tree, and should be directed to improvement of the fruit. An external cause of the shutting must therefore be found.

[c] *CP* 2 9. 7.

[d] The wild fig is not able to concoct its fruit (*CP* 1 18. 4).

αἰτίας καὶ διφοροῦσιν,[1] ἔνιοι δὲ καὶ τριφοροῦσιν αὐτῶν διὰ πλῆθος τροφῆς, ἀφαιρουμένων γὰρ τῶν πρώτων ῥᾳδίως ἕτερα φύουσιν, καὶ πάλιν τούτων ἕτερα, πληθύοντές τε τῇ ὑγρότητι καὶ μέχρι 15 τούτου κρατοῦντες, ἐπὶ πλεῖον δὲ οὐ δυνάμενοι διὰ 9. 14 τὴν ὑπερβολήν. μόνον γὰρ δὴ τοῦτο τῶν ἀγρίων, ἢ μετ᾽ ὀλίγων,[2] ἀτελῆ φέρει[3] τὸν καρπόν (εἰ μὴ ἄρα ἐνταῦθα αὐτῶν ἦν τὸ τέλος· φύεται γοῦν καὶ αὐτόματος· ἡ δὲ τῶν αὐτομάτων γένεσις ἐκ 5 σπέρματος, καὶ ὅσα δὴ διὰ σῆψίν τινα, μᾶλλον δ᾽ ἀλλοίωσιν, γίνεται τῆς γῆς)· ὥστε καὶ τοῦτο ὠλεσίκαρπον[4] ἄν τις ἔφη, καθάπερ ὁ ποιητής φησιν τὴν ἰτέαν· ἐπεὶ τά γ᾽ ἄλλα, κἂν ἀβρώτους ἔχῃ πᾶσι τοὺς καρπούς, ὅμως ἐκτελεῖ καὶ πεπαίνει 10 κατὰ τὴν τῆς φύσεως ὁρμήν.

9. 15 ἀλλὰ γὰρ αὕτη μὲν ἰδιότης ἄν τις εἴη πρὸς τὰ ἄλλα.

τὸ δ᾽ ἐπὶ τῶν φοινίκων συμβαῖνον οὐ ταὐτὸν μέν, ἔχει δέ τινα ὁμοιότητα τούτῳ (διὸ καλοῦσιν

[1] διφοροῦσιν U^r HP: δια- U^ar; δυσ- N.
[2] μετ᾽ ὀλίγων u HP: μετολίγον U (μετ᾽ ὀλίγον N).
[3] φέρει U^r N HP: -ειν U^ar.
[4] ὠλεσίκαρπον HP (and so U^ar at *HP* 3 1. 3): ὀλεσικαρπον U; ὀλεσίκαρπον u N (and so U^r at *HP* 3 1. 3).

---

[a] *Cf. CP* 1 1. 2. The wild fig-tree is too large to be the work of spontaneous generation by decomposition (*cf. CP* 1 5. 1, where the largest plant cited as so produced is silphium); it must therefore grow from seed, and the seed must be completed far enough to generate.

few get so much food that they even bear three: so on the removal of the first crop the tree easily grows a second, and when this is removed, a third, since the trees abound in fluid and can master it up to this point, but no further, because the amount is too great. For this tree is the only wild tree, or among the few, to bear imperfect fruit (unless after all perfection in this fruit lies in getting thus far; in any case the tree also grows without being planted, and such spontaneously produced plants come from seed, except for the cases due to some decomposition—or rather alteration [a]—of the earth). So one might have called this tree too, as Homer calls the willow, 9. 14

loser of its fruit,[b]

since the rest, even though their fruit may be inedible to every living thing, nevertheless complete and ripen it, so far as their natural initiative is concerned.[c]

But enough. We doubtless have here a peculiarity that marks fig-trees off from the rest.[d] 9. 15

The Date-Palm

What occurs in the date-palm, while not the same as caprification, nevertheless bears a certain re-

[b] *Odyssey* x. 510; *cf. HP* 3 1. 3.

[c] Ripening of the pericarpion, on the other hand, is often due to agricultural procedures.

[d] The fig-tree (cultivated and wild) is marked off by the use of caprification from the other trees in which fruit-drop occurs. Theophrastus favours the open fig theory, which stresses the difference between caprification and other procedures: *cf. CP* 2 9. 11, 2 9. 5 (first sentence).

5 ὀλυνθάζειν αὐτούς)· τὸ γὰρ ἀπὸ τοῦ ἄρρενος ἄνθος καὶ ὁ κονιορτὸς καὶ ὁ χνοῦς συγκαταπαττόμενος[1] ποιοῦσίν τινα τῇ θερμότητι καὶ τῇ ἄλλῃ δυνάμει ξηρότητα καὶ εὔπνοιαν, διὰ τούτων δὲ ἡ ἐπιμονή. φαίνεται δὲ τρόπον τινὰ ὅμοιον τούτῳ[2]
10 καὶ ἐπὶ τῶν ἰχθύων ξυμβαῖνον, ὅταν ὁ ἄρρην ἐπιρραίνῃ[3] τοῖς ᾠοῖς ἀποτικτομένοις τὸν θορόν. ἀλλὰ τὰς μὲν ὁμοιότητας καὶ ἐκ τῶν ἀπηρτημένων ἐστὶ λαμβάνειν.

10. 1 τῶν δὲ ὁμογενῶν ἐν οἷς τὰ μὲν ἄκαρπα, τὰ δὲ κάρπιμα τῶν ἀγρίων,[4] ἃ δὴ θήλεα, τὰ δ' ἄρρενα καλοῦσιν, ἐν ἐκείνῃ τῇ αἰτίᾳ περιλαμβάνεται τῇ καὶ πρότερον εἰρημένῃ περὶ τῶν ἀκάρπων, ὅτι διὰ
5 πυκνότητα καὶ ἰσχὺν καὶ εὐτροφίαν ἄκαρπα

[1] συγκαταπαττόμενος Gaza, Schneider: συγκαταπατουμενος U.
[2] τούτῳ u: τοῦτο U.
[3] ἐπιρραίνῃ u: -ειν U.
[4] τῶν ἀγρίων U: Schneider transposes after ὁμογενῶν.

---

[a] *Cf. HP* 2 8. 4: "In the date-palms there are the remedies that come from the males to the females; for it is the males that effect the retention and full concoction of the fruit, and some people call this process, from its similarity to the process in the figs, *olyntházein*." *Olyntházein* is derived from *ólynthos*, the edible wild fig fruit.

[b] The dust is the pollen, the down the stamens (anthers). *Cf. CP* 3 18. 1 and *HP* 2 8. 4: "When the male (*sc.* date-palm) blooms they cut off the blade from which the flower has come and shake the down, flower and dust directly from the severed blade over the fruit of the female; and if this is done to the female, she keeps the fruit and does not drop it."

[c] *Cf. HP* 2 8. 4: "It appears that in both (*sc.* the fig and the date-palm) a remedy comes to the females (*sc.* to the

semblance to it, which is why the procedure is called *olyntházein*.[a] For the flower and dust and down[b] from the male date-palm, when sprinkled on the fruit, effect by their heat and the rest of their power a certain dryness and ventilation, and by this means the fruit remains on the tree. Something similar in a way to this is seen to happen with fish, when the male sprinkles his milt on the eggs as they are laid. But resemblances can be found even in things widely separate.[c]

Bearing: " Males "

10. 1 Among wild trees[d] the case where some trees of the same kind bear but others do not, these trees being called " male " and " female,"[e] comes under that other cause that was given earlier[f] to account for trees that fail to bear: it is close texture, strength

cultivated fig-tree [*sykê* fem.] and the female date-palm) from the male (*sc. erineós* masc. " wild fig-tree " and male date-palm), for they call the tree that bears the fruit the female; but in the case of the date-palm this is as it were a mingling, whereas in the case of the figs it occurs in another way."

[d] Among cultivated trees the only case is the date-palm.

[e] *Cf. HP* 1 14. 5: " The distinction that is used only or mainly for wild trees is that of female and male; cultivated trees on the other hand are distinguished in a number of different ways; " *HP* 3 8. 1: " In all (*sc.* wild) trees . . ., taking them by kinds, several distinctions are found within each. One is common to all the kinds, the distinction whereby people differentiate a female and a male in each, of which the one bears fruit, whereas the other (in some) is fruitless; and where both bear fruit, the female bears a finer and more abundant crop (except that some call such females ' males,' there being some who use the terms in this way)."

[f] *CP* 1 16. 5.

γίνεται, συμβαίνει γὰρ ἅπαντα εἰς ἑαυτὰ καταναλίσκειν. ἡ δὲ καρποτοκία δεῖται μὲν τῆς φυσικῆς περιττώσεως, ἐκ ταύτης γὰρ ὁ καρπός, ὥσπερ καὶ τοῖς ζῴοις τὸ σπέρμα· τρεπομένης[1] δ' εἰς ἕτερον
10 ἀεὶ καὶ καταναλισκομένης, ἀφαιρεῖται τὴν γένεσιν, ἡ γὰρ φύσις οὐ διαρκὴς εἰς ἄμφω, μὴ λαμβάνουσα τὸ σύμμετρον.

10. 2 οἷς μὲν οὖν ἐνίοτε τοῦτο συμβαίνει, τότε ἄκαρπα γίνεται, οἷς δ' εὐθὺς ἐνταῦθα ἡ ὁρμὴ τῆς φύσεως, ὅλως ἄκαρπα,[2] διὸ δὴ καὶ τῶν ἀγρίων ἔνια τοιαῦτ' ἐν τοῖς ὁμογενέσιν, ἅπερ ἰσχυρότερα καὶ πυκνότερα
5 καὶ ὡς ἐπίπαν μείζω γίνεται διὰ τὴν εὐτροφίαν. ἐπεὶ καὶ τὰ μικρόκαρπα πάνθ' ὡς ἐπὶ τὸ πολὺ μείζω, καὶ τὰ εἰς μέγεθος ὡρμημένα [ἢ][3] μικροκαρπότερα, καθάπερ καὶ ἡ Ἰνδικὴ συκῆ καλουμένη. θαυμαστὴ γὰρ οὖσα τῷ μεγέθει,
10 μικρὸν [γὰρ][4] ἔχει φύσει σφόδρα τὸν καρπὸν καὶ ὀλίγον, ὡς εἰς τὴν βλάστησιν ἐξαναλίσκουσα πᾶσαν τὴν τροφήν (καὶ γὰρ σφόδρα μεγαλόφυλλος)· ἀφ' ὧν ἔοικεν διὰ τὴν εὐβοσίαν[5] καὶ ἡ τῶν ῥιζῶν

§ 2. 8–12. Athenaeus iii. 12 (77 F): πάλιν δὲ ὁ Θεόφραστος ἐν τῷ β' τῶν αἰτίων 'ἡ Ἰνδική' φησι 'συκῆ καλουμένη θαυμαστὴ οὖσα τῷ μεγέθει μικρὸν ἔχει τὸν καρπὸν καὶ ὀλίγον, ὡς ἂν εἰς τὴν βλάστησιν ἐξαναλίσκουσα ἅπασαν τὴν τροφήν.'

[1] τρεπομένης u HP: τρεφομένης U; πρεπομένης N.
[2] ἄκαρπα U HP: εὔκαρπα u N.
[3] [ἢ] u (erased): ἢ U.
[4] [γὰρ] Athenaeus HP (τε Wimmer): γὰρ U N.
[5] εὐβοσίαν u: εὐβοΐαν U (-ῗαν N HP).

---

[a] *Cf.* Aristotle, *On the Generation of Animals*, i. 18 (725 a 11–12): "Therefore semen is a part of the useful residue;" i. 18 (726 a 26–28): "So it is clear . . . that the semen is a

and good feeding that makes them barren, for the trees expend all their resources on themselves. But fruit production requires a natural residue, since from this comes the fruit, like the semen in animals.[a] Constant diversion and expenditure of this residue on something else robs the tree of procreation,[b] since its nature, not receiving an adequate provision, does not suffice for both tasks.

10. 2 Now trees to which this occasionally happens are barren on those occasions,[c] whereas trees whose nature takes this direction from the start are permanently barren. Hence among wild trees too of the same kind [d] some are permanently barren; and these are the ones that turn out stronger, closer in texture, and on the whole larger because of their good feeding. Indeed as a rule all trees with small fruit are larger, and trees whose drive has been to a large size have smaller fruit, as the so-called Indian fig.[e] For although its size is amazing the tree has fruit that is naturally very small in size and in amount. This suggests that the tree expends all its food on vegetative growth (the leaves in fact are extremely large); and it is from this growth, because (it seems) of the good feeding, that the roots too are produced

residue of useful food, and of food in its final state, whether all animals emit semen or not."

[b] *Cf.* Aristotle, *On the Parts of Animals*, ii. 5 (651 b 13–15): "Again fat animals are less fertile for the same reason: what should have gone from the blood to become generative fluid and semen is used up on fat and suet . . ."

[c] These are cultivated trees that are fed too much: so with the vine and almond (*cf. CP* 1 17. 9).

[d] *Cf. CP* 1 5. 5. In the wild trees the nature must be responsible; the result cannot be attributed to cultivation.

[e] The banyan.

τῶν καθιεμένων εἶναι γένεσις (ἴσως δ' ἀνάπαλιν,
15 ὅτι καὶ εἰς ταῦτα καταμερίζεται, διὰ τοῦτο οὖν
ἀσθενέστερός ἐστιν καὶ ἐλάττων ὁ καρπός)· ὑπὲρ
δὲ τῶν ῥιζῶν ἐν ἑτέροις εἴρηται.

10. 3 μόνα δὲ ᾗ [1] μάλιστα τῶν πολυκάρπων [2] αὔξησιν
λαμβάνει μεγέθους ἄμπελος καὶ συκῆ· δύναται γὰρ
ἡ μὲν ἐφ' ὁσονοῦν ἐφικνεῖσθαι κληματουμένη, καὶ
ἔδαφος εὔγειον [3] ἔχουσα καὶ εὔτροφον, ἡ δὲ καὶ
5 εἰς ὕψος αἵρεσθαι,[4] καὶ πολύκλαδος γινομένη
πολὺν καταλαμβάνειν [5] τόπον.

11. 1 εὐλόγως δὲ καὶ μακροβιώτερα τὰ ἄκαρπα τῶν
καρπίμων, καὶ τὰ ὀλιγόκαρπα τῶν πολυκάρπων,
ὅσα μὴ δι' ἀσθένειαν ἢ ὑγρότητα ἢ δι' ἄλλην τινὰ
αἰτίαν ἄκαρπα ἢ ὀλιγόκαρπα, καθάπερ τά τε
5 πάρυδρα καὶ ἀλσώδη καὶ ὅσα μανὰ [6] καὶ εὔσηπτα,
καθάπερ ἡ δάφνη. ταῦτα μὲν γὰρ καὶ εἴ τι τοιοῦτον
ἄλλο, διὰ τὰς εἰρημένας αἰτίας· ἡ δὲ καρποτοκία

[1] ᾗ Wimmer (ᾗ καὶ Schneider): καὶ U.
[2] πολυκάρπων Scaliger: μονοκάρπων U.
[3] εὔγειον u: εὐγείων U.
[4] αἵρεσθαι u: αἱρεῖσθαι U.
[5] καταλαμβάνειν Wimmer: -ει U.
[6] μανὰ u: μανὴ U.

---

[a] For similar reversals of the explanation *cf.* *CP* 2 4. 4, 6 12. 5.

[b] *HP* 1 7. 3: " The nature and power of the Indian fig is unique. For the tree sends its roots from its shoots until they touch the ground and get rooted . . .; " *HP* 4 4. 4: " The branches when they are in contact with the ground make a kind of fence all around the tree . . . The foliage above

that are let down to the ground. (But perhaps we should put this the other way round: because the food is apportioned to these parts as well, the fruit is weaker and scantier.)[a] We have spoken of the roots elsewhere.[b]

10. 3 Of trees with abundant fruit, on the other hand, the vine and the fig are the only or principal ones to attain to a great size. For the vine can branch out and cover any distance when it has ample and nutritious soil; and the fig can also grow in height, as well as fork into numerous branches and cover a great extent of ground.

Bearing and Longevity[c]

11. 1 It is also reasonable that non-bearers live longer than bearers, and bearers of little fruit than bearers of much, except where the tree bears none or little by reason of weakness or fluidity or some other cause, like trees growing by the water,[d] or in groves, or those that like the bay[e] are open in texture and decompose easily. Now these and the like are short-lived for the

ground is also abundant, and the whole tree is well-rounded in shape and very great in size . . . Its leaf is as large as a targe, but the fruit is extremely small, of the size of a chick-pea and resembling a fig . . . And the amount of it is amazingly small, even absolutely, let alone in comparison with the size of the tree . . ."

[c] Longevity in trees and plants is discussed in *HP* 4 13. 1–6.

[d] *Cf. HP* 4 13. 2: "Trees growing by the water are considered more short-lived than trees on drier ground; so the willow, white poplar, elder and black poplar."

[e] *Cf. HP* 4 13. 3: "Some trees, though ageing and decomposing rapidly, send up side growths from the same root, as the bay . . ."

πολὺ τῆς φύσεως ἀφαιρεῖ, καὶ τὸ κυριώτατον, ὅπερ καὶ ἐπὶ τῶν ζῴων συμβαίνει, τὰ γὰρ
10 πολυτοκώτατα καὶ γηράσκει τάχιστα καὶ ἀπόλλυται. φανερὸν δὲ ἀμέλει καὶ ἐπ' αὐτῶν ἐστι τῶν δένδρων· ὅσα γὰρ πολυφόρα[1] καὶ πολύκαρπα, ταῦτα καὶ ὅλως καταγηρᾷ θᾶττον, καὶ ἐν τοῖς ὁμογενέσιν (οἷον ἀμπέλοις συκαῖς τοῖς τ' ἄλλοις),
15 τὰ δὲ στέριφα καὶ ὀλιγόκαρπα χρονιώτερα ὡς εἰπεῖν.

11. 2 ⟨ἤ⟩δη ⟨δέ⟩[2] ποτε, μᾶλλον δὲ πολλάκις, ὑπερκαρπήσαντα τὰ δένδρα καὶ[3] δι' ἀσθένειαν ἀφαυάνθη,[4] καὶ μάλιστα τοῦθ' αἱ ἄμπελοι πάσχουσιν, καὶ τἆλλα δὲ τὰ πολύκαρπα, διὰ τὸ ἐξανηλῶ-
5 σθαι τὴν φύσιν εἰς τοὺς καρπούς· καὶ[5] τοῖς σιτηροῖς σπέρμασιν, καὶ ὅλως τοῖς ἐπετείοις, συμβαίνει[6] (διὸ καὶ ἐπέτεια)· συνεξαυαίνονται γὰρ εὐθὺς αἱ ῥίζαι τελειουμένων τῶν καρπῶν διὰ

[1] πολυφόρα Schneider: πολύφορα U.

[2] ⟨ἤ⟩δη ⟨δέ⟩ ego (*Fit ergo* Gaza; συμβαίνει μὲν οὖν Schneider; καί Wimmer): δή U (no punctuation precedes).

[3] καὶ U: N HP omit.

[4] ἀφαυάνθη HP: ἀφαυανθῆι U (-ῆ N); ἀφαυανθῆναι Schneider (*ut . . . arescant* Gaza).

[5] ⟨ὃ⟩ καὶ Wimmer.

[6] ⟨ταὐτὸ⟩ συμβαίνει Schneider (after Gaza).

---

[a] In the sentence that precedes: "weakness, fluidity or open texture." The riparian are short-lived because they are too fluid (which also makes them seek the water); the growers in groves because they are too weak (which is also why they seek company); and the bay because it is too open and easily decomposes.

reasons just mentioned;[a] but bearing takes much away from the tree's nature, and the most important part of it, and the result is the same for animals as well, for the most prolific are the quickest to age and die. For that matter the fact is evident when we merely consider the trees: the trees that bear many crops and abundant fruit are also the trees that age sooner, both absolutely and when compared to others of the same kind (as among vines,[b] figs and the rest); whereas the barren and the scanty bearers last longer as a rule.[c]

11. 2 It has sometimes (or rather, often) happened that on bearing too large a crop a tree has even withered away from weakness.[d] This occurs chiefly in the vine and other abundant bearers, because their nature has been expended on the crop. It also happens with cereals and annuals in general, indeed it is what makes them annual: the roots wither away as soon as the crop is matured, and the withering here is due to the

[b] *Cf. HP* 4 13. 2: " Short-lived too are a few varieties of the vine, especially the ones bearing abundant fruit . . ."

[c] *Cf. HP* 4 13. 1: " Broadly speaking, wild trees live longer than cultivated, both as a class and compared to their cultivated counterparts, as wild olive than olive, wild pear than pear and caprifig than fig, since they are stronger, of closer texture, and less generative of pericarpia."

[d] *Cf.* Aristotle, *On the Generation of Animals*, iii. 1 (750 a 20–29): " That in the prolific the food is diverted to the semen (seed) is evident from what happens. So most trees, after bearing an excessively large crop, wither after the yield, when no food is left for their bodies, and annuals appear to be affected in the same way, as legumes, grain and the rest. For they use up all their food on the seed, since this class of plant is many-seeded. And many hens, after being excessively prolific and laying up to two eggs a day, have died after such a display of fertility."

τὰς αὐτὰς ἀνάγκας, ἐκδιδομένης πάσης τῆς
10 φύσεως. τὰ δὲ δένδρα παραχρῆμα μὲν ἢ οὐ πάσχει τοῦτο, ἢ οὐ φανερὰ γίνεται, περιισταμένης δὲ τῆς ὥρας ἐξεδήλωσεν· ἐὰν δέ τι καὶ ἐπιγένηται τοιοῦτον ἐκ τοῦ ἀέρος ὥστε πιέσαι καὶ κακῶσαι, καὶ θᾶττον.

11. 3 οὐ μόνον[1] οὕτω τὰ ὑπερκαρπήσαντα· ἀλλὰ καὶ πολυκαρπήσαντα πονεῖ (πολλάκις δὲ[2] καὶ φθείρεται) κενωθέντα· ⟨κενωθὲν⟩[3] γὰρ ἅπαν ἀσθενές. ἐν ἀσθενεῖ δὲ καὶ τὸ μὴ ἰσχυρόν· διὰ τοῦτο καὶ τὰ
5 γεωργούμενα τῶν ἀγεωργήτων θᾶττον γηράσκει, καὶ τὰ βελτίω τῶν χειρόνων, καὶ τὰ ἥμερα δ' ὅλως τῶν ἀγρίων. καίτοι τάχ' ἂν δόξειεν ἄτοπον, εἰ τὰ μᾶλλον τυγχάνοντα θεραπείας· ἀλλ' ἡ θεραπεία πρὸς καρπογονίαν, οὐ πρὸς ἰσχύν, αὕτη δ' ἀναιρεῖ,
10 τὸ δὲ μακρόβιον ἐν τῷ ἰσχύειν, ἰσχυρὸν δὲ τὸ πυκνὸν καὶ στέριφον. ὅθεν καὶ Θάσιοι, τὰς γεωργίας ἀπομισθοῦντες, οὐ φροντίζουσι τῶν ἄλλων ἐτῶν, ἀλλὰ καὶ βούλονται κακουργεῖν, ὑπὲρ δὲ τοῦ τελευταίου συγγράφονται πρὸς τὴν αὐτῶν

11. 4 κατάληψιν.[4] συμβαίνει δὲ τοῦτο καὶ ἐπὶ τῶν ζῴων· καὶ γὰρ τὰ ὅλως εὔτοκα βραχύβια, καὶ τὰ ὑπερτο-

[1] μόνον ⟨δὲ⟩ Schneider.
[2] δὲ Wimmer: ἀεὶ U N; ἢ HP.
[3] κενωθέντα· ⟨κενωθὲν⟩ ego: κενωθέντα U; κενωθὲν u (-νο- N) HP.
[4] κατάληψιν U: κατάλειψιν Schneider.

---

[a] The necessary cause is an efficient or mechanical cause as opposed to a final cause, the final cause being aimed at a good.

same necessary causes,[a] the whole nature of the plant being given out. But in the trees this drying out does not occur at once or else does not at once become evident, but is manifested when the season returns; and if weather ensues that holds the tree back and subjects it to hardship the result is evident even sooner.

11. 3 Not only do trees that have borne to excess fare thus, but even when they have borne a large crop trees suffer (and often perish) from depletion, for everything depleted is weak. Lack of strength too comes under this head of weakness. For this reason trees under tendance age sooner than the untended, the better sort than the inferior, and indeed cultivated trees in general than wild.[b] Yet it might perhaps seem strange that trees receiving more care should age sooner. But the care is directed to fruit production, not to strength; and fruit production kills the tree, whereas longevity lies in strength, and a close-textured and barren tree is strong. So when the Thasians let out their orchards for cultivation they are unconcerned about any year of the lease but the last, and even welcome bad husbandry, but for the last year they stipulate that the lessor shall recover the very trees that he is renting out. 11. 4 This also happens in animals: not only are ready bearers short-

[a] *Cf.* for instance Aristotle, *On the Generation of Animals*, iv. 8 (776 b 32–33): ". . . for both the causes, . . . both for the sake of what is best and out of necessity . . ."

[b] *Cf. HP* 4 13. 1: "Now woodcutters say that there is practically speaking no distinction between the long-lived and short-lived in wild trees, all of them being long-lived and none short-lived, and perhaps they are so far right, for all these trees exceed by far the life of the rest."

κήσαντα πολλάκις ἀπόλλυται (καὶ μάλιστα εἴ τι ἀειδές, ὡς[1] τῶν γε συνανθρωπευομένων αἱ ὄρνιθες).
5 αὕτη τε δὴ βραχυβιότητος αἰτία, καὶ ἡ ἐναντία δῆλον ὅτι μακροβιότητος (ὥσπερ εἴρηται).

καὶ ὅσα δι' ἀσθένειαν εὔφθαρτα, μὴ πάντως ὄντα πολύκαρπα, ἀλλ' ἔνια καὶ ἄκαρπα, καθάπερ ἐν τοῖς ἐπετείοις (ὡς[2] ἡ σικύα,[3] περὶ ἧς καὶ
10 πρότερον ἐλέχθη)· ταῦτα γὰρ οὐδ' εἰς τελείωσιν ἀφικνεῖσθαι δύνανται τῆς φύσεως. ἔνια δὲ καὶ τῶν δένδρων εὔφθαρτα διὰ μανότητα καὶ ἀσθένειαν, οὐκ ὄντα πολύκαρπα (καθάπερ ἡ δάφνη· ταύτης γάρ ἐστιν ἣ[4] καὶ ἄκαρπος, ὥσπερ[5] ἡ
11. 5 βρυοφόρος). πλὴν οὐχ ἁπλῶς ταύτῃ γε, ἀλλὰ κατὰ μέρος καὶ ἡ φθορὰ καὶ τὸ γῆρας, ἀεὶ[6] γὰρ τὸ μάλιστα παχυνόμενον (ὡς εἰπεῖν) σήπεται καὶ φθείρεται. παραβλαστήσεις δὲ ἔχει πολλάς, ὡς ὅμοιον εἶναι
5 τρόπον τινὰ τοῖς ἀφαυαινομένοις ἀκρεμόσιν ἐπὶ τῶν δένδρων· ἀλλ' ἡ διαφορὰ τοῖς μὲν ὅτι τὸ κυριώτατον, τοῖς δὲ τῶν ἀπηρτημένων τι μορίων.[7] ἀλλ' ὅτι ἡ παραβλάστησις[8] ἀπὸ τῶν αὐτῶν, διὰ τοῦτο καὶ ταὐτὸ δοκεῖ καὶ τὸ δένδρον εἶναι, περὶ
10 οὗ καὶ διηπορήθη πρότερον.

[1] εἴ τι ἀειδές, ὡς ego (Schneider deletes; εἴ τι ἀεὶ ὡς Wimmer): ἐπι ἀεὶδος U; ἐπὶ ἀειδῶς N HP.
[2] [ὡς] Schneider.
[3] σικύα Gaza, Scaliger: σκιᾶ U.
[4] ἣ Wimmer: ἡ U.
[5] ὥσπερ U: ὅλως Wimmer.
[6] ἀεὶ Itali: εἰ U.
[7] τι μορίων u (τιμοριῶν U): τιμωριῶν (τι μωριῶν N) HP.
[8] παραβλάστησις u: παραβλάστησεις U.

lived, but those also that have borne to excess often perish, especially any animal of undifferentiated shape, as among domestic animals with the hen.[a]

Here then is a cause of brevity of life, and the opposite is evidently a cause of longevity, as we said.[b]

Further there are the plants that perish readily from weakness, although not necessarily abundant bearers, some in fact being non-bearers, as among annuals (such as the gourd mentioned earlier);[c] for these are not even able to bring their nature to its full development. A few trees too perish readily because of open texture and weakness, although they are not abundant bearers, as the bay; for there is in fact a non-bearing bay, as the one that produces catkins.[d] Except that the bay at least does not merely age and perish but does both piecemeal, for whatever part is stoutest at the moment, so to speak, decomposes and perishes. But the bay has many side-shoots, so that in a way the case is like that of branches withering on a tree. But there is a difference: in the clump of bay it is the most important stem that perishes, but among the others it is some peripheral part. Nevertheless since the side-shoots in the bay come from the same roots it is considered that the tree too is the same. (This difficulty has been explored before.)[e]

11. 5

[a] *Cf. CP* 1 22. 1 with note *f*; Aristotle, *On the Generation of Animals*, iii. 1 (750 a 27–29) [cited in note *d* on *CP* 2 11. 2] and iii. 1 (749 b 30–32): "And the baser breeds are more prolific than the better, since their bodies are more fluid and bulky . . ."

[b] *CP* 2 11. 1, 3.

[c] *CP* 2 8. 4 (where the plant is called *kolokýnthē*).

[d] *Cf. HP* 3 7. 3; 3 11. 4.

[e] *HP* 4 13. 3–4.

11. 6 ἀρχαὶ δὲ φθορᾶς τοῖς ἀσθενέσιν καὶ ἀπὸ πληγῆς, καὶ ἀπὸ πνευμάτων μεγέθους, καὶ ἀφ' ἑτέρων τινῶν τοιούτων· ἐκ πολλῶν γὰρ εὐκίνητον τὸ ἀσθενές. ἐὰν δὲ πρὸς τούτῳ καὶ πολύκαρπον ᾖ, 5 καθάπερ ἡ ῥόα καὶ μηλέα ἡ ἐαρινή, καὶ μᾶλλον· ἐξ ἀμφοτέρων γάρ, τάχα δ' ἐκ πολλῶν, αἱ ἀρχαί· καὶ αἵ γε μηλέαι καὶ σκωληκοῦνται τάχιστα, θᾶττον δὲ αἱ γλυκεῖαι. καὶ τοῦτο πάσχουσιν, καὶ ὅλως γηράσκουσιν, οὐχ αἱ μηλέαι μόνον,[1] ἀλλὰ 10 καὶ αἱ ῥόαι, τὸ γὰρ ποτιμώτατον ἀφαιρεῖται τῆς φύσεως· ἅμα δὲ καὶ αἱ μὲν ὥσπερ ἄγριαί τινες, αἱ

11. 7 δ' ἥμεροι τυγχάνουσιν. ἀνὰ λόγον[2] δὲ καὶ αἱ ἀπύρηνοι τῶν πυρηνωδῶν, καὶ αἱ μαλακοπύρηνοι τῶν σκληρῶν, καὶ ἐπὶ μυρρίνων δὲ καὶ τῶν ἄλλων ὡσαύτως· καὶ γὰρ πλεῖον[3] ἀφαιρεῖται τῆς φύσεως, 5 καὶ τὸ ὅλον ἀσθενέστερα καὶ μανότερα ταῦτα· διὸ καὶ πρωϊκαρπότερα, θᾶττον γὰρ καὶ μᾶλλον ὑπακούει τῷ ἀέρι τὸ ἀσθενές.

ὡς δὲ Δημόκριτος αἰτιᾶται, τὰ εὐθέα τῶν σκολιῶν βραχυβιώτερα καὶ πρωϊβλαστότερα διὰ τὰς 10 αὐτὰς ἀνάγκας εἶναι—τοῖς μὲν γὰρ ταχὺ διαπέμ-

§ 7. 8. Democritus, Frag. A 162, Diels-Kranz, *Die Fragmente der Vorsokratiker*, vol. ii[8], p. 128.

[1] μόνον N HP: -αι U.
[2] ἀνὰ λόγον Wimmer: ἀνάλογον U.
[3] πλεῖον U[r] N HP: -ων U[ar].

---

[a] *Cf. HP* 4 13. 2: ". . . trees of the following type are short-lived, as the pomegranate, fig, apple, and of the last the spring apple is more so than other apples, and the sweet apple than the sour, just as among pomegranates the stoneless kind are more short-lived."

[b] And therefore shorter-lived: *cf. CP* 2 11. 3.

When trees perish from weakness the process can begin with blows or great winds or other similar occurrences, for weakness is susceptible to many influences. And if the tree is not only weak but a good bearer as well, like the pomegranate and spring apple,[a] it is still more susceptible, since the source may lie in both characters, perhaps in many. Apple-trees moreover also breed worms soonest, the sweet ones sooner than the others. And this early worminess, and early old age in general, occurs not only in the sweet apples but in the sweet pomegranates as well, since by the production of the sweet fruit their nature is deprived of its most purified part; then too the inferior kinds are (as it were) of a wild character, the sweet kinds of a cultivated one.[b] The same holds for the pomegranates that have no stones[c] compared to the ones that have them, and for the soft-stoned ones compared to the hard-stoned, and so too with myrtles and the rest; for in the first kinds more of the nature of the tree is removed, and these trees are in general the weaker and more open in texture. This moreover is why they fruit earlier, for the weak tree has a quicker and greater response to the weather.

11. 6

11. 7

The explanation given by Democritus—that the shorter life and earlier sprouting of straight as compared with crooked trees are due to the same necessary causes,[d] since in the straight trees the food is

[c] *Cf. HP* 4 13. 2, cited in note *a* (p. 292).

[d] *Cf.* Aristotle, *On the Generation of Animals*, v. 8 (789 b 2–5): "But Democritus neglects to speak of the final cause, and traces to necessity all the means that nature employs. It is true that the means are necessary, but they nevertheless have an end, and the end is what in each case is better."

πεσθαι τὴν τροφήν, ἀφ' ἧς ἡ βλάστησις καὶ οἱ καρποί, τοῖς δὲ βραδέως, διὰ τὸ μὴ εὔρουν εἶναι τὸ ὑπὲρ γῆς, ἀλλ' αὐτὰς τὰς ῥίζας ἀπολαύειν, καὶ γὰρ μακρόρριζα ταῦτα εἶναι καὶ παχύρριζα—

11. 8 δόξειεν ἂν οὐ καλῶς λέγειν. καὶ γὰρ τὰς ῥίζας ἀσθενεῖς φησιν εἶναι τῶν εὐθέων, ἐξ ὧν ἀμφοτέρων θᾶττον γίνεσθαι τὴν φθοράν, ταχὺ γὰρ ἐκ τοῦ ἄνω διιέναι καὶ τὸ ψῦχος καὶ τὴν ἀλέαν ἐπὶ τὰς ῥίζας διὰ 5 τὴν εὐθυπορίαν, ἀσθενεῖς δ' οὔσας, οὐχ ὑπομένειν· ὅλως δὲ τὰ πολλὰ τῶν τοιούτων κάτωθεν ἄρχεσθαι γηράσκειν διὰ τὴν ἀσθένειαν τῶν ῥιζῶν. ἔτι δὲ τὰ ὑπὲρ γῆς, διὰ τὴν λεπτότητα καμπτόμενα ὑπὸ τῶν πνευμάτων, κινεῖν τὰς ῥίζας, τούτου δὲ 10 συμβαίνοντος ἀπορρήγνυσθαι καὶ πηροῦσθαι, καὶ ἀπὸ τούτων τῷ ὅλῳ δένδρῳ γίγνεσθαι [1] τὴν φθοράν.

ἃ μὲν οὖν λέγει ταῦτά ἐστιν.

11. 9 οὐ μὴν δόξειέ γ' ἂν (ὥσπερ εἴρηται) καλῶς λέγειν. οὔτε γὰρ τὸ τῶν ῥιζῶν ἐστιν ἀληθές (ὥσπερ καὶ πρότερον ἐλέχθη, τὸ [2] τῶν μακροβίων εἶναι μακρὰς καὶ παχείας), οὐ γὰρ μακρόβιον 5 οὔθ' ἡ συκῆ οὔθ' ἕτερα τῶν μακρορρίζων καὶ παχυρρίζων· οὔτε τὰ εὐθέα καὶ τὰ ὀρθὰ βραχύβια, οἷον ἐλάτη φοῖνιξ κυπάριττος. οὐδ' εὐαξῆ δὲ ταῦτα, οὐδὲ πρωΐκαρπα [οὐδὲ πρωϊκαρπῆ]· [3]

[1] γίγνεσθαι Wimmer (ἐπιγίνεσθαι Schneider): πήγνυσθαι U.
[2] τὸ u N HP^{c}: τῷ U P^{ac}(?).
[3] [οὐδὲ πρωϊκαρπῆ] Scaliger.

---

[a] (1) Straightness of the upper parts and (2) weakness of the roots.

quickly distributed (and from the food comes the sprouting and the fruit), whereas in the crooked the distribution is slow because the part of the tree above ground offers no easy channel for the food, and the roots instead consume it by themselves (crooked trees having roots that are long and thick)—would not appear to be well taken. So he also asserts that in straight trees the roots are weak, and both of these causes[a] have the result that straight trees perish sooner, since both cold and heat pass rapidly from the part above ground to the roots owing to the straightness of the passage, and the roots on their part, being weak, offer no resistance; indeed most straight trees (he says) begin to age from below because of the weakness of their roots. Again the parts above ground (he says) are so thin that they are bent by the winds and move the roots, and when this occurs the roots are broken off and crippled, and death, starting from these, spreads to the whole tree. 11. 8

These then are Democritus' arguments.

Nevertheless it would not appear (as we said)[b] that he is right. For (1) neither is his point about the roots true (the one mentioned before,[c] that they are long and thick in the short-lived), since neither the fig nor others with long and thick roots are long-lived; (2) nor are trees that are straight and erect short-lived, for example the silver-fir, date-palm and cypress. These again are no rapid growers[d] either, 11. 9

[b] *CP* 2 11. 7.
[c] *CP* 2 11. 7.
[d] The date-palm and cypress are slow growers (*CP* 1 8. 4), the silver-fir a rapid one (*HP* 3 6. 1).

καίτοι καὶ ταῦτα ἐχρῆν εὐθύς, τοιούτων γε τῶν
10 πόρων ὄντων, καὶ τῶν ῥιζῶν μὴ μακρῶν· ὅσα
γὰρ ἀπὸ τῆς αὐτῆς αἰτίας, ἅπαντα δεῖ[1] συνακολουθεῖν τοῖς αὐτοῖς.

11. 10 ἀλλὰ μή ποτε οὐ τοῦτο ᾖ τὸ αἴτιον, ἀλλὰ τὰ
προειρημένα πρότερον, ἐν αἷς[2] καὶ τὸ δυσφυὲς
ὅλως καὶ μικρόκαρπον καὶ ὀλιγόκαρπον, καὶ ὅλως
τὸ ἰσχυρόν ἐστιν· ἡ γὰρ πυκνότης καὶ ἡ ξηρότης
5 καὶ ἡ στερεότης καὶ ἡ λιπαρότης, ἐν οἷς ὑπάρχει,
καὶ μακροβιότητος, καὶ πάντων τῶν τοιούτων
αἴτια, τὰ δ' ἐναντία τῶν ἐναντίων.

ὅσα δὲ πολυκαρποῦντα μὴ βραχύβια, μηδὲ ταχὺ
γηράσκει, καθάπερ ἄπιος ἀμυγδαλῆ δρῦς, καὶ
10 πρεσβύτερα γιγνόμενα καρπιμώτερα, καθάπερ
ἐλέχθη· παραιρουμένης γὰρ τῆς ἰσχύος, παραιρεῖται
τὸ πλῆθος τῆς τροφῆς, ὥστε[3] ῥᾴδιον[4] καταπέττειν

11. 11 τὴν λοιπήν. ἅμα δ' ἴσως συμβάλλεται καὶ τὸ μὴ
ἐνδελεχές, ἢ πᾶσιν ἤ τισιν· ἡ γὰρ ἄπιος καὶ ἔτι
μᾶλλον ἡ ἀμυγδαλῆ προφαίνουσιν πολύν, ⟨οὐ μὴν⟩[5]
ἐκτρέφουσίν γε[6] πολλάκις τοῦτον· ἔστιν δὲ ἐν τῇ
5 τελειώσει ὁ πόνος καὶ ἡ ἀπέρασις. ἡ δὲ συκάμινος
ἐλαφρόν τινα καὶ ὑδατώδη καὶ μικρὸν ὡς πρὸς τὸ
μέγεθος ἔχει τοῦ δένδρου τὸν καρπόν.

[1] δεῖ u: δὴ U.
[2] αἷς (αἶς U) u: οἷς Gaza, Heinsius.
[3] ὥστε Gaza (*atque ita*), Basle ed. of 1541: ὥσπερ U.
[4] ῥᾴδιον U: ῥᾷον Hindenlang.
[5] ⟨οὔ μην⟩ Basle ed. of 1541 (*sed* Gaza).
[6] γε Scaliger (Basle ed. of 1541 omits): γὰρ U.

---

[a] The silver-fir fruits late (*HP* 3 4. 5). Theophrastus does not tell us specifically about the fruiting of the date-palm or the

or early fruiters;[a] yet these characters should follow at once, since the food passages are of the sort to bring this about and the roots are not long, for all consequences of the same cause should appear in the same group of trees.

11.10 But perhaps the cause of longevity is not this, but the characters mentioned earlier,[b] which are the causes both of slow growth in general with smallness and scantiness of fruit, and to sum up, of strength: it is closeness of texture, dryness and oiliness (where it is present) that are responsible not only for longevity but for all such characters as these, and their opposites are causes of the opposites.

(As for abundant bearers that are not short-lived or quick to age, such as the pear, almond and oak, these in fact get to be more fruitful as they grow older,[c] as we said:[d] with the reduction of their strength goes a reduction in the amount of food taken, so that it is easy to concoct what remains.)[e] 11.11 Then too in all or some of these the failure to finish what they have begun perhaps contributes to their longevity: so the pear and still more the almond promise an abundant crop but often fail to rear it; and the hardship and depletion attend on the perfecting of the fruit. The mulberry[f] on the other hand has fruit that is light, watery and small for the size of the tree.

cypress. But he says that evergreens are generally late-fruiting (*CP* 1 10. 7).

[b] *CP* 1 8. 2, 4.

[c] *Cf. CP* 5 9. 2.

[d] *CP* 1 13. 8.

[e] *Cf. CP* 2 11. 1–4; *CP* 1 17. 9–10.

[f] A long-lived tree: *cf.* Pliny, *N. H.* 16. 119: morus tardissime senescit.

ἀλλὰ γὰρ περὶ μὲν μακροβιότητος ἐν τούτοις ἔστωσαν αἱ αἰτίαι· περὶ δὲ πολυκαρπίας τῶν 10 δένδρων εἴρηται πρότερον, ὅτι τὰ θερμὰ καὶ μανὰ καὶ ὑγρά.

12. 1 τῶν δὲ σπερμάτων ὡς ἁπλῶς εἰπεῖν τὰ ἐλάττω πολυχούστερα· καὶ γὰρ τὰ ἐλάχιστα μάλιστα, καθάπερ κέγχρος σήσαμον ἐρύσιμον μήκων κύμινον. αἴτιον δὲ δοκεῖ, καθόλου μὲν καὶ κοινῶς 5 εἰπεῖν, ὅτι τὰ ἐλάττω ῥᾷον ἐπιτελεῖν, ἐν δὲ τῷ ῥᾳδίῳ τὸ πλῆθος· ὡς δ'[1] ἐγγυτέρως, ὅτι πάντα τὰ τοιαῦτα εὐβλαστότερα καὶ θᾶττον ὑπακούει τῷ ἀέρι, σημεῖον δὲ καὶ ⟨ἡ⟩[2] ὀλιγοχρονιότης τῆς τελειώσεως, ἐπεὶ καὶ τὰ σπέρματα διὰ τοῦθ' 10 ὡς ἁπλῶς εἰπεῖν πολυκαρπότερα τῶν δένδρων, καὶ

[1] ὡς δὲ Schneider (*vel ut* Gaza): ὥστ' U.
[2] ⟨ἡ⟩ u.

---

[a] *CP* 1 15. 4; 1 16. 7.
[b] " Seed-crops " renders *spérmata*, literally " seeds," as the Greeks called all plants especially valued for their " seeds " as distinguished from fruit or root or leaves or juice or flower.

Theophrastus distinguishes four classes of plants: trees, shrubs, undershrubs (literally " fire-wood ") and herbaceous plants. Coronaries (plants used in making crowns) are valued for their flowers or fragrance or leaves; they come partly under undershrubs and partly under herbaceous plants. The rest of the herbaceous plants are divided into vegetables (valued for the culinary use of leaf, root, stalk, bulb) and *spérmata* (" seed-crops " or grains); these in turn are divided into legumes (*chedropá*, " gathered by the hand," as opposed to gathered by the sickle), cereals, and " summer seeds," this

As for longevity, then, we shall take the causes to lie in these features; as for abundant fruiting, it was said earlier [a] that hot, open-textured and fluid trees are the heavy bearers.

Fruitfulness: Seed-Crops [b]

Of seed-crops the smaller seeds have (broadly speaking) the greater yield; indeed the smallest have the greatest yield of all, as millet, sesame, hedge-mustard, poppy, cummin.[c] The reason, put very generally and loosely, is considered to be that it is easier to finish things that are smaller, and abundance results from ease of production. To put it more particularly, the reason is that such smaller seeds all sprout more readily and respond more quickly to the air. Proof of this is the shortness of the time taken to mature; indeed on this account seed-crops not only yield more abundantly than trees, but also among seed-crops themselves legumes [d] yield more 12. 1

last a term imposed by Theophrastus for want of a current term, because unlike other seed-crops they were exclusively sown in the April planting for harvesting in summer.

[c] *Cf. HP* 7 3. 3 (of vegetables): "All yield a large crop and have many shoots, but cummin yields the largest." (At *HP* 8 6. 1 cummin is one of the " summer seeds.")

[d] *Cf. HP* 8 3. 4–5: " In general legumes are more productive and prolific (*sc.* than cereals), but the summer seeds millet and sesame are still more prolific than these, and in the class of legumes lentil (*sc.* where the seed is smallest) is most so. And broadly speaking the plants with the smaller seeds are more prolific (one might say), as among vegetables cummin, though all of these have many seeds."

αὐτῶν τούτων τὰ χεδροπὰ τῶν σιτωδῶν, θάττων[1] γὰρ ἡ τελέωσις καὶ ἡ ἅδρυνσις.

12. 2 ἔτι δ' οἱ καυλοὶ τῶν πλείστων ἰσχυροὶ καὶ οὐ ξυλώδεις καὶ οὐ μονοφυεῖς, ἀλλὰ πλείους καὶ ἀκρεμονικοί· πολλῶν δὲ καὶ ἰσχυρῶν ὄντων, εὔλογον ἤδη καὶ τὸ τοῦ καρποῦ πλῆθος, ἄλλως τε 5 κἂν αἱ ῥίζαι μὴ ἀντισπῶσιν, ὥσπερ τῶν ναρθηκωδῶν καὶ κεφαλορρίζων, ἀλλ' αἰεὶ διαδιδῶσιν[2] ὧν[3] ἂν λαμβάνωσιν· ὅπερ καὶ ἐπὶ τῶν χεδροπῶν ἐστιν καὶ ἐπὶ τῶν ἄλλων τῶν πλείστων· μονόρριζα γὰρ καὶ οὐ παχύρριζα, ἀλλ' εἰς τἄνω φερόμενα τῇ 10 ὁρμῇ, καὶ μάλιστα (ὡς εἰπεῖν) τὸ κύμινον, μικρὰ

12. 3 γάρ τις ἡ ῥίζα τούτου πάμπαν. (ὅσα δὲ ἰσχυρόρριζα τούτων—ἔνια γὰρ τοιαῦτα, καθάπερ ὁ κέγχρος καὶ ὁ μέλινος[4]—, ταῦτα δὲ κατὰ λόγον ἀποδίδωσι τοὺς καυλούς, ἀπ' ἰσχυρῶν ἰσχυροὺς 5 καὶ πολυσχιδεῖς.) ὥσθ', ὅταν ἀπὸ μικρᾶς[5] ἀρχῆς πολλοὶ γένωνται, κατὰ λόγον ἤδη καὶ τὸν καρπὸν εἶναι πολύν, ὥσπερ καὶ ἐπὶ τῶν

[1] θάττων Schneider: θᾶττον U.
[2] διαδιδῶσιν HP: διαδίδωσιν U N.
[3] ὧν N HP (ὃ Schneider): ὂν U.
[4] μέλινος Schneider: μειλινος U (μί- u; μή- N HP).
[5] μικρᾶς U: μιᾶς Schneider.

---

[a] *Cf.* Aristotle, *On the Generation of Animals*, i. 18 (726 a 9–11): "Similarly (*sc.* to the case with animals) some (*sc.* plants) are prolific in their yield, producing many seeds, because of power, others because of lack of power."

[b] *Cf. HP* 8 2. 3: "There is as it were an opposition between the two groups: legumes have single roots but many branch-

abundantly than cereals, since they mature and get sturdy sooner.[a]

12. 2 Further, the stalks of most of these prolific crops are strong and not woody, and are not single but grow several to a plant and have many branchings;[b] and in view of their number and strength, the abundance of the crop becomes reasonable, especially if the roots do not pull the food the other way (as they do in the fennel-like and bulbous plants),[c] but constantly portion out some of what they take in. Such in fact is the case both with legumes and with most of the rest; for they have a single root and this not thick, the impetus of growth tending upward instead, above all (one might say) cummin, since its root is quite small.

12. 3 (On the other hand all of these plants with strong roots—there being a few such, as millet and Italian millet—produce from their strong roots correspondingly strong and many-branched stalks.)[d] And so, with multiple stalks coming from a small base, it becomes reasonable that the crop too should be

ings above ground from their stalks (except for bean); whereas cereals have many roots and send up many shoots, but these do not branch, except for a kind of wheat of this description . . ."

[c] The larger (and stronger) the root, the greater its pull. *Cf. HP* 1 6. 10 (on whether the bulbs of bulbous plants are roots): " Indeed it is evident that the nature of all such plants tends rather downward; for the stalks and the upper parts in general are short and weak, whereas the lower parts are large and numerous and strong . . . Also in the fennel-like plants the roots are large and fleshy."

[d] *Cf. CP* 4 15. 1, *HP* 8 9. 3: " Of the crops planted in the summer-seed time sesame is held to be worst for the land and to exhaust it most, and yet millet has more numerous and thicker stems and more numerous roots."

πυρῶν καὶ κριθῶν ὅταν ἐξ ἑνὸς πλείονες [1] ἐκβλαστῶσιν κάλαμοι, πλείους γὰρ οἱ στάχυες.

12. 4 ταὐτὸ δὲ τοῦτο καὶ ἐπὶ τῶν λαχανηρῶν ἐστιν, ὅτι πολύσπερμα καὶ ἰσχυρόρριζα τυγχάνει· τὰ μὲν γὰρ πλείστους καυλοὺς ἀφιᾶσι, τὰ δ' ἀπὸ τοῦ ἑνὸς ἀκρεμόνας [2] πλείους, ἅπαντα γὰρ ἀποδενδροῦται 5 τῇ ὄψει διακαυλήσαντα,[3] πανταχόθεν δὲ καὶ ἐκ πολλῶν πολὺς ὁ καρπός.

ὅταν οὖν καὶ τὰ [4] τοῦ ἀέρος εὐμενῆ τυγχάνῃ, βραχὺς [5] γὰρ ἤδη καὶ ἀχείμαντος τοῖς ὀψισπόροις, καὶ ἡ φύσις εἰς τὰ ἄνω μᾶλλον ὁρμᾷ, καὶ μὴ 10 ἀντισπᾷ ⟨τὰ⟩ [6] τῶν ῥιζῶν, ἀλλὰ συνεργῇ, καὶ προσέτι μανὰ [7] σπείρηται, κατὰ λόγον ἤδη ταῦτα πολυκαρπότερα γίνεται τῶν ἄλλων.

12. 5 ὁ δὴ [8] κύαμος καὶ εἴ τι τῶν χεδροπῶν ἄλλο [9] μὴ πολύκαρπον, δι' ἀσθένειαν ὀλιγόκαρπον, διὸ δὴ

[1] πλείονες U^c from -ος.
[2] ἀκρέμονας u: ἀκρεμόνες U.
[3] διακαυλήσαντα Schneider: -ίσαντα U.
[4] καὶ τὰ u: κατὰ (κατα U) N HP.
[5] βραχὺς U (*sc.* ὁ χειμών): πραῢς Wimmer.
[6] ⟨τὰ⟩ Schneider.
[7] μανὰ Wimmer: -ῆ U (-ὴ u).
[8] δὴ U HP: δὲ N.
[9] ἄλλο HP: ἀλλ' ὃ U N.

[a] For differences in the number of haulms of wheat see *CP* 4 11. 3–4 and *HP* 8 4. 3: "And one kind of wheat has a single haulm, another many, and there are degrees of this."

[b] *Cf. HP* 1 3. 4: "For of undershrubs and vegetables some have but one stem and grow up possessing (as it were) the nature of a tree, for example cabbage and rue, and for this reason some people call plants of this description 'tree-vegetables;' and all or most plants of the vegetable kind when

multiplied, as it is in wheat and barley when several haulms come out of a single seed,[a] there being then more ears.

12. 4 The same combination is also found in vegetables, because they are plants with many seeds and strong roots: for some send out very many stalks, whereas others send out a number of branches from their single stalk, since all of them become tree-like in appearance when they have run to stalk; [b] and when the product comes from all sides and many branches it is plentiful.

So when the air too happens to be clement (since for the plants that are sown late [c] the part of winter that remains is short and without storms), and when the tendency of the plant's nature is to grow upward rather than down, and the roots do not pull the other way but do their share, and when further the plants are sown thin,[d] it becomes reasonable that they bear (as they do) a more plentiful crop than the rest.

12. 5 The bean (and any other legume producing no large crop) is a small producer through weakness.

they have been long in the ground acquire ' branches ' (as it were) and the whole plant takes on the appearance of a tree, except that it is shorter-lived."

[c] There are three sowings of vegetables, the late one in Gamelion (January) after the winter solstice; one sowing of summer seeds, in Munychion (April); and two sowings of cereals and legumes, the later one after the winter solstice. So the late sowing for vegetables, cereals and legumes—that is, for all the seed-crops Theophrastus has been discussing except summer seeds—would be in January. *Cf. HP* 7 1. 1–2, 8 1. 2–3. The terms " early " and " late " are used with reference to the Attic year, which began with the summer solstice.

[d] This (like sowing in January: *cf. CP* 2 12. 5) shows that the bean is not included (*cf. CP* 4 14. 2).

καὶ πρωϊσπορεῖται, προλαμβανόντων τοὺς χειμῶνας εἰς τὴν ῥίζωσιν· ἔτι[1] δὲ ἐπίκηρον ἐν πολλοῖς καὶ
5 πολλάκις, ὥστ' ἐὰν[2] καὶ προφάνῃ,[3] μὴ δύνασθαι τελεοῦν. ἡ δὲ ἀσθένεια καὶ τῇ αἰσθήσει φανερά· μανὸν γὰρ καὶ κενὸν[4] καὶ οὐ πολύρριζον, ὥστε εὐδίοδον εἶναι τῷ κακοποιοῦντι.

12. 6 τοῦ μὲν οὖν πολυκαρπεῖν ἐν τούτοις αἱ αἰτίαι καὶ τοῖς ἐπετείοις καὶ τοῖς χρονιωτέροις. ἐπεὶ καὶ αἱ διακαθάρσεις τῶν δένδρων καὶ αἱ κατακοπαὶ ποιοῦσι πολυκαρπεῖν, ὅτι τὰ μὲν αὐτὰ[5] κωλύει,
5 τὰ δ' αὐτὰ λαμβάνει τὰς τροφάς· τούτων οὖν ἀφαιρεθέντων εἰς τὸν καρπὸν ἡ ὁρμή. καὶ διὰ τοῦτο τὰς Ἡρακλεωτικὰς καρύας θαμνώδεις ποιοῦσι κατακόπτοντες, ἐξαναλίσκουσιν γὰρ ἀποδενδρούμεναι πᾶσαν[6] τὴν τροφήν.

13. 1 αἱ δὲ μεταβολαὶ τῶν καρπῶν κατὰ τὰς χώρας, δῆλον ὅτι κατὰ τὸν ἀέρα καὶ τὰ ἐδάφη γίνονται, διὰ τούτων γὰρ καὶ ἐκ τούτων αἱ τροφαὶ πᾶσιν.

[1] ἔτι U: ἔστι Itali.
[2] ὥστ' ἐὰν ego: ὥστε ἂν U.
[3] προφάνῃ Schneider: προφανῆ U.
[4] καὶ κενὸν u: κεκαινόν U.
[5] αὐτὰ U: Schneider deletes; ἄλλα Wimmer.
[6] πᾶσαν Gaza (*totum*): πᾶσαι U.

---

[a] *Cf. HP* 8 1. 3 (of plants sown early, after the setting of the Pleiades): "Among legumes bean and bird's pease are sown earliest (one may say), for owing to their weakness they like to get rooted before the cold weather."

[b] *Cf. HP* 2 7. 2: "All trees require the clearing away of deadwood, for they improve with its removal, as if it were a foreign body that interferes with growth and feeding."

That is why it is sown early,[a] the farmers wishing it to get roots before the cold weather sets in. Again, in many matters and at many points in its development it is liable to injury, so that even if it promises a large crop it is unable to mature it. Its weakness is also evident on inspection, for the plant is open in texture, hollow, and has few roots, thus affording an easy passage to injurious influences.

12. 6 To conclude: the causes of abundant bearing, both in annuals and plants of longer duration, lie in the points mentioned. Indeed in trees not only pruning[b] but also cutting back of stems[c] makes for abundant bearing, because some of the parts removed are a hindrance to the bearing parts themselves and others themselves take their food, and so when these interferences are removed the impetus of the tree is to fruit production. And the prevention of this diversion is the reason why growers leave the filbert with the habit of a shrub when they prune away the stems,[d] since when filberts assume the habit of a tree they expend all their food on growth.

Mutation of Fruits and Crops: Natural Mutation

13. 1 When the crop changes with the country the mutation evidently follows the air and the soil, since all plants get their food through these and from these,

[c] *Cf. HP* 3 15. 1 (of the filbert, treated as a wild tree): "It always becomes more fruitful when the canes are cut off."

[d] *Cf. HP* 1 3. 3: "When the myrtle is not pruned it takes on the character of a shrub and so too does the filbert. The latter is considered to bear better and more plentiful fruit if one leaves behind a number of its canes on the assumption that the nature of the plant is that of a shrub."

ἰσχυρὸν δ' ἡ τροφὴ πρὸς ὁμοίωσιν, εἴ γε καὶ ἐν τοῖς
5 ζῴοις οὕτως ἡ [ἐν][1] τοῖς θήλεσιν ὁμοιότης
φαίνεται δὲ οὐ μόνον τὰ σπέρματα καὶ τὰ φυτὰ
καὶ τὰ δένδρα μεταβάλλειν, ἀλλὰ καὶ τὰ ζῷα, κα
τρόπον τινὰ μᾶλλον ταῦτα· καὶ γὰρ τὰς μορφὰς
ἐξομοιοῦνται κατὰ τοὺς τόπους, ἐν δὲ τοῖς καρποῖς

13. 2 οὐχ ὁμοίως τοῦτο ἐπίδηλον. οὐ μὴν ἀλλὰ κα
ἐνταῦθα γίνεται, καὶ μάλιστα ἐκδήλως κατὰ[2] τ
χρώματα καὶ τὰ μεγέθη καὶ τοὺς χυλούς· χρώματ
μέν, οἷον τὰ σπέρματα, λευκὰ γὰρ ἐκ μελάνων
5 καὶ μέλανα ἐκ λευκῶν μεταβάλλει· χυλοὺς δέ, κα
οἱ καρποί, τὸ δ' ἐκ[4] τῶν χρωμάτων ἢ οὐκ ἐμφανές
ἢ οὐχ ὁμοίως συμβαίνει, πλὴν ἐὰν ὅλον μεταστῇ
τὸ δένδρον, ὥστε ἐκ μέλανος γενέσθαι λευκόν
ὅπερ ἐνίοτε συμβαίνει τοῖς ἐκ σπέρματος φυομένοις

13. 3 κοινοῦ δὲ τοῦ πάθους ὄντος ὁμοίως ζῴων τε κα
φυτῶν, κοινήν τινα δεῖ καὶ τὴν αἰτίαν ζητεῖν
ἐπεὶ κἀκεῖνο ὅμοιον ἔν τε τοῖς σπέρμασιν καὶ ἐ
τοῖς ζῴοις, ὥστε μὴ εὐθύς, ἀλλὰ τριγονήσαντ

[1] [ἐν] Gaza, Schneider.
[2] κατὰ Schneider (*in* Gaza): καὶ U.
[3] μελάνων HP: μελανῶν U N.
[4] δ' ἐκ U: δὲ Schneider (after Gaza).

---

[a] *Cf.* Aristotle, *On the Generation of Animals*, ii. 4 (738 25–36): "The body comes from the female, the soul from th male . . . And for this reason where male and female o different kinds of animal unite . . . at first the offspring in th matter of resemblance shares in both kinds, as with mixture of fox and dog and of partridge and chicken; but as time goe on and new generations are produced, the offspring end b resembling the female in form, just as imported grains end b taking on the character of the country, for it is the country tha gives the grains their material and body."

[b] *Cf.* Aristotle, *On the History of Animals*, viii. 28 (606 a 13

and the food has a strong effect in producing similarity, seeing that in animals too it is through the food that similarity to the females comes about.[a] Not only grains, slips and trees are observed to change but also animals, and animals in a way even more, since assimilation to the regional character affects even their shapes,[b] whereas this is not so noticeable in crops. Still the change occurs in crops too, most noticeably in colour, size and flavour: in colour, as grains, from black to white [c] and from white to black, and in flavour, fruit too,[d] whereas a change in its colour is either not noticeable or not so frequent, except in the instances where the whole tree is changed, with the result that a white variety comes from a black, a thing that occasionally occurs in trees growing from seed.[e] 13. 2

Since animals as well as plants are affected, we must look for a cause that applies to both; indeed there is a further point of resemblance between grains and animals,[f] that the change does not occur at the start but only in the third generation,[g] both 13. 3

b 3): " In Syria the sheep have tails a cubit broad, and the goats have ears a span and a palm long and some reach the ground with them, and the cattle, like camels, have humps on their shoulders . . . The cause that is given for this is the food . . . But in some places the climate is also responsible . . .; " *On Length and Brevity of Life*, chap. v (466 b 16–28).

[c] *Cf. CP* 3 21. 3: a certain kind of ground makes barley whiter by concocting the food more thoroughly.

[d] So with the change of the pomegranate in Egypt: *cf. HP* 2 2. 7; *CP* 1 9. 2; 2 13. 4.

[e] Changes in the colour of a tree are mentioned in *HP* 2 2. 4; 2 2. 6; 2 3. 1, but none appears to be the change referred to here, since no change of country appears to be involved. *Cf.* perhaps Aristotle, *On the Generation of Animals*, v. 6 (786 a 5).

[f] *Cf. CP* 2 13. 1, note *a*.

[g] *Cf. CP* 1 9. 3 with note *a*.

5 μεταβάλλειν, ἅπαντα γὰρ κατὰ μικρὸν ἐξαλλοιοῦται καὶ μεθίσταται. τὸ δ' ἐπὶ τῶν δένδρων ἀνάπαλιν εὐλόγως· ἀσθενέστατα γὰρ ἐν ἀρχῇ, καὶ μάλισθ' ὅταν ἐκ σπέρματος.

13. 4 ἀτοπώτατον δέ, καὶ θαυμάζεται μάλιστα ἐπὶ τῶν δένδρων, ἡ εἰς τὸ βέλτιον μεταβολή, καθάπερ ἐν Αἰγύπτῳ τε καὶ ἔτι μᾶλλον ἐν Κιλικίᾳ τῶν ῥοῶν, ἔτι δ' ἡ τῆς μυρρίνης περὶ Αἴγυπτον 5 εὐωδία· τὰς γὰρ ἐπὶ τὸ χεῖρον καὶ πολλὰς ὁρῶμεν καὶ πανταχοῦ, διὸ καὶ θαυμάζομεν.

ἔστιν οὖν δῆλον ὅτι καὶ τὸ ἐπὶ τῶν σπερμάτων συμβαῖνον, ὅταν εἰς τὸ βέλτιον ᾖ, σχεδὸν ὅμοιον· ἐν τούτῳ γὰρ ἡ διαφορά, τῷ τότε μὲν ἀεὶ τοῦτο 10 μένειν ὅταν φυτευθῇ, τὸ δ' ἄλλο καὶ ἄλλο μεταβάλ-13. 5 λειν· ἐπεὶ τό γε γινόμενον ταὐτό, παραιρεῖται γὰρ ἀεὶ τῆς[1] φύσεως, ἐπικρατῶν[2] ὡσαύτως ἀμφοῖν. ξυμβαίνει γὰρ τοῦτο καὶ ἐπὶ τῶν ζῴων, ἐκ μελάνων[3] γὰρ καὶ λευκὰ γίνεται, καὶ ἐκ τραχέων

[1] ⟨τι⟩ τῆς Wimmer.
[2] ἐπικρατῶν U N HP: -οῦν u; τὸ ἐπικρατοῦν Schneider.
[3] μελάνων u: μελανῶν U.

[a] *Cf. HP* 2 2. 7; *CP* 1 9. 2; 2 14. 2; 5 3. 3; 6 18. 6–7.
[b] *Cf. HP* 6 8. 5; *CP* 6 18. 4–10.
[c] *Cf.* Aristotle, *History of Animals*, iii. 12 (519 a 3–19): ". . . because of abnormal occurrences in the seasons, as when there is severer cold, some (*sc.* birds) of uniform colour change from black or blacker to white, as raven, sparrow and swallow; but of the white kinds none has been observed to change to black. Again, most birds change their colour with the seasons, so that one not familiar with the bird would not recognize it. And some animals change the colour of their hair with the change of water: in one locality the animals become white, in another the same animals become black. And at mating time

the grains and the animals being altered in quality and shifted a little at a time. That the change in trees, on the other hand, should come at the start and not as a conclusion happens reasonably enough; since trees are weakest when they begin, and most of all when grown from seed.

13. 4 But what does appear highly odd in the case of trees and is most wondered at is mutation for the better, as the mutation in Egypt and still more in Cilicia of the pomegranates;[a] and there is further the fragrance of the myrtle in Egypt.[b] As for mutations for the worse, we see them everywhere and in all countries. Hence our wonderment.

Now what occurs in grains when the mutation is for the better is evidently (one may say) similar, since the difference lies in this: the tree, once planted in the new country, continues as the same individual, whereas the grain changes through a succession of different individuals. But the process in the second case is the same as in the first: 13. 5 for the region keeps taking away from the nature of the plant, prevailing over trees and grain alike. For this gradual taking away from the nature happens in animals too, since starting as black some end as white,[c] and starting as rough some

there are in many places waters of such a kind that when the rams have drunk them and then cover the ewes they beget black lambs, as was the effect of the so-called ‘Cold’ river in the Assyritis district of the Thracian Chalcidice. And in the territory of Antandros there are two rivers, of which one makes sheep white, the other black. Again, the Scamander river is believed to make sheep yellow . . .;” *On the Generation of Animals*, v. 6 (786 a 2–5): “Animals that have by nature a uniform colour, but are of a kind that allows a number of uniform colours, change most because of the water: warm water makes their hair white, cold water black, just as with plants;” *cf.* [Aristotle], *Problems*, x. 7 (891 b 13–20).

5 μαλακά, καὶ ἄλλας τοιαύτας ἔχοντα μεταβολάς. αἴτιον δὲ ἐνταῦθα δοκεῖ τῶν μέν, φανερὸν[1] εἶναι, τὸ ὕδωρ, τῶν δέ, καὶ ὅλως αἱ τροφαὶ καὶ ὁ ἀήρ· ὥστε κἀκεῖ χρὴ νομίζειν, καὶ εἴ που ἄλλοθι τοιοῦτόν τι συμβαίνει, τὰς αὐτὰς εἶναι καὶ 10 παραπλησίας ἀνάγκας. τὰ δὲ καθ' ἕκαστα μᾶλλον, ἴσως δὲ καὶ μόνως, ἄν τις ἀποδοίη[2] τὴν ἐμπειρίαν προσλαβὼν χώρας καὶ τόπου διὰ τῆς ἱστορίας.

14. 1 ὅμοιον γοῦν τούτῳ καὶ παραπλήσιον φαίνεται καὶ ὅσα διὰ τῆς θεραπείας ἀλλοιοῦται, πρῶτον μὲν καὶ καθόλου πᾶσιν εἰπεῖν ἡμερούμενα, δεύτερον δὲ ἐν αὐτῷ τούτῳ βέλτιον, καὶ χαίρει γεωργούμενα· καὶ 5 γὰρ τοὺς πυρῆνας (ὥσπερ εἴπομεν) ἐλάττους ἔχει,[3] τὰ δὲ βελτίονα· ἔτι δ' ἡ πολυυδρία[4] γλυκαίνει τε τὰς ῥόας καὶ μαλακωτέρας ποιεῖ, δεῖται γὰρ ἡ στρυφνότης τοιαύτης καὶ ἡ σκληρότης

[1] φανερὸν ego (-ῶς Schneider): -ῶν U.
[2] ἀποδῴη Schneider: ἀποδοθείη U.
[3] ἔχει Heinsius: ἔχειν U.
[4] ἔτι δ' ἡ πολυυδρία ego (ἐστι. ἡ δὲ πολυΰδρία Wimmer): ἔτι δήπου ὑδρία U.

---

[a] *Cf.* Aristotle, *On the Generation of Animals*, v. 3 (783 a 12–18): " Sheep in cold climates are affected in the opposite way from man: the Scythians have soft hair, but the Sauromatian sheep rough hair. The cause of this applies also to all wild animals. For the cold hardens because it makes dry when it congeals; for as the heat is pressed out it carries the fluid with it, and both the hair and the skin become earthy and hard."

[b] See the passages cited in note *c* (p. 308), and *cf.* Aristotle,

end up soft,[a] and some are found to have undergone other changes of the sort. The cause of some of these changes is regarded as evident, namely the water,[b] and of others as being both the food in general and the air;[c] so that we must believe that in those parts too and in any country where the like occurs, the necessary[d] causes are the same and similar. But the particular cases could be better explained, or perhaps only explained, if one has gone further and become acquainted with the country and region concerned through the collection of information.

Improvement by Tendance

14. 1 Similar in any case and close to this improvement by a new country are the instances where trees are altered in quality through tendance, first and speaking generally of all, when they are brought under cultivation, and second when the cultivation itself is improved and the trees delight in it: thus they get smaller stones (as we said)[e] and the rest of the fruit is better;[f] again the copious watering makes pomegranates sweeter[g] and softer, their astringency and hardness requiring such succour, since deficiency re-

On the Generation of Animals, iv. 2 (767 a 28–35) for an explanation of the great effect of water.

[c] For the food *cf. CP* 2 13. 1, note *b*; for the cold of the air *cf.* Aristotle, *History of Animals*, iii. 12 (519 a 3–7), cited in note *c* (p. 308), Aristotle, *On the Generation of Animals*, v. 3 (783 a 12–18), cited in note *a*.

[d] That is, mechanical: *cf.* note *a* on *CP* 2 11. 2.

[e] *CP* 1 16. 2.

[f] The pericarpion: *cf. CP* 1 16. 2 (the plant becomes more fluid, devotes more food to the pericarpion, and concocts the fruit in a way that makes it suitable for human consumption).

[g] *Cf. CP* 6 18. 7.

ἐπικουρίας, καὶ γὰρ τὸ ἐνδεὲς δεῖται[1] βοηθείας.[2]
10 μόνον[3] γε δεῖ[4] τοῦτο διχῶς, ἢ πρόσθεσίν τινα
λαμβάνον, ἢ ἀφαιρέσεως γινομένης θατέρου[5]
(καθάπερ καὶ ἐπὶ τῶν ἀμυγδαλῶν ἐλέχθη τῶν
κολαζομένων καὶ περιαιρουμένων τὰς ῥίζας, καὶ
ἐπὶ τῶν συκῶν τῶν κατασχαζομένων·[6] ἀφαιρε-
15 θέντος γὰρ τοῦ πλήθους τῆς τροφῆς ἰσχύει μᾶλλον
τὸ σύμφυτον θερμὸν εἰς τὸ[7] κατάλοιπον).

14. 2 ἰδιωτάτη δ' ἂν δόξειεν ἡ ἀπὸ τῆς τῶν ῥιζῶν
εἶναι θεραπείας, τῆς τε κόπρου τῆς ὑείας ταῖς
ῥόαις[8] παραβαλλομένης, καὶ εἴ τις ἄλλη τοιαύτη
τινὶ δίδοται τροφὴ δι' ἧς γλυκαίνει τὸν χυλόν.
5 λέγεται δὲ καὶ ὡς ἡ πολυυδρία καὶ ἡ ψυχροϋδρία
ποιεῖ τινα μεταβολήν, καὶ ἐμφανέστατα δὴ καὶ
μάλιστα αἱ[9] τῶν ἐδαφῶν[10] ἐνίων καὶ τοῦ ἀέρος
φύσεις,[11] ὥσπερ καὶ ἐν Αἰγύπτῳ καὶ ἐν Κιλικίᾳ·
ἐν τούτοις γὰρ καὶ ζητεῖν ὅλως δεῖ τὰς ἀλλοιώσεις
10 καὶ μεταβολάς· ἐδάφει, καὶ ὕδατι, καὶ ἀέρι, καὶ

[1] δεῖται u: δεῖ τὸ U.
[2] βοηθείας u: βοηθοῦν U.
[3] ⟨οὐ⟩ μόνον HP.
[4] γε δεῖ ego: γὰρ δὴ U.
[5] γινομένης θατέρου ego (θατέρου γινομένης Scaliger): θατέρου γινομενου U.
[6] κατασχαζομένων Schneider: κατασχιζομενων U.
[7] τὸ U^r N HP: τὸν U^{ar}.
[8] ῥόαις u: ῥοαῖς U N HP.
[9] αἱ U: ἡ Schneider.
[10] ἐδαφῶν Schneider: ἐδάφων u; ἐλάφων U.
[11] φύσεις a: φύσις U N HP.

---

[a] *Cf. CP* 2 1. 1 (intelligence wishes to help nature).
[b] *Cf. CP* 5 9. 11.
[c] *Cf. CP* 1 17. 9–10; *HP* 2 2. 11; and *HP* 2 7. 6: "If a tree

quires help.[a] The help however must come in two forms: either the tree obtains a supplement, or the surplus that creates the deficiency is removed[b] (as we said[c] of the almond trees that are chastised and get their roots cut off and of the fig trees that are slashed, since with the removal of the excess food the native heat has greater strength for acting on the remainder).

14. 2 Of most special application would appear to be the remedy for treatment of the roots by the use of swine manure[d] or where a tree is given some other such food[e] whereby it turns its flavour to sweet. Plentiful watering and the use of cold water are also said to produce a mutation, and most evidently and greatly the natures of certain soils and kinds of air, as in Egypt and Cilicia.[f] These in fact are the sources where we must look for all alterations of quality and mutations: soil, water, air and tendance; indeed

will not bear fruit but turns to vegetative growth, they split the part of the trunk that is underground and insert a stone to make the split open up, and say that the tree then bears. So too if one cuts some of the roots around the trunk, and this is why they do this to the surface roots of the vine when it gets goaty. In figs they not only cut the roots but plaster ashes around and slash the trunk and say that the tree bears better. In the almond they also hammer an iron peg in and after making a hole replace the iron peg with a wooden one and cover it up with earth, and some call this 'chastisement' . . ."

[d] Swine manure was called the next strongest after that of man (*cf. HP* 2 7. 4); for its application to the pomegranate *cf. HP* 2 2. 11: "Tendance changes the pomegranate and almond, the pomegranate changing when it gets swine manure and plenty of flowing water . . .;" *cf.* also *CP* 3 9. 3.

[e] The strongest manure was not considered suitable for trees (*CP* 3 9. 2); *cf.* however the application of urine and of tanner's manure (*CP* 3 9. 3; 3 17. 5).

[f] *Cf. HP* 2 2. 7; *CP* 1 9. 3; 2 13. 4; 5 3. 3; 6 18. 6–7.

ἐργασίᾳ (καὶ γὰρ ἁπλῶς ἡ γεωργία μεθίστησιν, ἐξημεροῦσα τὰ δένδρα καὶ τοὺς καρπούς).

14. 3 ὑπὲρ μὲν οὖν τῶν λοιπῶν ἕτεραί τινες αἰτίαι· ὑπὲρ δὲ τῆς ἀπὸ τῶν ῥιζῶν μεταβολῆς, ὑπὲρ ἧς τὰ νῦν ὁ λόγος, ἐκεῖνο δεῖ λαβεῖν· ὅτι καθάπερ ἀρχαί τινες αἱ ῥίζαι τῶν δένδρων· ἀκολουθεῖν δὲ 5 φιλεῖ ταῖς ἀρχαῖς τὰ ἄλλα (διὸ καὶ ἐπὶ τῶν σικύων ἐλέχθη πρότερον ὅτι βρεχομένων ἐν γάλακτι τῶν σπερμάτων ἢ ἐν μελικράτῳ γλυκύτεροι γίνονται καὶ ἐπ' ἄλλων). αὗται δὲ καὶ τὴν τροφὴν πεπεμμένην μᾶλλον λαμβάνουσαι [1] καὶ 10 αὐταὶ μεταβάλλουσαι, συμμεταβάλλειν ποιοῦσι καὶ τὸ δένδρον· ἀπὸ γὰρ τούτων ἡ διάδοσις. ἐπεὶ [2] καὶ αἱ πολυυδρίαι καὶ τὰ ἐδάφη καὶ αἱ κατεργασίαι περὶ ταύτας [3] πρῶτον καὶ ἀπὸ τούτων ἀρχόμεναι [4] τὰ ἄλλα συναλλοιοῦσιν.

14. 4 πρὸς ἕτερα δ' ἴσως καὶ ἕτεραι βοήθειαι [5] συνεργοῦσιν, οἷον [αἱ] σχάσεις [6] συκῶν καὶ κλάσεις τῶν

5–8. Athenaeus iii. 5 (74 A–B): *Θεόφραστος δέ φησι* (*HP* 7 4. 6) *σικυῶν τρία εἶναι γένη . . . 'γίνονται δέ' φησι 'καὶ εὐχυλότεροι οἱ σικυοί, ἐὰν τὸ σπέρμα ἐν γάλακτι βραχὲν σπαρῇ ἢ ἐν μελικράτῳ' ἱστορεῖ δὲ ταῦτα ἐν Φυτικοῖς Αἰτίοις. θᾶττον* (*HP* 7 1. 6) (⟨*δὲ*⟩ ego) *αὔξεσθαι, κἂν ἐν ὕδατι κἂν ἐν γάλακτι πρότερον ἢ εἰς τὴν γῆν κατατεθῆναι βραχῇ*.

[1] λαμβάνουσαι Schneider: -σιν U.
[2] ἐπεὶ u: ἐπι U.
[3] ταύτας Schneider: ταῦτης U.
[4] ἀρχόμεναι u: -οι U.
[5] ἕτεραι βοήθειαι u: ἑτέρα βοήθειᾳ U.
[6] [αἱ] σχάσεις ego: αἱ σχάσεις ⟨τῶν⟩ Schneider.

husbandry brings about a total shift in kind by turning trees and fruit from wild to cultivated.[a]

Causes of Mutation From the Roots

14. 3 Now these other mutations have other causes, but to explain the mutation proceeding from the roots, which is our present theme, we must rest our argument on the point that the roots are a kind of starting-point (as it were) of trees, and that everything else tends to follow the starting-points.[b] (This is why it was said earlier[c] of cucumber that the plant comes up sweeter when the seeds[d] have been soaked in milk or hydromel; so too of others.) As for the roots, both when they receive their food in a better concocted state[e] and when they are themselves changed,[f] they bring about an accompanying change in the tree as well, since it is from them that the food is distributed. Indeed plentiful watering, good soil and tillage deal first with the roots and begin with them in bringing about changes of quality in the rest of the tree.

14. 4 It is for another set of results that other remedies than this collaborate, as slashing the fig,[g] pruning

[a] *Cf. CP* 2 14. 1 *init.*

[b] *Cf. CP* 2 16. 3; 3 17. 7; 3 24. 4; 5 17. 5.

[c] *HP* 7 3. 5: "Again when the seeds of some plants have been treated in advance the plants grow up different in their flavours, as when the seed of cucumber is soaked in milk before sowing;" *cf. CP* 3 9. 4; 3 24. 4; 5 6. 12.

[d] The seed is the starting-point in the strictest sense.

[e] No doubt from the heat of the manure.

[f] By being made to attract more vigorously (*CP* 3 17. 3–5); manure does this too.

[g] *Cf. HP* 2 7. 6, cited in note *c* on *CP* 2 14. 1.

ἀμπέλων καὶ αἱ κολούσεις τῶν ἀμυγδαλῶν ἢ[1] πληγαῖς ἢ διειρόντων τοὺς παττάλους· ἀπορρεούσης γὰρ
5 τῆς ὑγρότητος ἡ καταλειπομένη ῥᾷον ἐκπέττεται.

γλυκύτητας[2] δὲ ἐνοφθαλμιζομένας[3] καὶ[4] ἐγκεντριζομένας οὐ πέφυκε μεταβάλλειν, ὅτι καθάπερ ἀρχή τις ἑτέρα τούτων ἐστίν, ἧς οὐκέτι δύναται κατακρατεῖν· ὥσπερ γὰρ γῇ χρῆται τῷ ὑποκειμένῳ τὸ
10 ἐμφυτευόμενον ἢ ἐνοφθαλμιζόμενον (ὥσπερ ἐλέχθη).
14. 5 καί εἰσιν αὗται δύο μεταβολαὶ καὶ ἀπὸ δυοῖν ἥ τε ἀπὸ τῶν ῥιζῶν καὶ ἡ ἀπὸ τῶν ἐνοφθαλμιζομένων[5] ἢ ἐγκεντριζομένων ἢ ἐμφυτευομένων· ἡ μέν, αὐτῶν τῶν ὑποκειμένων ἀλλοιουμένων, ἡ δέ,
5 ἑτέρων τινῶν ἐμβαλλομένων, διὸ καὶ ἧττον ἐπὶ ταύτης τὸ θαυμαστόν.

διὰ τί δ' ἡ ῥόα μάλιστα [μὲν][6] μεταβάλλει, ⟨μάλιστα μὲν⟩ τὴν[7] μανότητα καὶ τὴν ἀσθένειαν αἰτιάσαιτ' ἄν τις, εὐηκοώτατα[8] γὰρ τὰ τοιαῦτα
10 πρὸς μεταβολήν· ἐπισκεπτέον δὲ καὶ εἴ τις ἄλλη τῆς φύσεως ἰδιότης.

15. 1 εἰς δὲ τὸ χεῖρον μεταβολῆς[9] δῆλον ὡς ἐναντίαι, καὶ ἐμφανεστάτη γε ⟨καὶ⟩[10] κοινοτάτη πᾶσιν

[1] ἢ N HP: ἡ U.
[2] γλυκύτητας ego: γλυκύτης. τὰς U.
[3] ⟨ἐν⟩ο- Wimmer: ὀ- U.
[4] καὶ U: ἢ Schneider.
[5] ⟨ἐν⟩ο- Wimmer: ὀ- U.
[6] [μὲν] Wimmer.
[7] ⟨μάλιστα μὲν⟩ τὴν ego: τὴν ⟨μὲν⟩ Wimmer.
[8] εὐηκοότατα u (ο² ss.): εὐηκότατα U N HP.
[9] εἰς—μεταβολῆς ego (τῶν δ' εἰς τὸ χ. μεταβολῶν Schneider [*Mutationum autem in partem deteriorem* Gaza]; τῆς δ' εἰς τὸ χ. μεταβολῆς Wimmer): εἰς δὲ τὸ χ. μεταβολαὶ U.
[10] ⟨καὶ⟩ Gaza (*et*), Basle ed. of 1541.

the vine [a] and checking the almond by blows [b] or by inserting pegs,[c] since with the consequent flowing away of fluid the remainder is more easily brought to full concoction.

As for sweetness that comes from bud grafts and cleft grafts, the root is not of a nature to change it, because the grafts have another starting-point (as it were) than the root, and the root cannot master it, as it masters its own progeny, since the grafted twig or bud treats the stock as if the stock were the soil (as we said).[d] Here then are two distinct mutations of a tree, and they have two distinct origins: the mutation of a tree from its roots, and the mutation of a tree from grafts inserted into it as buds or wedges or twigs. In the one mutation the very bases and underlying parts are altered in quality; in the other, a different set of parts is inserted into it, which is why there is less to marvel at here. 14. 5

Why most of all trees the pomegranate undergoes mutation for the better one would preferably explain by its open texture and weakness, such trees being most responsive to the influences that induce mutation; but we must investigate to see whether it also has some further peculiarity in its nature.

Changes for the Worse:
(1) *From Lack of Tendance*

Change for the worse has evidently the opposite causes, and most noticeable of these and most com- 15. 1

[a] *Cf. CP* 3 14. 1.
[b] Presumably inflicted in pruning the roots: *cf. CP* 2 14. 1.
[c] *Cf.* note *c* on *CP* 2 14. 1.
[d] *CP* 1 6. 1. For the scion the root of the stock counts as earth, not as a root.

ὡς ἀγεωργησία· πάντα γὰρ (ὡς εἰπεῖν) ἀπαγριοῦται.

5 ἐνίοτε δὲ καὶ οἱονεὶ πηρώσει τινὶ μεταβάλλουσιν εἰς τὸ χεῖρον κολουόμενα κατὰ τὴν πρώτην γένεσιν τὰ φυτά, καθάπερ ἡ ἀμυγδαλῆ, πικρὰ γὰρ ἐκ γλυκείας γίγνεται καὶ ἐκ μαλακῆς σκληρά· τὰ δ' ἄλλα οὐκ ἔστιν ἐπίδηλα μεταβάλλοντα (καίτοι τά 10 γε τῆς ἀμπέλου φυτὰ καὶ ἀπόλλυται πονοῦντα· τὰ δὲ μᾶλλον, καὶ μεταβάλλειν εἰκὸς μᾶλλον ἦν· τῇ γὰρ ἀμυγδαλῇ καὶ τὸ ὅλον φαίνεται παράλογον, ἰσχυρότερον γὰρ τῶν δένδρων).[1] ἔνια δὲ καὶ βελτίω κολουόμενά φασι γίνεσθαι, καθάπερ οἱ Χῖοι τὴν ἄπιον τὴν Φωκίδα.

15. 2 τὸ μὲν οὖν ἰσχυρὸν τῆς ἀμυγδαλῆς οὐκ ἐν τῷ 5 αὐτῷ λαμβάνεται· δενδρωθεῖσα[2] μὲν γὰρ ἰσχυρά, φυομένη δὲ ἀσθενής, ἄλλως τε καὶ ἀπὸ σπέρματος. ἐπεὶ καὶ μὴ κολουσθεῖσαι[3] πικραὶ γίνονται καὶ σκληραί, καθάπερ ἐν τοῖς πρότερον ἐλέχθη (πᾶν γὰρ ὅλως ἀπὸ σπέρματος ἐξαλλοιοῦται πρὸς τὸ 10 χεῖρον)· κινηθείσης ⟨δὲ⟩[4] τῆς φυσικῆς ὁρμῆς εἰκὸς ἔτι μᾶλλον, οἷον γὰρ ἄλλη καὶ ἀσθενεστέρα

[1] τῶν δένδρων U: τὸ δένδρον Gaza (*arbor*), Schneider.
[2] δενδρωθεῖσα Wimmer: δένδρω. κολουσθεῖσα (-ουθ- U[r]) U[ar].
[3] κολουσθεῖσαι N P[ac] (-εί- U): -ουθεῖσαι u HP[c].
[4] ⟨δὲ⟩ HP.

---

[a] *Cf. HP* 2 2. 9: "Trees also change because of their food and other forms of care, which are the means whereby wild trees are turned into cultivated ones and even among cultivated trees themselves some become more cultivated, as the [καθάπερ ἥ ego; καὶ ἀπορῇ U] pomegranate and almond."
[b] *Cf. CP* 5 17. 5. [c] *Cf. CP* 5 17. 5.
[d] As shown by the need for "chastisement."

mon to all trees is lack of husbandry, since then all (one might say) turn wild.[a]

(2) *From Cutting Back in the Young Almond*

Occasionally the young slips also change for the worse by a sort of mutilation (as it were) when they are cut back in the course of their first growth, as the almond, for it changes from the sweet kind to the bitter and from the soft kind to the hard;[b] other young trees do not change noticeably. Yet the young vines suffer such hardship when they are cut back that they are even killed;[c] and one would expect that plants more apt to suffer should also be more apt to change. Indeed it appears anomalous that the almond should change at all, for it is a stronger tree than the rest.[d] Some trees are even said to improve when cut back, as the Chians say of the Phocian pear.

15. 2

As for the point about the almond's being "strong,"[e] it comes from taking the almond at a different stage: the almond is strong when it has grown to a tree, but weak when it is growing up, especially when it comes from seed. Indeed it needs no cutting back then to become bitter and hard (as we said earlier),[f] since in general every tree grown from seed alters for the worse; and it is likely to do this even more when its natural impulse has been interfered with,[g] for this turns it into a different (as it were) and weaker almond. Indeed things

[e] Made in the last sentence of *CP* 2 15. 1.
[f] *HP* 2 2. 5.
[g] The cutting back compels the natural movement of feeding and growth to take another course.

γέγονεν. ἐπεὶ καὶ τὰ ἑψόμενα μωλύεται,[1] μὴ ἐν καιρῷ κινούμενα, καὶ ἡ γῆ δυσδιάτηκτος ἡ βρεχομένη· ὃ [2] δὲ ἐν ἀρχῇ μέγα, καὶ διατενὲς πρὸς
15 τὴν τελείωσιν.

15. 3 ἐπὶ μόνης δὲ ταύτης ⟨ἣ⟩ [3] μάλιστ' ἔνδηλον εἶναι τὴν μεταβολὴν οὐκ ἄτοπον, καὶ διὰ τὰς προειρημένας αἰτίας καὶ διότι τῆς μὲν ἀμπέλου καὶ τῶν ἄλλων, οὐθενὸς ὁ καρπὸς τελεουμένου [4]
5 πικρὸς ἢ ὀξύς, ἀλλ' ἤτοι γλυκὺς ἢ οὐκ ἐπέφθη· ταύτης δὲ οὐχ [5] ὑπάρχει τοιοῦτος εὐθὺς ἐν τοῖς τελείοις καὶ εὐκαρποῦσιν. τάχα δὲ καὶ τῶν ἄλλων ἐστί τις καὶ εἰς τὸ στρυφνότερον ἢ ὑδαρέστερον ἢ μὴ ὁμοίως γλυκύ, λανθάνει δὲ τὴν
10 ἡμετέραν αἴσθησιν· καὶ γὰρ αὐτῶν τῶν ὁμογενῶν αἱ μὲν μᾶλλον, αἱ δὲ ἧττον, τοιαῦται.[6] συμβαίνει δὲ ταῖς κολουσθείσαις,[7] ἂν πρεσβύτεραι γενόμεναι πάλιν ἐπικόπτωνται καὶ διακαθαίρωνται, γλυκυτέραις [8] γίνεσθαι, καὶ τέλος ἀποκαθίστασθαι πρὸς
15 τὴν φύσιν.

15. 4 αἴτιον δὲ ἔτι πρὸς τῷ εἰρημένῳ διότι τὸ μὲν ἡ κόλουσις κωλύσασα τὴν εἰς τὸν ὄγκον [9] βλάστην κατέμιξε καὶ χείρω τὴν ὑγρότητα τὴν εἰς τὸν

[1] μωλύεται ego (μολύνεται Coray): κωλυεται U.
[2] ὃ u: ὁ U.
[3] ⟨ἣ⟩ ego.
[4] τελεουμένου ego: τελεοῦμενος U.
[5] [οὐχ] Schneider.
[6] αἱ μὲν—τοιαῦται U: τὰ μὲν μᾶλλον, τὰ δ' ἧττον τοιαῦτα Schneider.
[7] κολουσθείσαις U^c (-λοσθ- U^ac): -λουθ- U^r N HP.
[8] γλυκυτέραις Schneider: γλυκύτεραι U.
[9] ὄγκον u: οἶκον U.

that are being boiled become only half-cooked if the process is interfered with at the wrong moment, and the earthy material that is immersed does not then dissolve properly.[a] And what is done at a beginning is of great weight, and its effects carry over to the maturity of the tree.

15. 3 It is not strange that the mutation should only occur in the almond, or be most detectable in it. The reasons are not only the ones mentioned, but also this: in the vine and the rest the fruit is never bitter or sour when the tree is becoming full grown, but is either sweet or did not ripen; whereas in the almond the fruit lacks that character even when we begin with the full-grown and properly bearing tree. Perhaps in the rest too there is a departure in the direction of more astringency or wateriness or else of less sweetness, but it is not detected by our senses. Indeed even among individual almonds there are some with these characters to a greater or less degree; and it happens that almonds that have been cut back, if on growing older they are cut back and pruned again, become sweeter and eventually recover their natural state.

15. 4 A causation in addition to the one mentioned[b] is this: on the one hand the pruning prevented the sprouting out of mere unproductive bulk and thus mixed the food for this into the fluid destined for the fruit, thus making the fluid inferior, and with this

[a] *Cf.* Aristotle, *Meteorology*, iv. 3 (381 a 12–22) [Boiling and ripening are two forms of concoction]. In the tree the improperly dissolved earthy material is the bitter and hard pericarpion.

[b] *CP* 2 15. 3.

καρπὸν ἐποίησε, πλείονος δὲ οὔσης ἄπεπτος, ὥστε
5 πικρός· καὶ κατακοπτομένη δὲ λαμβάνει τινὰ
ἀποπνοὴν καὶ ἀφαίρεσιν, ὥσπερ καὶ ὅταν οἱ
σφῆνες διακρουσθῶσιν· ἐλάττονος δὲ γινομένης,
καὶ αὐτὸ [1] μᾶλλον ἰσχῦον διὰ τὴν εὐθένειαν,[2] καὶ
ἐκπέττει τε μᾶλλον καὶ ἀποκαθίσταται.

15. 5 ταύτης μὲν οὖν ὥσπερ ἀνασῴζεται πάλιν ἡ
φύσις.

ἔνια δέ, ἐὰν μὴ κολουσθῇ,[3] τὸν καρπὸν οὐ
πέττει, καθάπερ ἡ ἄμπελος ἡ κανθάρεως [4] καλου-
5 μένη· διὸ καὶ κολούουσιν ἄκρον τὸν βότρυν· εἰ
δὲ μή, σήπει καὶ διαφθείρει. δῆλον οὖν (ὡς
ἁπλῶς εἰπεῖν) ὅτι ἀφαιρέσεως δεῖται τῆς ὑγρότητος.

ἡ δὲ Φωκὶς κολουομένη [5] βελτίων πρὸς δένδρω-
10 σιν, οὐ πρὸς εὐκαρπίαν· ἐκτρέχει [6] γὰρ ἄγαν μὴ
κολουσθεῖσα [7] καὶ γίνεται μονόκαυλος [8] καὶ ἀσθε-
νής· εἰ δὲ μή, παραβλαστάνουσα δενδροῦται.
15. 6 τάχα δ' ἄν τι [9] συμβάλλοιτο τοῦτο πρὸς εὐκαρπίαν,
ἰσχυροτέρας γὰρ γινομένης ἡ πέψις καλλίων.

§ 5. 4. Hesychius *s. v.*: κανθαριος· ἀμπέλου εἶδος.

[1] καὶ αὐτό u: και αὖ | τῶ U[c] (και αὖ in an illegible erasure).
[2] εὐθένειαν ego (εὐσθένειαν Heinsius): ἀσθένειαν U.
[3] κολοῦσθηι U[ac]: κολουθῆι U[c+r].
[4] κανθάρεως Schneider: κανθαρεως U.
[5] κολουομένη Wimmer (καλουμένη <ἄπιος κολουομένη> Schneider after Gaza): καλουμένη U.
[6] ἐκτρέχει Schneider: ἐκτρεφει U.
[7] κολουσθεῖσα U[ar]: -ουθ- U[r] N HP.
[8] μονοκαυλος U: μονόκωλος u (-κολος N HP).
[9] τι Gaza: τις U.

increase in the fluid the fruit fails to get concocted and is therefore bitter; on the other hand the tree, when also cut back later, obtains a certain relief and removal of a burden, just as it does when the wedges are knocked out,[a] and the amount of fluid being diminished, and the tree itself having greater strength for the task because of its well-being, it not only succeeds better in fully concocting the fruit but also recovers its natural state.

In this tree, then, the nature is (as it were) recovered. 15. 5

Changes Due to Pruning in Other Trees and Plants

Some trees, if not pruned, fail to concoct their fruit, as the so-called scarabaeus vine; this is why growers prune away the tips of the clusters, since otherwise the tip causes the cluster to decompose and so destroys it. So it is clear (speaking broadly) that these trees require removal of their fluid.

With the Phocian pear,[b] on the other hand, pruning improves its habit as a tree, but not its crop, since without pruning the tree shoots up too high and fails to branch and is weak; with pruning it sends out branches and acquires the habit of a tree. Perhaps however this habit would contribute to good 15. 6 fruit production, since as the tree gets stronger concoction is better carried out.

[a] The " wedges " appear only here; they are the " pegs " spoken of elsewhere (*HP* 2 2. 11; *HP* 2 7. 6; *CP* 1 17. 10; *CP* 2 14. 4).

[b] *Cf. CP* 2 15. 2.

τῶν δὲ λαχανωδῶν ἢ ποιωδῶν ὅσα κολουόμενα [1] καὶ [2] κειρόμενα βελτίω (καθάπερ τά τε πράσα καὶ
5 ἡ ῥάφανος ἡ παλιμβλαστὴς καὶ ἡ μηδίκη [3] καὶ ἡ θρῖδαξ καὶ τὸ ὤκιμον), ἅπαντα ταῦτα τῇ ἁπαλότητι [4] καὶ εὐτροφίᾳ βελτίω καὶ εὐχυλότερα γίνεται. παραιρεῖται γὰρ ἡ δριμύτης καὶ ἡ ξηρότης καὶ εἴ ἔν τινι τὸ ὀπῶδες ἀπὸ [5] τούτων, ἁπαλὰ δὲ καὶ
10 εὐτροφῆ [6] γίνεται διὰ τὸ τὰς ῥίζας ἰσχυροτέρας εἶναι (καὶ ὅλως αὐξανομένης [7] ἔτι μᾶλλον, κολουομένων, ⟨τῆς⟩ [8] ἰσχύος)· αἱ δὲ καὶ ἐπισπῶνται πλείω καὶ καταπέττουσιν μᾶλλον. ἔτι δέ, ἀφαιρουμένων τῶν ξυλωδῶν καὶ σκληρῶν, ἡ ἐπίδοσις
15 πλείων, μηθενὸς ἐμποδίζοντος. ἡ δὲ πρώτη βλάστησις, ἐξ ἀσθενοῦς ὡρμημένη μᾶλλον, χείρων.

αὗται μὲν οὖν ἐν τοῖς χυλοῖς αἱ μεταβολαί.

16. 1 γίνονται δὲ καὶ κατὰ τὰς ὀσμάς, μάλιστα μὲν αὐτομάτως διὰ τὸν ἀέρα [9] καὶ τὴν χώραν· εὐοσμότερα [10] γὰρ (ὡς ἐπὶ πᾶσιν) αἱ ξηραὶ ποιοῦσιν καὶ ὁ ἀὴρ ὁ τοιοῦτος· διὸ καὶ τὰ ἄγρια εὐοσμότερα

[1] κολουόμενα Gaza, Schneider: καλούμενα U N HP; καυλούμενα u; κακούμενα a.
[2] καὶ U: ἢ Schneider.
[3] ἡ παλιμβλαστὴς (ἣν παλιμβλαστήσῃ Wimmer) καὶ ἡ μηδίκη ego: ἣν πάλιν βλαστήσηι καὶ ἡ μηδικὴ U.
[4] ἁπαλότητι u a: ἀπλότητι (ἁ- N HP) U.
[5] ἀπὸ ego (Gaza omits, Schneider deletes): ὑπὸ U.
[6] ευτροφῆ U: εὐτραφῆ u.
[7] αὐξανομένης Gaza, Schneider: -ων U.
[8] ⟨τῆς⟩ Schneider.
[9] ἀέρα u: αἶρα U.
[10] εὐοσμότερα U^r: -αι U^ar.

---

[a] *Cf. CP* 3 19. 1–2 and *HP* 7 2. 4 (of vegetables): "When the stems are cut back practically all except the stemless (?)

Among vegetables and herbaceous plants those that are better when cut back and clipped, like leek, cabbage (of the kind that sprouts again), lucerne, lettuce and basil,[a] all improve in tenderness and plumpness and flavour-juice, since the pungency and dryness and rennet-like quality (where it exists) of these parts is removed, and the parts come up tender and plump because the roots are stronger (in fact there is an absolute gain in strength when the plant is clipped); and the roots not only attract more food but concoct it better. Again, with the removal of the parts that were woody and hard, growth is greater, since there is now nothing to impede it. The first sprouting, on the other hand, had come from a weaker source,[b] and was inferior.

These changes, then, occur in the flavours.

Changes in Odour

16. 1 Changes also occur in odour; they mostly arise spontaneously because of the air and the country, since it is on the whole dry countries and dry weather that make plants more fragrant, and this is why wild plants are the more fragrant. But perhaps they

ones sprout again, and most evidently (as if to serve our needs) basil, lettuce and cabbage. In lettuce they say that the stems that come up again are better eating, since the first stem is rennet-like in quality and bitter, as being unconcocted, whereas others say that on the contrary the second stems are more rennet-like in quality but appear sweeter so long as they are tender. But about the cabbage there is agreement on this point, that it is better eating when it sprouts again, so long as the leaves are gathered before it runs to stalk."

[b] That is, the roots as they were then.

5 (τάχα δὲ οὐ πάντως, [οὐδ'] ὀδμωδέστερα ⟨δέ⟩,[1] δριμύτερα γάρ· ἀρίστη δ' ἡ μέση καὶ ὁ μέσος, ἐπεὶ καὶ τὰ ἐν τῷ [γῆι] ὕδαρεῖ[2] καὶ ἄοσμα).

καὶ περὶ μὲν ὀσμῶν καὶ χυλῶν αὐτὰ καθ' αὑτὰ δεῖ θεωρεῖν ἐπὶ πλέον ἐν τοῖς ὕστερον. αἱ δὲ 10 μεταβολαὶ διότι καὶ ἐν τούτοις γίνονται καὶ φυσικῶς καὶ ἐκ θεραπείας φανερὸν ἐκ τῶν εἰρημένων.

16. 2 ἔνιαι δὲ δοκοῦσιν ὅλων τῶν δένδρων καὶ φυτῶν αὐτόματοί τινες εἶναι μεταβολαί, καθάπερ τὴν λεύκην ἐξαιγειροῦσθαί φασιν καὶ φύλλοις καὶ τῇ ὅλῃ προσόψει· καὶ τὸ σισύμβριον εἰς μίνθαν[3] 5 μεταβάλλειν μὴ κατεχόμενον ταῖς ἐργασίαις[4] καὶ μεταφυτευόμενον πολλάκις· ἔτι δὲ καὶ τὸν πυρὸν ἐξαιροῦσθαι καὶ τὸ λίνον.

16. 3 αὕτη μὲν οὖν, εἴπερ ἀληθής, ὥσπερ φθορά τις ἔοικεν εἶναι διὰ πλῆθος ὑγροῦ (γίνεται γὰρ δι' ἐπομβρίαν)· ἀλλοιωθείσης δὲ τῆς ἀρχῆς ἀλλοῖον τὸ ἀναβλαστάνον· ἡ δ' αἶρα φίλυδρον.

[1] [οὐδ'] ὀδμωδέστερα ⟨δέ⟩ ego: οὐδ' ὀσμωδέστερα U.
[2] τῷ [γῆι] ὕδαρεῖ ego (τῇ γῇ ⟨τῇ καθύγρῳ⟩ ὑδαρῇ Schneider after Gaza): τῆι γῆι ὑδαρῇ U.
[3] μίνθαν Wimmer: μίνθον U.
[4] ταῖς ἐργασίαις U^c: ταὶς ἐργασίας U^ac.

---

[a] As in Egypt: *CP* 6 18. 3.
[b] In *CP* books vi–vii.
[c] As well as in entire plants.
[d] *Cf.* *CP* 4 5. 7.
[e] *Cf.* *HP* 2 4. 1: ". . . bergamot mint is held to change to green mint if not held back by tendance, which is why they transplant it frequently, and wheat to darnel."
[f] *Cf.* the end of the preceding note and *HP* 8 7. 1: "Now

are not exactly more fragrant, but more odorous, since they are too pungent; and the best country and air are those intermediate in dryness and humidity; indeed plants where the air is watery lack any odour at all.[a]

But odours and flavours must be studied by themselves at greater length later.[b] Meanwhile it is clear from the preceding that mutations occur also[c] in odour and flavour, not only naturally but as a result of tendance.

Mutations of the Entire Plant

16. 2 There are held to be a few spontaneous mutations of the entire tree or plant. So it is said that white poplar changes to black poplar[d] not only in its leaves but in its entire appearance, that bergamot mint changes to green mint[e] if not held back by agricultural procedures and frequently transplanted, and further that wheat (and flax) change to darnel.[f]

16. 3 Now this mutation to darnel (if true) would appear to be a kind of extinction (so to say) due to too much water, since it is brought about by rainy weather,[g] and once the starting-point has suffered alteration, what sprouts from it is altered too;[h] and darnel is fond of water.

it is not in the nature of other grains to lose their identity and change to something else, but people assert that wheat and barley change to darnel, wheat doing this more . . . This then is a peculiarity of these plants, and furthermore of flax, for people assert that darnel also comes from it."

[g] *Cf. HP* 8 7. 1: ". . . they say that wheat and barley (and wheat more than barley) change into darnel, and that this occurs during rainy spells and especially in well-watered and rainy places."

[h] *Cf. CP* 2 14. 3; 3 17. 7; 3 24. 4; 5 17. 5.

5 ἡ δὲ[1] τῆς λεύκης, εἰ ἄρα ἐστὶ μεταβολή, γινομένη τις ἂν εἴη παχυνομένου τοῦ δένδρου μᾶλλον, ὃ συμβαίνει διὰ τὴν ἡλικίαν· εἰς βάθος γὰρ ἡ αὔξησις ἀπογηρασκόντων, ἐν ᾧπερ οἵ τε φλοιοὶ παχύτεροι[2] καὶ οἱ ἀκρεμόνες μείζους[3] καὶ
16. 4 πλείους. ὁ δὲ τῶν φύλλων μετασχηματισμὸς καὶ ἑτέρων κοινός· ἐπεὶ καὶ τὰ τοῦ κρότωνος νέα,[4] περιφερῆ φυόμενα τὴν ἀρχήν, ὕστερον ἀπογωνιοῦται, καθαπερανεὶ διαρθρούμενα· τοῦτο δ' ὅτι ῥᾷον[5]
5 τὸ ἁπλοῦν ἢ τὸ πολυειδές, ἀσθενὴς δ' ἡ ἀρχή.

τὸ δὲ σισύμβριον εἰς μίνθαν,[6] κατὰ τὴν ὀσμήν, εἴπερ ἄρα μεταβάλλει, μόνον, ἀπολλύον τὴν οἰκείαν, ἐξαμαυρούμενον διὰ τὴν ἀργίαν, ἐκείνην δ'[7] οὐ λαμβάνον, ἀλλ' ὥσπερ[8] ὅμοιον ταῖς καλα-
10 μίνθαις γινόμενον· ἡ γὰρ μεταβολὴ πᾶσιν εἰς ὅμοιόν τι, καὶ οὐ πόρρω τελέως φθειρομένων.
16. 5 ἡ δὲ θεραπεία καὶ ἡ μεταφυτεία κατέχει, καὶ σῴζει τὴν φύσιν· σημεῖον δέ, ὅτι καὶ τὸ ἄγριον τοιοῦτον τῇ ὀσμῇ. καὶ γὰρ δὴ κἀκεῖνο[9] τὸ τῆς φύσεως ἐναντίον· ἡ μὲν γὰρ μίνθα βαθύρριζον, τὸ δὲ
5 σισύμβριον ἐπιπολῆς καὶ οὐχ ὁμοίως πολύρριζον. ὥστε μᾶλλον ἔοικεν, ἐπί γε τῶν τοιούτων, κατὰ

[1] ἡ δὲ Schneider: ἤ τε U.
[2] παχύτεροι U: τραχύτεροι Schneider.
[3] μείζους u: μεί | ους U.
[4] νέα Schneider: ἔνια U.
[5] ῥᾷον Gaza, Schneider: ῥᾴδιον U.
[6] μίνθαν U^c: μίθαν U^ac.
[7] ἐκείνην δ' U: *speciem vero* Gaza; ἐκείνης δὲ τὴν μορφὴν Schneider.
[8] ὥσπερ U^c: ὥσπερ τε U^ac.
[9] κἀκεῖνο N HP: κακείνωι U.

The mutation of the white poplar (if it is a mutation) would come about rather with the thickening of the tree, which occurs with age. For as trees grow old their growth is lateral, and this involves thicker bark and longer and more numerous branches. The re- 16. 4 shaping of the leaf is common to other plants as well. Indeed in the castor bean the young leaves come out round at first and later become angular, as if they were being more precisely formed.[a] The reason is that it is easier to form the simple than the complex, and the plant is weak at the beginning.

The mutation of bergamot mint to green mint (supposing there is a mutation) is only in the odour, the bergamot mint losing its own, when the plant becomes dulled by lack of tendance, without acquiring the odour of green mint, but coming instead to resemble (as it were) calamint. For mutation in all plants is to something similar, and is not a complete extinction into something remote. The care and the 16. 5 transplanting restrain the plant and preserve its nature; that this is its nature is shown by the similar odour in the wild bergamot mint. In fact there is another point in the nature of bergamot mint that is in opposition to that of green mint: green mint has deep roots, whereas bergamot mint has shallow roots, and they are not so numerous as in green mint. So it would rather seem, at all events in cases such as these,

[a] *Cf. HP* 1 10. 1: " The leaves of other trees are alike to each other in all, but in the white poplar and the so-called *krótōn* (that is, castor bean; literally " tick ") they are unlike and have two different shapes; for the young leaves are round and the older ones angular, and all end up as angular; " *HP* 3 18. 7: " It is a rare phenomenon and occurs in few, that the leaf changes with age, as it does in the white poplar and the castor bean."

φαντασίαν ἡ μεταβολὴ γίνεσθαι, καὶ ὥσπερ εἰ τὸ ἥμερον [1] εἰς τὸ ἄγριον. οὐδετέρως δ' ἄτοπον, ἐπεί γε καὶ οἱ τόποι μεταβάλλουσιν.

16. 6 εἰ δὲ καὶ ἐπὶ τῶν ζῴων τοῦτο συμβαίνει, καθάπερ φασὶν ἐπὶ τῶν ὀρνίθων καὶ χρώμασι καὶ σχήμασι καὶ δυνάμεσι, καὶ τοῦτο καθ' ἕκαστον ἐνιαυτόν, οὐκ ἐν πλήθει χρόνου πλείονι, κἂν 5 θαυμάσειεν ἄν τις μᾶλλον εἰ μή τι συμβαίνει καὶ ἐνταῦθα τοιοῦτον, ἀτακτοτέραν ⟨γὰρ⟩ [2] καὶ μᾶλλον ξυγκεχυμένην εἰκὸς ταύτην εἶναι τὴν φύσιν· τάχα δ' ἰσχυροτέραν, διὸ καὶ μεταβάλλειν οὔτε χρώμασιν οὔτε ἐν τοῖς [3] ἄλλοις φθειρομένην, καὶ τοῦτο ἐν 10 ὀλίγοις ὥστε εἰς ἄλλο γε μεταλλάττεσθαι φυτόν.

16. 7 (οὔτε [4] γὰρ κατὰ τὴν γένεσιν [5] οὐδὲν μεταβάλλει τὰς μορφάς, ὡς [6] ἔνια τῶν ζῴων, ἀλλ' ἁπλῆ τις ἡ [7] φύσις πάντων.) ἀλλ' αἱ μεταβολαὶ γίνονται (καθάπερ πολλάκις λέγεται) τοῖς τε χυλοῖς 5 μάλιστα καὶ ταῖς ὀσμαῖς καὶ τοῖς μεγέθεσιν αὐτῶν

[1] τὸ ἥμερον u: τὸν μερον U. [2] ⟨γὰρ⟩ HP.
[3] τοῖς N HP: τοῖ U.
[4] οὔτε U: οὐδὲ Wimmer.
[5] γένεσιν N HP: γεννεσιν U.
[6] ὡς u: ὥστ' U (ὥστ' N); ὥσπερ HP.
[7] ἡ Uc: Uac omits.

---

[a] *Cf. CP* 2 13. 1–5.

[b] *Cf. HP* 2 4. 4: "It might appear more surprising that natural changes of the kind should be even more numerous in animals. For some animals are held to change with the seasons, like the hawk and hoopoe and other similar birds, and again when localities undergo an alteration, as the water-snake turns into a viper when the streams dry up. Most obviously again some animals change in the course of generation, and change through a number of animals: so a caterpillar turns into a chrysalis and this into a butterfly; and this sort

that the mutation is merely apparent, and as if a cultivated plant lapsed into a wild. But there is no absurdity in supposing true mutation either, since even localities bring mutations about.[a]

16. 6 If change also occurs in animals, as it is said of birds that they change in colour, in shape and in power, and do it every year, taking no longer,[b] one would be actually more astonished if the like did not also occur in plants, since it is likely that nature as we have it in plants is more irregular and confused.[c] But it is perhaps likely that it is stronger, and therefore that it changes without total loss of colour or the other characters;[d] and only in a few cases is the change such that there is transition to a different plant. 16. 7 (For neither does a plant change its shape in the process of generation, as do some animals;[e] instead the nature of all plants is a simple one.) But the changes occur mainly (as we keep saying)[f] in flavour, odour and

of change is found in a number of other cases. But perhaps there is no absurdity, and what we are trying to explain is not similar to it." *Cf.* Aristotle, *History of Animals*, ix. 49 (632 b 14–633 a 28) for annual changes in the colour and singing [Theophrastus' " power "] of birds, especially 633 a 18: " The hoopoe also changes both its colour and appearance [Theophrastus' " shape "] . . ." *Cf. ibid.*, iii. 12 (519 a 7–9), cited at 2 13. 5, note *c*.

[c] *Cf.* Aristotle, *Physics*, ii. 8 (199 b 9–10): " Further the final cause is found in plants too, although it is there less articulated; " *cf. On the Parts of Animals*, ii. 10 (655 b 37–656 a 2): " Now the nature of plants, being stationary, has no rich variety of anhomoeomerous parts, for since it has few actions to perform, it has use for but few organs . . ."

[d] That is, shape or power.

[e] *Cf. HP* 2 4. 4 (cited in note *b* on *CP* 2 16. 6).

[f] *CP* 2 13. 2; 2 14. 1, 2, 4; 2 15. 1, 2, 3, 4, 5, 6 (changes in flavour); *CP* 2 13. 4; 2 16. 1 (in odour); *CP* 2 14. 1 (smaller stones); *CP* 2 16. 4 (shape of leaves; but size is implicit).

τε τῶν καρπῶν καὶ τῶν φύλλων (καὶ γὰρ τὰ στενόφυλλα πλατυφυλλότερα γίνεται) καὶ ὅλων τῶν δένδρων τούτων.

διὰ τοῦτο καὶ ζητεῖ τόπον ἕκαστον οἰκεῖον,
16. 8 οἰκεῖος δὲ ἐν ᾧπερ εὐθενεῖ.[1] διόπερ καὶ οὐ πᾶσιν ὁ[2] ἄριστος, ἀλλ' ἔνια λεπτὴν καὶ λυπρὰν χώραν φιλεῖ, τὰ δ' ὕφαμμον, ἔνια δὲ καὶ ἁλμώδη[3] τινά, καθάπερ ἡ ῥάφανος.
5 διττῶς δὲ καὶ τὸ τῆς χώρας πρόσφορον· ἢ γὰρ τὸ οἰκεῖον τῆς φύσεως, ἢ τὸ πρὸς ἰσχὺν καὶ δύναμιν ἁρμόττον, οἷον ταῖς ἀμυγδαλαῖς ἡ λεπτή, βαθείας γὰρ οὔσης καὶ πιείρας, ἐξυβρίσασαι[4] διὰ τὴν εὐτροφίαν, ἀκαρποῦσιν.
10 καὶ καθόλου περὶ τῶν δένδρων εἴρηται πρότερον· ἀλλὰ δὴ τὰ μὲν περὶ τὰς ἀλλοιώσεις καὶ μεταβολὰς ἄχρι τούτων διωρίσθω.

17. 1 θαυμασιώτατον δ' ἂν δόξειεν, καὶ ὅλως ἄτοπόν τι καὶ παράδοξον εἶναι, τὸ ἔνια μὴ δύνασθαι βλαστάνειν ἐν τῇ γῇ καὶ σπέρματα καὶ φυτά, καθάπερ ἡ

§ 1. 2–4. Pliny, *N.H.* 16. 244: quaedam enim in terra gigni non possunt et in arboribus nascuntur. namque cum suam sedem non habeant, in aliena vivunt, sicut viscum . . .

3–11. Pliny, *N.H.* 16. 245: visci tria genera. namque in abiete, larice stelin dicit Euboea nasci, hyp⟨h⟩ear Arcadia, viscum autem in quercu, robore, ilice, piro silvestri, terebintho, nec ⟨non et⟩ aliis arboribus adgnasci pleri⟨s⟩que, copiosissimum in quercu, †adhasphear (quod hyphear *Mayhoff*) vocant.

[1] εὐθενεῖ U: εὐσθενεῖ u.
[2] ὁ U N HP: a omits; ὁ αὐτὸς Gaza (*idem*), Schneider.
[3] ἁλμώδη Schneider: αμμώδη U.
[4] ἐξυβρίσασαι u HP: ἐξυβρίσαι U; ἐξύβρίσας N.

size of the fruit alone and of the leaves (thus plants with narrower leaves get broader ones),[a] and of the entire trees in the case mentioned.[b]

This is why each tree seeks out its appropriate locality, that locality being appropriate in which it is at its ease. It is for this reason that the best locality is not best for all,[c] but some plants like thin and poor country, some sandy, and some even (as cabbage) country with a certain salinity. 16. 8

Again a country can be good for a plant in two ways: either the good may be its appropriateness to the plant's nature, or it may be its comporting with the plant's strength and power, as thin soil comports with almonds, since when the soil is deep and fat the trees get out of hand because of the rich feeding and fail to bear.

We have earlier[d] treated of the best locality with general reference to trees; now that the discussion of qualitative alterations and of mutations has been brought to this point, we leave it.

The Case of the Mistletoe: The Problem

17. 1 It might appear most amazing and a thing quite strange and unexpected that some plants are unable

[a] *Cf. CP* 6 18. 4, of the Egyptian myrtle transplanted to Cyprus and Rhodes.
[b] *CP* 2 16. 2–3 (white poplar to black).
[c] *Cf. CP* 1 18. 1 (the best land is not the best for trees).
[d] *CP* 1 18. 1–2; 2 4. 1–12.

ἰξία καὶ ἡ στελὶς καὶ τὸ ὑφέαρ, ὧν τὴν μὲν
5 καλοῦσιν Εὐβοεῖς, τὸ δ' ὑφέαρ Ἀρκάδες, ἡ δὲ ἰξία κοινή.[1]

φασὶ δ'[2] οἱ μὲν εἶναι πάντα μίαν τινὰ φύσιν, τῷ δὲ ἐν ἑτέροις φύεσθαι ⟨καὶ⟩[3] διαφέρειν [καὶ][4] δοκεῖν· τὸ γὰρ ὑφέαρ ἐν ταῖς ἐλάταις καὶ πεύκαις
10 γίνεται καὶ ἡ στελίς, ἡ δ' ἰξία καὶ ἐν δρυῒ καὶ ἐν τερμίνθῳ καὶ ἐν ἑτέροις πλείοσιν.

17. 2 οἱ δὲ διαφέρειν, καὶ σημεῖον λέγουσιν οὐ μικρόν (εἰ γὰρ[5] ἀληθές), ὡς οὐ μόνον ἐν τοῖς ὁμογενέσιν ἕκαστον ἐμφύεται τούτων (οἷον ἐλάταις καὶ πεύκαις), ἀλλὰ καὶ ἐν τῷ αὐτῷ πλείω καθ'
5 ἑκάτερον τῶν μερῶν, ἔνθεν μὲν [στελις ἢ][6] ἰξία, ἔνθεν δὲ[7] ὑφέαρ. ἔτι δ' οὐ μόνον τὰς μορφάς· ⟨ἀλλ'⟩[8] οὐδὲ καρποὺς ὁμοίους ἔχειν φασίν· καίτοι τοῦτό γε πανταχόθεν διατηρεῖται[9] καὶ ἐν τοῖς[10] πλεῖστον διαφέρουσιν[11] χώραις. ἐπεί τό γε

§ 1. 4–5. Hesychius *s.v.* στελίς· . . . περὶ φυτῶν τὴν ἰξίαν ὑπὸ Εὐβοέων; *cf.* ἀστυλίς· φυτὸν ὅθεν ὁ ἰξός.

9–10. Hesychius *s.v.* ὑφαίαρ· τὸ ἐπιφυόμενον ταῖς πεύκαις καὶ ἐλάταις.

§ 2. 2–7. Pliny, *N.H.* 16. 245 (continued): in omni arbore, excepta ilice et quercu, differentiam facit (⟨acini⟩ *Mayhoff*) odor virusque, et folium non iucundi odoris, utroque visci amaro et lento.

9–12. Pliny, *N.H.* 16. 246: adiciunt discrimen: visco in iis quae folia mittant et ipsi decidere, contra inhaerere nato in aeterna fronde.

[1] κοινὴ u HP (-ῆ a): κοινῆν U (-ὴν N).

[2] φασὶ δ' ego (φασὶν οὖν Wimmer): φασιν. U (the point is now erased).

[3] ⟨καὶ⟩ ego.

[4] [καὶ] Gaza, Schneider.

[5] εἰ γὰρ U: εἴπερ Gaza (*si quidem*); εἴ γ' Schneider.

[6] [στελις ἢ] Schneider.

[7] δὲ ⟨στελὶς ἢ⟩ Schneider.

to sprout—either the seeds or the plants—in the ground, as the *ixía* (mistletoe),[a] the *stelís*[b] and the *hyphéar*,[c] *stelís* being the Euboean word, *hyphéar* the Arcadian, and *ixía* the word in general use.

Some assert that all of them are a single natural entity, but because they grow on different plants they are also considered to be different; so the *hyphéar* occurs on silver-fir and pine, and so too the *stelís*, whereas the *ixía* (mistletoe) occurs on oak, terebinth, and a number of other trees.[d]

17. 2 Others assert that the plants are different and cite in proof a circumstance of no small weight (for such it is, if true): not merely does each of them grow on different individuals of the same kind (such as silver-firs or pines), but several grow on the same individual tree, with a distinction between the two sides, the *ixía* always being on the one side, the *hyphéar* on the other.[e] They further assert that these plants do not even have similar fruits, let alone shapes; and yet we see everywhere that similarity of fruit is retained even in plants differing widely in country. As

[a] *Loranthus europaeus.*
[b] *Viscum album.*
[c] *Viscum.*
[d] *Cf. HP* 3 7. 6 (*ixía* on the oak and other trees); *HP* 3 16. 1 (*ixía* and *hyphéar* on the kermes-oak).
[e] *Cf. HP* 3 16. 1 (of the kermes-oak): "It bears in addition to its acorn a sort of scarlet berry, and also gets both *ixía* and *hyphéar*, with the result that it sometimes has four sets of fruit, two of its own and two others, one that of the *ixía* and the other that of the *hyphéar*. And it bears the *ixía* on the north side, the *hyphéar* on the south."

---

[8] ⟨ἀλλ'⟩ Gaza, Itali.
[9] διατηρεῖται U: διατηρεῖσθαι Schneider.
[10] τοῖς ego: ταῖς U.
[11] διαφαίρουσιν U: διαφερούσαις u.

10 τὴν μὲν ἀείφυλλον εἶναι[1] τῶν ἰξιῶν[2] οὐθὲν ἄτοπον, κἂν ἡ μὲν[3] ἀειφύλλοις, ἡ δὲ ἐν φυλλοβόλοις ἐμβιῷ·[4] συμβαίνει γὰρ ἔνθα μὲν ἔχειν, ἔνθα δὲ μὴ ἔχειν διαρκῆ τὴν τροφήν, αἰτία δὲ αὕτη τῆς ἀειφυλλίας καὶ μή (καθάπερ εἴπομεν).

17. 3 ἀλλὰ τοῦτο μὲν ὁποτέρως ποτ' ἔχει, πρὸς τὸ νῦν ἀπορούμενον οὐθὲν διαφέρει.

τὸ δὲ μὴ φύεσθαι χαμαὶ μηδαμῶς ἄτοπον, ἄλλως τε καὶ οὕτω πολὺν[5] καὶ ἰσχυρὸν ἔχουσαν 5 καρπόν.

εἰ δὲ καὶ σπέρματα [τα][6] τοιαῦτά ἐστιν, οἷον τὸ περὶ Βαβυλῶνα τῇ ἀκάνθῃ περὶ τὸ ἄστρον ἐπισπειρόμενόν φασιν αὐθημερὸν ἀναβλαστάνειν καὶ ταχὺ περιλαμβάνειν καὶ τὴν ἄκανθαν, ἔτι δὲ τὸ Συριακὸν 10 βοτάνιον, ὁ καλούμενος κασύτας,[7] ⟨ὃ⟩[8] καὶ δένδροις καὶ ἀκάνθαις ἐμφύεται καὶ ἄλλοις τισίν, τῇ μὲν ἔλαττον ⟨ἂν⟩[9] εἴη, τῇ δὲ πλεῖον[10] τὸ

6–9. Pliny *N.H.* 13. 129: non omittendum est et quod Babylone seritur in spinis, quoniam non aliubi vivit, sicut et viscum in arboribus, sed illud in spina tantum quae regia vocatur. mirum quod eodem die germinat quo iniectum est (inicitur autem ipso canis ortu) et celerrime arborem occupat.

9–10. Hesychius: κασύτας· Συριακὸν βοτάνιον.

9–11. Pliny *N.H.* 16. 244: namque cum suam sedem non habeant, in aliena vivunt, sicut viscum et in Syria herba quae vocatur casytas (cassitas *or* castas MSS) non tantum arboribus, sed ipsis etiam spinis circumvolvens sese . . .

[1] εἶναι u: ἔνιαι U.
[2] ἰξιῶν (ἰξίων U) ⟨τὴν δὲ φυλλοβόλον⟩ Gaza, Itali.
[3] μὲν ⟨ἐν⟩ Schneider.
[4] ἐμβιῶι U: ἐμβιώιη u (-ιώη N HP).
[5] πολὺν u: -ὺ U.
[6] [τα] Schneider.

for there being an evergreen *ixia*, there is nothing strange here, or in the circumstances that it grows on evergreens, and the other on deciduous trees, since it turns out that in the one case it has a constant supply of food, in the other, that it does not; and the constancy of the supply (as we said)[a] accounts for a plant's being evergreen or not. But whether the three are the same or not makes no difference for our present problem. 17. 3

That the plant under no circumstances grows on the ground[b] is strange, especially when its fruit is so plentiful and strong.

If there are grains of the sort too,[c] such as the one reported sown in the dog days on the thorn-bushes in Babylonia that sprouts the same day and then speedily envelops the bush, and again the small Syrian weed called *kasytas*[d] that grows on trees, thorn-bushes and certain other plants, the strangeness in one way is diminished, in another increased, since in things paradoxical the multiplication of instances has

[a] *CP* 1 10. 7; 1 11. 6.

[b] Modern experiments have demonstrated that the seeds will germinate on the ground and even on glass plates.

The strength is also literal: bird-lime was made from the mistletoe berry.

[c] That is, that (1) always grow on another plant and that (2) are strong.

[d] Dodder (*Cuscuta* var.); in Syriac and Aramaic keṣ̌ūṯā, in Mishnaic Hebrew keš̌ūṯ, in Arabic kašūṯ: *cf.* I. Löw, *Die Flora der Juden*, vol. i (Vienna and Leipzig, 1928; photographic reprint Hildesheim, 1967), pp. 453–458.

---

7 κασύτας Hesychius: καδύτας U.

8 ⟨ὃ⟩ ego.

9 ⟨ἂν⟩ Wimmer.

10 πλεῖον u: πλείω U (-ων N HP).

θαυμαστόν· ἀμφότερα γὰρ ποιεῖ τὸ πλῆθος ἐν τοῖς [1] παραδόξοις, ὁτὲ μὲν ὡς πεφυκὸς οὕτως μὴ
15 θαυμάζειν, ὁτὲ δὲ μᾶλλον θαυμάζειν διὰ τὸ πλῆθος.

17. 4 ἐπεὶ τό γε ἐμφύεσθαι καὶ ἐν δένδροις καὶ ἐν φυτοῖς ἑτέροις τὸ καὶ ἐν τῇ γῇ φυόμενον οὐκ ἄτοπον, ἀλλὰ καὶ γινόμενον, ὥσπερ ὁ κιττὸς ἐν πολλοῖς. (ἔτι γὰρ τοῦτο παραδοξότερον, ὅτι καὶ
5 ⟨ἐν⟩ [2] ἐλάφου κέρασιν ὦπται· καὶ ἡ τέρμινθος δὲ ἐν ἐλαίᾳ, καὶ τὸ πολυπόδιον καλούμενον ἐπί τισι δένδροις, καὶ ὅσα δὴ σπανιώτερα καὶ τερατωδέστερα φαίνεται, καθάπερ ἡ δάφνη ποτὲ ἐν πλατάνῳ καὶ ἐν δρυΐ, καὶ τὰ ἄλλα ὅσα ὡς τέρατα προφαί-
10 νουσιν. ὅταν γὰρ εἰς γεῶδες γεγενημένον διὰ σῆψιν ἐμπέσῃ τὸ σπέρμα, διεβλάστησεν, εἶτα ζῇ τὴν τροφὴν τὴν ἐκ τοῦ δένδρου λαμβάνον, ὃ καὶ ἐπὶ τοῦ κιττοῦ τοῦ περὶ τὰ κέρατα βλαστοῦντος [3] —εἴπερ ἦν—οὐκ ἄλογον.)

17. 5 ἀλλὰ τὸ [4] ἐν ἑτέρῳ μόνον φύεσθαι, χαμαὶ δὲ μή, τοῦτ' ἄτοπον. προσφιλῆ μὲν γὰρ δὴ ἀλλήλοις καὶ σύμβια, καθάπερ καὶ τὰ ζῷα, καὶ τὰ φυτὰ τάχ' ἂν εἴη· τὸ δ' ὅλως ἐπὶ τῆς γῆς μὴ φύεσθαι

6–7. Pliny, *N.H.* 16. 244 (continuing passage cited on *CP* 2 17. 3): item circa Tempe Thessalica quae polypodion vocatur.

[1] *τοῖς* Heinsius: *ταῖς* U.
[2] ⟨*ἐν*⟩ Schneider.
[3] *βλαστοῦντος* U: *βλαστόντος* Wimmer.
[4] *τὸ* u: *τῶι* U; *τὰ* N P; *τὰ μὲν* H.

both effects: it sometimes makes us feel no surprise, since we take the thing to be normal, and it sometimes makes us all the more surprised because of the numbers.

17. 4 As for a plant's growing on trees and other plants when it also grows in the earth, there is no oddity; the thing is instead of common occurrence, as ivy grows on many plants. (For the instance observed of its even growing on the horns of a stag[a] is of a more unexpected sort, and so too that of a terebinth observed growing on an olive, and those of the so-called octopus-plant[b] growing on certain trees, and all the instances that strike people as having rather the character of rarities and portents, as the bay that grew on a plane tree,[c] and another that grew on an oak, and all the other instances displayed to us as portents.[d] For when the seed falls on some spot that has become earth-like through decomposition, it sprouts and then lives by taking the food that belongs to the tree; and it is not unreasonable that this is also what happened with the ivy that was growing on the horns, supposing the report true.)

17. 5 No; the oddity is that a plant grows exclusively on another plant, and not on the ground. Now it may perhaps be that like animals plants are fond of one another and live together; but that a plant

[a] *Cf.* Aristotle, *History of Animals*, ix. 5 (611 b 17–20) [*cf. Mir. Ausc.*, chap. v (831 a 2–3), Antigonus, *Mir.*, chap. xxix, Pliny, *N. H.* 8. 117, Athenaeus viii. 48 (353 A)]: "An *achaínes* stag has been known to be caught with a lot of fresh ivy growing on its horns. This would come from the ivy having grown on the horns when they were tender, as on green wood."

[b] Polypody (*Polypodium vulgare*), a fern.

[c] *Cf. CP* 5 4. 5.

[d] Perhaps a reminiscence of θεοὶ τέραα προὔφαινον (*Odyssey* xii 394): the gods "display portents" to Odysseus' crew, who have eaten the cattle of the Sun.

5 θαυμαστόν, ἄλλως τε καὶ καρπὸν ἔχον καὶ σπέρμα καὶ ἀπὸ τούτου βλαστάνον. εἰ γὰρ ἦν ἐκ διαφθορᾶς [1] τινος τῶν ἐν τοῖς δένδροις ἡ γένεσις, ὥσπερ ἐν τοῖς ζῴοις ἐγγίνεται τοιαῦτα ζῷα, λόγον τιν' ἂν [2] εἶχεν· ἀλλ' οὐκ ἔστιν οὐδὲ γίνεται πλὴν ἀπὸ
10 σπέρματος, ὅταν οἱ ὄρνιθες, ἐσθίοντες τὸν καρπόν, προϊῶνται [3] τὴν περίττωσιν ⟨ἐπὶ⟩ [4] τῶν δένδρων· τότε γὰρ αὐτὸς ὁ καρπὸς σῳζόμενος καὶ ἐπιμείνας διεβλάστησεν.

17. 6 τὸ μὲν οὖν θαυμαστὸν πολὺ καὶ ἐκ πολλῶν.

ἔοικεν δ' οὖν ὅμοιόν τι συμβαίνειν ταῖς ἐμφυτείαις καὶ τοῖς ἐνοφθαλμισμοῖς· ἑτοιμοτέραν γὰρ λαμβάνει τροφὴν καὶ ὥσπερ κατειργασμένην καὶ πεπεμ-
5 μένην σχεδόν, ὃ καὶ ἡ ἰξία ζητεῖν φαίνεται. τὸ δὲ τοιαύτης δεόμενον ἀσθενὲς ἂν εἴη τῇ φύσει.

9–11. Pliny, *N.H.* 16. 247: omnino autem satum (*sc.* viscum) nullo modo nascitur nec nisi per alvum avium redditum, maxime palumbis et turdi. haec est natura, ut nisi maturatum in ventre avium non proveniat.

9–11. Aelian, *N. A.* ix. 37: φυτοῦ ἑτέρου κλάδος ἐπιφύεται πρέμνῳ, προσήκων (-ῆκόν ego) οἱ μηδὲ ἓν πολλάκις. τὸ δὲ αἴτιον Θεόφραστος λέγει, φυσικώτατα ἀνιχνεύσας· ὅτι τὰ ὀρνύφια τὴν ἄνθην τῶν δένδρων σιτούμενα εἶτα ἐπὶ τοῖς φυτοῖς καθήμενα τὰ περιττὰ ἀποκρίνει· οὐκοῦν τὸ σπέρμα ἐν ταῖς κοιλάσι καὶ ταῖς ὀπαῖς αὐτῶν καὶ τοῖς σηραγγώδεσιν ἐμπῖπτον καὶ ἐπαρδόμενον τοῖς ὄμβροις τοῖς ἐξ οὐρανοῦ, εἶτα ἀναφύει ἐκεῖνο ἐξ ὧν ἐβλάστησεν ἀναπείθει, οὕτω τοι καὶ ἐν ἐλαίᾳ συκῆν (terebinth *CP* 2 17. 4) κατανοήσεις, καὶ ἐν ἄλλῳ ἄλλο.

[1] διαφθορᾶς Gaza, Itali: -φο- U.

[2] τιν' ἂν ego: τινὰ U.

[3] προϊῶνται (*egesserunt* Gaza) Scaliger: προορῶνται U; προαιρῶνται u.

[4] ⟨ἐπὶ⟩ Gaza (*in*), Itali.

[a] *Cf.* Aristotle, *History of Animals*, v. 31 (556 b 21–28): "Those insects that are not carnivorous but live on the flavours

should push this to the point of not growing on the ground at all is astonishing, especially when the plant bears fruit and seed and is produced from it; since if it came from a corruption of something in the host tree, as animals that can reproduce[a] arise in other animals,[b] there would be some accounting for this exclusive preference. But it does not come from that, and does not arise except from seed, when the birds eat its fruit and let their droppings fall on the host tree, since the fruit proper is then left intact and remains on the host and sprouts.

17. 6 So the oddity is great and appears in many features.

Still, it seems that something similar takes place in grafted twigs and buds: the scion gets food that is more readily available and that has been (as it were)[c] prepared and practically concocted, and this is what the *ixía* (mistletoe) appears to seek. A plant seeking such food would be weak in its nature.

of living flesh, as lice . . . and bedbugs, all generate by copulation the so-called nits, but from these nothing is generated in turn. Of this class . . . the bedbugs (*sc.* are produced) from the moisture from animals when it sets on the outside of the body, and the lice from the flesh." *Cf. CP* 2 9. 6.

[b] *Cf.* Aristotle, *On the Generation of Animals*, i. 1 (715 b 25–30, cited on *CP* 1 1. 2 note *e*). *Cf. ibid.*, iii. 9 (759 a 3–7), iii. 11 (762 b 18–21).

[c] The cautious language is no doubt due to Aristotle's statements: *cf. On the Parts of Animals*, ii. 3 (650 a 20–23): ". . . plants take with their roots their food already prepared from the earth (which is why plants have no excrement, since they use the earth and the heat in it as a stomach . . .);" ii. 10 (655 b 32–36): "Now plants . . . have no place for the useless residue, since they get their food concocted from the earth, and in place of residue put forth their seeds and fruits."

τοῦτο δὲ πάλιν οὐκ ἔοικεν, ἀλλ' ἰσχυρὸν εἶναι καὶ τρόφιμον καὶ ἡ ἰξία καὶ ἡ στελὶς καὶ τὸ ὑφέαρ·[1] τούτοις γὰρ δὴ καὶ τοὺς βοῦς καὶ τὰ
10 ὑποζύγια χιλεύουσιν καὶ ἀνατρέφουσιν μετὰ τοὺς θερισμούς. ἔτι δὲ καὶ αὐτὸς ὁ καρπὸς τῆς ἰξίας[2] μηνύει τὴν ἰσχύν.

17. 7 ἀλλὰ μὴν εἴ γε ἰσχυρὰ καὶ μὴ ἀσθενῆ, διὰ τί ποτ'[3] οὐ βλαστάνει καθ' αὑτὰ καὶ φύεται; διαβιάσασθαι γὰρ τὴν γῆν τῶν ἰσχυόντων ἐστίν, ὅπερ ποιεῖ καὶ ὁ θέρμος.[4]

5 εἰ δ' αὖ ψυχρὸν ἔχει τὸ σπέρμα καὶ δύσπεπτον, ἀλλὰ χρονιωτέραν ἐχρῆν εἶναι τὴν ἔκφυσιν, ὥσπερ καὶ ἑτέρων· ἐπεὶ καὶ τῶν τευτλίων[5] ἔνιά φασιν τῷ ὕστερον ἔτει διαφύεσθαι καὶ διαβλαστάνειν· οὐδὲ γὰρ οὐδὲ ταύτῃ κίνδυνος ὥστε σαπῆναι,
10 διαμένον γὰρ καὶ τοῦτο φαίνεται καὶ ἕτερα πολλῷ τούτων ἀσθενέστερα.

ταῦτα μὲν οὖν οὐ λύει τὴν ἀπορίαν, ἀλλ' ἐπιξυνδεῖ μᾶλλον.

17. 8 ἡ δ' ἀρχὴ ληπτέα φυσικῶς, ἀκολουθοῦσι κατὰ τὸ γινόμενον, ὅτι πέφυκεν ἐν ἑτέρῳ μόνον ταῦτα

§ 6. 9–11. Pliny, *N.H.* 16. 246: hyphear ad saginanda pecora utilius.

[1] ὑφέαρ Schneider: ὕφεαρ U.
[2] ἴξύας U.
[3] ποτ' U^r N HP: πουτ' U^ar.
[4] θέρμος Gaza (*lupinum*): θερισμος U.
[5] τευτλίων Basle ed. of 1541: σευτλιων U.

---

[a] From the fruit was made *ixós*, bird-lime, a tenacious substance.

[b] *Cf.* *HP* 1 7. 3 (of lupine): ". . . if sown in deep vegetation it is so strong that it threads its root through to the ground and sprouts;" *HP* 8 11. 8 (of lupine): ". . . and often when

But this again does not appear to be the case: mistletoe (*ixía*), *stelís* and *hyphéar* all appear to be strong plants and nutritious as food, since they are fodder on which oxen and mules are kept after the harvest. In the mistletoe (*ixía*) moreover the fruit itself indicates the strength of the plant.[a]

17. 7 But if the plants are strong and not weaklings, why do they not sprout and grow by themselves? For strong plants can force themselves into the soil, which is what lupine does.[b]

If again it is objected that the seed is cold and ill-concocted,[c] we reply that it should simply take longer to come up, just as the seeds of other plants do (so some beet seed is said to push through and come up a year later),[d] since in the mistletoe too the seed is in no danger of decomposing, for not only is this seed observed to survive but many much weaker seeds are observed to do so as well.

These considerations then do not solve our difficulty but render it more acute.

The Case of the Mistletoe: The Solution

17. 8 We must rest our explanation on nature and be guided by the event: it is the nature of these plants

the seed falls on some shrub or herbaceous plant it pushes it aside and connects its root with the ground and sprouts."

[c] *Cf. CP* 4 7. 2–3 (the seed of lupine is unconcocted as it were and requires a great deal of heat).

[d] *Cf. HP* 7 1. 6: "They say that something distinctive happens with beet: not all the seed comes up at first, but some of it much later, and some even one or two years later, and this is why few beets come up out of many seeded." For explanations *cf. CP* 4 3. 2; 4 13. 1.

γίνεσθαι, καθάπερ καὶ ζῷα ἐν ζῴοις, οἷον τά τε ἐν ταῖς πίνναις ἢ[1] καὶ ὅσα ἄλλα ζῳοτροφεῖ.[2] πλὴν τῶν
5 μὲν οὐκ ἔχομεν τὴν γένεσιν, τῶν δ' ἔχομεν λέγειν. τὸ δὲ ἀπὸ τῆς τῶν ὀρνίθων προσφορᾶς εἶναι τὴν ἀρχὴν ὥσπερ συμβεβηκός ἐστι πρὸς τὴν γένεσιν (ὅπερ καὶ ἐπ' ἄλλων γίνεται· κατορύττει γὰρ ἡ κίττα θησαυριζομένη τὰς βαλάνους, καὶ ἄλλα τῶν
10 ὀρνέων)· περιαιρεθέντος δὲ τοῦ ἰξοῦ καὶ κατεργασθέντος ἐν ταῖς κοιλίαις, ὅπερ ἐστὶ ψυχρότατον, ἅμα[3] τῷ περιττώματι καταπῖπτον τὸ σπέρμα καθαρὸν (καὶ τοῦ δένδρου λαμβάνοντός τινα μεταβολὴν ὑπὸ τῆς κόπρου) διαβλαστάνει καὶ
15 φύεται.

17. 9 τὴν δ' ἰσχὺν εὐλόγως ἔχει, καὶ διὰ τὴν εὐτροφίαν, καὶ μάλιστ' ἴσως διὰ τὴν αὐτοῦ φύσιν. ἰσχυρότερα δὲ εἰκὸς τὰ ἐν ταῖς ἐλάταις καὶ πεύκαις εἶναι· πλείων γὰρ ἡ τροφὴ καὶ λιπαρωτέρα.
5 πολλὰ δ' ἡ φύσις φαίνεται καὶ ἐν τοῖς ζῴοις τοιαῦτα ποιεῖν ὥσθ' ἕτερον ἑτέρῳ χρήσιμον εἶναι πρὸς σωτηρίαν καὶ γένεσιν, ἅπερ ἐν ταῖς ἱστορίαις ταῖς περὶ τούτων εἴρηται· διὸ καὶ ἐνταῦθα ἴσως οὐ θαυμαστέον τὸ τῶν ὀρνίθων, εἴτε ἐπίτηδες εἴτε
10 κατὰ συμβεβηκὸς γέγονεν, οὐδὲ λεκτέον ὅτι οὐκ

[1] ἢ U: ἐστὶ Wimmer.
[2] ζῳοτροφεῖ U^r N HP: ζῷα τροφεῖ U^ar.
[3] ἅμα Schneider: μία U^ar N; βία U^r; σὺν HP.

[a] *Cf.* Aristotle, *History of Animals*, v. 15 (547 b 25–31): "In some testacea grow white crabs, very small in size, most of them in the mussels with large cavities, but there also grow in the pinnae the ones called pinnoterae [literally "guardians of the pinna"]. They are also found in scallops and lagoon oysters. These do not grow in size in any way that can be noticed, and

to be generated only in another plant, just as there are animals that are only found in other animals, like those in the pinnae [a] and in others that support animals.[b] Only we cannot tell how the animals are generated, whereas we know about the generation of these plants. That it should all begin with the birds' eating the berry is accidental (as it were) to the generation, and such accidents occur in the generation of other plants as well: so the jay [c] buries a provision of acorns in the ground, and so too do other birds. When the flesh has been stripped from the berry and digested in the birds' digestive tract, and the flesh is the coldest part of the fruit, the seed, now rid of it, is dropped with the excrement (the host too undergoing a certain change because of the dung) and sprouts and grows.

17.9 The plant comes by its strength reasonably, both because it feeds well and perhaps mainly because of its own nature. It is likely that the ones growing on silver fir and pine are stronger, since the food there is more abundant and fatty.

Nature is seen to produce many such instances in animals too of one serving the other for preservation and generation, and these have been mentioned in the History of Animals.[d] So here too we must perhaps not be surprised at the role played by the birds, whether purposeful or accidental, and argue that

the fishermen say that they are generated with the shellfish that harbour them."

[b] *Cf.* also *ibid.*, v. 16 (548 b 15) [of the sponge]: "It supports animals inside itself."

[c] *Cf. ibid.*, ix. 13 (615 b 22–23) [of the jay]: ". . . when acorns run short it hides them away and stores them up."

[d] See notes *a–c* below.

ἂν ἦν ἡ γένεσις εἰ μὴ διὰ τούτους. οὐδὲ[1] γὰρ ἴσως ταῖς πίνναις βίος εἰ μὴ διὰ τὸν καρκίνον, οὐδ'[2] ἡ τῶν μελιττῶν φύσις ⟨εἰ μὴ διὰ⟩[3] τὸν γόνον (ὥς φασί τινες), οὐδ'[2] ἡ τοῦ κόκκυγος εἰ μὴ
15 ἦν ἡ ὑπολαῒς ἧς[4] εἰς τὴν νεοττιὰν[5] τὰ ᾠὰ τίθησιν. ἀλλ' ὥσπερ καὶ φθοραί, καὶ σωτηρίαι τινὲς γίνονται, καὶ[6] ἀλλήλων καὶ εἰς τοὺς βίους καὶ εἰς τὰς γενέσεις, οὕτω καὶ πρὸς τὰ φυτὰ διήκειν οὐθὲν κωλύει παρὰ τῶν ζῴων. ὥστε ταύτην ἢ
20 τοιαύτην αἰτίαν ὑποληπτέον εἶναι τῶν ἀπορηθέντων.

17. 10 ἐν δὲ τοῖς ἐπισπειρομένοις ἀφῄρηται καὶ τὸ ξυμβεβηκός, προαιρέσει γὰρ δρῶσιν, ἀλλὰ τὸν καιρὸν δῆλον ὅτι λαμβάνουσιν τῆς ἐπισπορᾶς ὅταν ὀργᾷ τὸ ὑποκείμενον, ὥσπερ ἡ ἄκανθα φαίνεται
5 περὶ τὸ ἄστρον· ἔνικμον γὰρ δεῖ καὶ εὐαφὲς εἶναι πρὸς τὴν διαβλάστησιν. ⟨τὸ⟩[7] δὲ ταχὺ τῆς ἐκφύσεως οὐ μόνον διὰ τοῦτο καὶ τὴν ὥραν, ἀλλὰ δῆλον ὅτι καὶ διὰ τὴν ἰδίαν γίνεται φύσιν.

καὶ περὶ μὲν τούτων ἅλις.

[1] οὐδὲ ego: οὔτε U.
[2] οὐδ'—οὐδ' U: οὔθ'—οὔτε Schneider.
[3] ⟨εἰ μὴ διὰ⟩ u.
[4] ἧς Moldenhawer (ᾗ Wimmer): καὶ U.
[5] νεοττιὰν ego: νεοττίαν U.
[6] καὶ U N: HP omit; δι' Wimmer.
[7] ⟨τὸ⟩ Schneider.

---

[a] Aristotle, *History of Animals*, v. 15 (547 b 25–31) [cited in note *a* on *CP* 2 17. 8].

[b] Aristotle, *History of Animals*, v. 21 (553 a 18–19): "For some assert that the bees do not bear and do not copulate, but fetch their grubs (τὸν γόνον) . . .;" *cf.* also Aristotle, *On the*

there would have been no generation of mistletoe but for them. For there would perhaps have been no life for the pinnae either but for the crab,[a] and no such thing as bees but for the grubs (as some assert),[b] and no such thing as cuckoo but for the wheatear, in whose nest it deposits its eggs.[c] No; just as in animals there are cases not only of killing but of saving, and not only of killing and saving of one by the other, but of destroying and preserving ways of life and ways of propagation, so there is nothing to prevent this from carrying through from animals to plants as well. And so we must suppose that the explanation of the difficulties that have been discussed is either this or the like.

17. 10 Where one plant is sown on another[d] the accidental character is also eliminated, since the sowers act intentionally. Still they evidently choose the moment when the host is prepared to receive the seed (as the thorn-bush is evidently ready to do in the dog days), since the host must contain moisture and be of the right consistency if the seed is to come up. The rapidity with which it sprouts is due not only to this and to the season, but also evidently to its own distinctive nature.

So much for these cases.

Generation of Animals, iii. 10 (759 a 11–13) [of the generation of bees]: "For they must either fetch their grubs from a distance, as some assert, and the grubs must either grow spontaneously or be laid by some other animal . . ."

[c] Aristotle, *History of Animals*, vi. 7 (564 a 2) [of the cuckoo]: "It also lays its eggs in the nest of the wheatear . . .;" ix. 29 (618 a 8–11): "The cuckoo . . . does not make a nest, but lays its eggs in the nests of others, chiefly on the ground in those of the . . . wheatear . . ."

[d] *Cf. CP* 2 17. 3.

18. 1 ὅτι δὲ καὶ ἐν τοῖς φυτοῖς ἔνια συνεργεῖ πρὸς τὴν ἀλλήλων σωτηρίαν καὶ γένεσιν, καὶ ἐκ τῶνδε φανερόν· ἐν μὲν γὰρ [1] τοῖς ἀγρίοις τὰ φυλλοβόλα [2] τοῖς ἀειφύλλοις, ὅτι σηπομένων [3] ξυμβαίνει καθά-
5 περ κοπρίζεσθαι τὴν γῆν, ὃ καὶ πρὸς εὐτροφίαν καὶ πρὸς τὴν βλάστησιν τῶν σπερμάτων χρήσιμον· ἐν δὲ τοῖς ἡμέροις ὅσα τοῖς φυτοῖς ἐπισπείρουσιν τῶν ἀμπέλων, ἀφαιρεῖν βουλόμενοι τὸ πλῆθος τῆς ὑγρότητος, καὶ τοῖς λαχάνοις ἢ τούτου χάριν, ἢ
10 τῶν γινομένων θηρίων (οἷον [4] ταῖς ῥαφανῖσι [5] τοὺς ὀρόβους πρὸς τὰς ψύλλας,[6] καὶ εἴ τι τοιοῦτον ἕτερον
18. 2 ἑτέροις). οἴεσθαι γὰρ χρὴ τοιαῦτα καὶ ἐν τοῖς αὐτομάτοις, τῆς φύσεως ὑπάρχειν,[7] ἄλλως τε εἰ καὶ [8] ἡ τέχνη μιμεῖται τὴν φύσιν. ἔτι δὲ ὅσα πρόσδενδρα καὶ περιαλλόκαυλα [9] τυγχάνει, ταῦτα γὰρ τὸ [10] πρὸς
5 ἑτέρῳ διώκει, καθάπερ ὅ τε κιττὸς καὶ ἡ σμῖλαξ καὶ ἡ σικύα καὶ ἄλλ' ἄττα, καὶ τῶν ἐλαττόνων ἕρπυλλος [11]

§ 2. 6. Pliny, *N.H.*16. 244 (continuing the passage cited on *CP* 2 17. 4): et quae dolichos ac serpyllum.

[1] γὰρ U^cm (with no index): U^t omits.
[2] φυλλοβόλα u HP: φυλλόβολα U N.
[3] σηπομένων ⟨τῶν φύλλων⟩ Heinsius after Gaza.
[4] οἷον Schneider (οἷον ἐν Gaza, Itali): ἐν U.
[5] ῥαφανίσι u: ῥαφανησι U.
[6] ψύλλας Schneider: ψυχὰς U N HP (ψύχας u).
[7] ὑπάρχειν u: ὑπάρχει U.
[8] εἰ καὶ U: καὶ εἰ Schneider.
[9] περιαλλόκαυλα Gaza (*amplexicaulia*), Schneider: περὶ ἄλλον καυλὸν U.
[10] τὸ Schneider: τὰ U.
[11] ἕρπυλος Schneider: ἑρπυλλον U.

That among plants too some collaborate to preserve and propagate others can also be seen from the following: among the wild the deciduous help the evergreen, since it happens that the earth is manured (as it were) by the decomposing leaves, and this is useful for good feeding and making the seeds sprout; among the cultivated there are the plants sown among the young vines when the growers wish to reduce their excess of fluid,[a] and the plants sown among vegetables either to do this or to keep them free of the pests that arise, as bitter vetch is sown among radish to help against the flea-spider,[b] and any similar case where a plant of this kind is sown with others.[c] For we must take it that such relations as these, in plants of spontaneous growth as well, belong to the nature of the plants, especially if art imitates nature.[d] Again there are the plants that rest against trees and have twining stems,[e] since these seek the support of another, as ivy, smilax, *sikýa*[f] and some others, and of smaller plants tufted

18. 1

18. 2

[a] *Cf. CP* 3 10. 3; 3 15. 4.

[b] *Cf. HP* 7 5. 4: "The pests found in radish are flea-spiders . . . It helps radishes against the flea-spiders to sow bitter vetch among the crop. But people say that there is no specific that will keep them from being produced;" *cf.* also *CP* 3 10. 3.

[c] *Cf. CP* 6 19. 1.

[d] *Cf.* Aristotle, *Meteorologica*, iv. 3 (381 b 6): ". . . for art imitates nature . . ."

[e] *Cf. HP* 7 8. 1 (of herbaceous plants): "A few have twining stems, and if they have nothing to rest on are prostrate-stemmed, like *epetínē* (corrupt; read ἰασιώνη 'bind-weed'?), bedstraw and in short those with a stem that is thin, soft and long, and this is why they mostly grow on other plants."

[f] *Cf. CP* 1 10. 4; 2 11. 4.

καὶ ἰασιώνῃ·[1] πάντα γὰρ ταῦτα ζῇ πρὸς ἕτερον,[2] μὴ ἔχοντα δέ, χαμαίκαυλα γίνεται, πλὴν ὅσα καὶ δενδροῦσθαι πέφυκεν, καθάπερ ὁ κιττός. ἔστι δὴ
10 καὶ ⟨ἡ⟩[3] ἄμπελος τοιοῦτον· οὐ γὰρ ἂν δύναιτο φέρειν τὰ κλήματα καὶ τὴν βλάστησιν (ἢ οὐχ ὁμοίως) καὶ μὴ[4] ἔχουσα τὸ ὑπερεῖσον, ἀλλὰ καὶ τὴν ἕλικα δοκεῖ τούτου χάριν ἔχειν, ὅπως εὐθὺς ἀντίληψις γίνηται καὶ οἷον δεσμός, ὥσπερ ὁ κιττὸς
15 τὰ ῥιζία[5] τὰ ἐκ τῶν βλαστῶν.

18. 3 ὅσα δὲ κοῦφα καὶ λεπτὰ ῥᾳδίως ἀναβαίνει καὶ ἐπιμένει[6] (καθάπερ ἡ ἰασιώνη[7] καὶ ὁ θέρμος καὶ ὁ δόλιχος). πάντα δὲ ταῦτα βλάπτει τὰ δένδρα, καταπνίγοντά τε καὶ ἐπισκιάζοντα, καὶ τὴν τροφὴν
5 ἀφαιρούμενα τὰ μὲν τῷ κωλύειν, ὁ δὲ κιττὸς καὶ τῷ ἐμφυόμενος ἐξαιρεῖσθαι (διὸ χαλεπώτατος τοῖς δένδροις οὗτος· πρὸς γὰρ τοῖς ἄλλοις, ἀείφυλλος ὢν καὶ ἰσχυρός, ἀεὶ δεῖται τροφῆς καὶ πολλῆς)· ἀφαυαίνεται[8] δὲ τελέως ὅταν ἐπὶ τὰ ἄκρα συναυ-
10 ξηθῇ· τότε γὰρ καταπέττει καὶ ἐξαιρεῖται πανταχόθεν τὴν τροφήν.

§ 3. 2–3. *Cf.* Pliny, *N.H.* 16. 244: namque cum suam sedem non habeant, in aliena vivunt, sicut viscum et . . . (*sc.* herba) quae (*sc.* vocatur) dolichos ac serpyllum.

9–10. *Cf.* Pliny, *N.H.* 16. 243: hedera necari arbores certum est.

[1] ἰασιώνη Gaza (*volucrum*), Scaliger (-όνη): ἡ ασινῆ U.

[2] πρὸς ἕτερον Gaza (*apud aliud*), Itali (πρὸς ἑτέρῳ Schneider): πρότερον U.

[3] ⟨ἡ⟩ Schneider.

[4] καὶ μὴ ego (*nisi* Gaza, μὴ Itali, εἰ μὴ Scaliger; καλὰ μὴ Moldenhawer): καὶ ακμὴν U.

[5] ῥιζία Gaza (*radiculas*), Itali (*cf.* *HP* 3 18. 10 ῥίζας): ῥιχεια U.

thyme and bindweed. For all live in dependence on another, and when they lack the other grow with their stems along the ground, except where the plant (like ivy) is also capable of assuming a tree-like habit.[a] Such too is the vine, since it would not be able to sustain its branches and foliage, or to do this so well, without something to lean on; in fact this is why it is considered to have its tendrils, to enable it at once to lay hold of its support and bind itself (as it were) to it, just as the ivy has the rootlets coming from its shoots.

18. 3 The light and slender plants easily climb to the top and stay there, as bindweed, lupine and longweed. But all these climbers injure the trees by choking and shading them and taking away their food, some by preventing it from reaching them, whereas ivy also extracts it by growing into them, which is why this climber does trees the most harm,[b] since in addition it is evergreen and strong and so requires food constantly, and food in plenty. The trees wither completely when ivy grows to the top, since it then concocts its fruit [c] and extracts its food from all parts of the tree.

[a] *Cf. HP* 1 3. 2 (of chaste-tree, Christ's thorn and ivy): ". . . these admittedly become trees; and yet they are of the class of shrubs."

[b] *Cf. CP* 5 15. 4 and *HP* 4 16. 5: "Trees also kill one another by taking their food away and interfering with them in other ways. Ivy too is harmful when it grows alongside . . ."

[c] Ivy does not bear fruit until it has outgrown the climbing stage.

---

[6] ἐπιμένει Schneider: ἐπιμάνει U; πημαίνει u.

[7] ἡ ἰασιώνη Wimmer (*volucrum* Gaza, ἰασιώνη Schneider): ἡ ἀσινη U.

[8] ἀφαναινεται U: ἀφαναίνει Wimmer (after Schneider; *exsiccat* Gaza).

18. 4 οὐ χαλεπὸν δ' ἴσως οὐδὲ ἐν ἄλλοις γε [1] πλείοσι λαβεῖν τὰς βλάβας (πολλῷ γὰρ πλείους εἰσὶ τῶν ὠφελίμων, ὥσπερ καὶ τοῖς ζῴοις)· ἐπεὶ καὶ ταῖς ὀσμαῖς ἔνια βλάπτεται, καθάπερ ἡ ἄμπελος τῇ τῆς
5 δάφνης καὶ τῇ τῆς ῥαφάνου, καὶ τοῦτο εὐθὺς ἐκδηλοῖ κατὰ τὴν βλάστησιν· ὅταν γὰρ πλησίον ᾖ τῆς ῥαφάνου καὶ τῆς δάφνης, ὁ βλαστὸς ἐπιστρέφεται [2] τὸ ἄκρον αὐτοῦ καὶ ὥσπερ ἀνακάμπτει διὰ δριμύτητα τῆς ὀσμῆς (ὀσφραντικὸν γὰρ ἡ
10 ἄμπελος, ὥσπερ καὶ ὁ οἶνος δεινὸς ἑλκύσαι τὰς ἐκ τῶν παρακειμένων ὀσμάς, καὶ μᾶλλον καὶ θᾶττον ὁ κατεσταμνισμένος [3] διὰ τὴν ὀλιγότητα καὶ τὸ γυμνόν). ἀλλὰ τὸ μὲν τοιοῦτον γένος ῥᾴδιον ἐν πολλοῖς (ὥσπερ ἐλέχθη) συνιδεῖν.

19. 1 ὅσα δὲ κοινὰ γένους τινός, ἢ καὶ πλειόνων μὴ ὁμογενῶν—οἷον τὸ στρέφειν τὰ φύλλα τὴν φίλυραν καὶ τὴν ⟨λεύκην καὶ τὴν⟩ [4] ἐλάαν καὶ τὴν πτελέαν

6–8. *Cf.* Varro, *R.R.* i. 16. 6 (olive trees so detest the oak) ". . . ut introrsum in fundum se reclinent, ut vitis adsita ad holus facere solet."

10–11. Theophrastus, *On Odours*, chap, iii. 11: ὁ γὰρ οἶνος, ὥσπερ καὶ πρότερον ἐλέχθη (a reference to the present passage or to *CP* 6 19. 2), δεινὸς δέξασθαι τὰς ὀσμάς.

[1] γε ego ([τε] or τε ⟨καὶ⟩ Schneider; καὶ Wimmer): τε U.
[2] ἐπιστρέφεται U: ἀποστρέφεται Wimmer.
[3] κατεσταμνισμένος Scaliger: καταεσταμνισμένος U.
[4] ⟨λεύκην καὶ τὴν⟩ ego.

---

[a] *Cf. HP* 4 16. 6: "Some plants do not kill but cause deterioration by the powers of their flavours and their odours, as cabbage and bay do to the vine. For people assert that the

It would doubtless not be difficult to set down injuries in many other encounters, since injuries are far more numerous than benefits, as in animals. Indeed a few plants are even injured by odours, as the vine by the odours of bay and cabbage,[a] and it shows that this is so from the moment it sends out shoots. For when the vine is near cabbage or bay its shoot curves its tip and (as it were) turns back because of the pungency of the odour. For the vine is sensitive to smell, just as wine too is apt to attract the odours of objects placed near it,[b] wine drawn off in jars doing this more and faster because of its small quantity and of its exposure. But (as we said) [c] effects of this sort are easily seen in many instances. 18. 4

Movements in Plants

With movements common to a certain kind of plant, or else to a number of plants of different kinds [d]—such as the curling of the leaves in lime, white poplar, 19. 1

vine smells and absorbs odours. This is why the shoot, when it gets close to them, turns back and faces the other way, treating the odour as hostile. Androcydes even used this behaviour as an example to show that cabbage is a good antidote to wine, ridding us of drunkenness, since (he said) even when alive the vine avoids the odour."

[b] *Cf. CP* 6 19. 2; *On Odours*, chap. iii. 11.

[c] In the first sentence of the paragraph.

[d] The rising and sinking of the stalk of the water-lily in the Euphrates is a movement common to plants of the same kind; the opening and shutting of the flowers and the curling of the leaves are each common to a number of plants of different kinds.

ταῖς τροπαῖς ταῖς θεριναῖς, κα�ὶ ὡς ἔνια τῶν ἀνθῶν
5 νύκτωρ μὲν συμμύει, μεθ' ἡμέραν δ' ἐκπετάννυται,
καὶ ὡς ἐν τῷ [1] Εὐφράτῃ λέγουσιν οὐ μόνον τοῦ
λωτοῦ τὸ ἄνθος ἀνοίγεσθαι καὶ συμμύειν, ἀλλὰ καὶ
τὸν καυλὸν ὁτὲ μὲν ἀναβαίνειν, ὁτὲ δὲ δύεσθαι, καὶ
καταβαίνειν [2] ἀπὸ δυσμῶν μέχρι μέσων νυκτῶν,
10 ὡσαύτως δὲ καὶ εἴ τι ἄλλο τοιοῦτον—ἐν ἅπασιν καὶ
πανταχοῦ κοινήν τινα αἰτίαν ὑποληπτέον εἶναι·
ῥᾷον [3] δὲ ἴσως ἐν τοῖς ὁμογενέσιν ἰδεῖν.

19. 2 ἐπὶ δ' οὖν τῶν εἰρημένων ἡ μὲν τῶν φύλλων
στροφὴ γίνεται διὰ τὸ περὶ τοῦτον τὸν καιρὸν
μάλιστά πως ἀεὶ φυλλορροεῖν· ἡ δ' αἰτία πρότερον
εἴρηται περὶ πάντων τούτων. γινομένης οὖν τῆς
5 φυλλορροίας, ἀνάγκη μάλιστα πάσχειν τι τὰ φύλλα,
καὶ τὰ μὲν ἀσθενῆ καὶ ὥσπερ γεγηρακότα κατα-

[1] τῷ H(τῶ P): τῆι U (τῇ N).
[2] καταβαίνειν Schneider (*cf.* *HP* 4 8. 10 ὑποκαταβαίνειν): ἀποβαίνειν U.
[3] ῥᾷον Gaza (*facilius*): ῥᾴδιον U.

[a] *Cf.* *HP* 1 10. 1 (of leaves): "Peculiar too is what occurs in olive, lime, elm and white poplar: they are held to turn the upper surface after the summer solstice, and it is by this that people know that the solstice has occurred."

[b] *Cf.* *HP* 4 7. 8 (reports about the island of Tylos): ". . . there is another tree with a many-petalled flower like the rose, and this shuts up at night and opens with the rising sun and at noon is completely unfolded, but again in the afternoon is gradually gathered in and at night is shut . . .;" *HP* 4 8. 9 (of the Egyptian water-lily): "When the sun sets the petals shut and conceal the head, but at sunrise open up and rise above the water."

[c] *Cf.* *HP* 4 8. 10 (of the Egyptian water-lily): "They say that in the Euphrates the head and flowers sink into the water and keep sinking lower from evening until midnight, going

olive and elm at the summer solstice,[a] and the cases of some flowers that shut at night and open by day,[b] and the case reported that in the Euphrates [c] not only the flower of the water-lily opens and shuts [d] but the stalk too rises above water at one time and sinks below it at another, moving downward from sunset to midnight, and any other such case—in all the instances and wherever they occur we must take the cause to be a common one; [e] but perhaps it is easier to see in plants of the same kind.[f]

19. 2 Be that as it may, in the cases mentioned [g] the curling of the leaves occurs because at about this time there is always some shedding of leaves; [h] the cause that accounts for leaf-drop in all these trees [i] has been given earlier.[j] And so, since shedding of leaves is taking place, the leaves most of all must be affected in some way, and the weak and (as it were) aged

very deep . . . Then with the first light they come up again and rise even further at daybreak, the plant appearing above the water with the rising sun and opening its flower, and continuing to rise after the flower opens, a considerable part of it emerging above the surface."

[d] As it does in Egypt: *cf. HP* 4 8. 9 (cited in note *b*).

[e] Loss of fluid to the sun with ensuing contraction.

[f] The connection with the sun is most obvious in the Euphrates water-lily.

[g] *CP* 2 19. 1.

[h] *Cf. HP* 1 9. 7: "The loss and withering of leaves in evergreens is staggered, since the same leaves do not always remain on the tree, but new ones come out as the others wither away. This withering away happens mainly at the summer solstice."

[i] In the deciduous ones as well as in the evergreen (olive).

[j] *CP* 1 11. 6 (the cause of the character evergreen is continuity of food supply). Dropping the leaf would then be due to interruption of the food supply, and this happens to a certain extent with the beginning of the dry season.

ξηρανθέντα πίπτειν, τὰ δ' ἄλλα τὴν ἐπιστροφὴν μόνον λαμβάνειν. ἅπασιν μὲν οὖν τοῦτο συμβαίνειν [1] ἢ μᾶλλον ἢ ἧττον, ἔνδηλον δὲ μάλιστα ἐπὶ τούτων,
10 ὅτι μεγίστη διαφορὰ τῶν χρωμάτων τοῖς πρανέσιν πρὸς τὰ ὕπτια, τὰ μὲν ⟨γὰρ⟩ [2] χλωρά, τὰ δὲ ὑγρὰ καὶ ὑδατώδη· μᾶλλον δ' ἐπὶ τῆς φιλύρας, μείζω γὰρ καὶ ἐκλευκότερα (δεῖ δ' ἀνομοίως) [3] καὶ ἐπὶ τῆς

19. 3 λεύκης.[4] τῶν δ' ἀειφύλλων οἷς μὴ συμβαίνει τοῦτο διὰ τὸ κατ' ἐκεῖνον τὸν καιρὸν ἀκμήν τινα καὶ ὥσπερ ἰσχὺν [τινα] [5] εἶναι τῆς βλαστήσεως, διὰ τοῦτο οὐκ ἂν συμβαίνοι· [6] τῶν [7] γὰρ ἰσχυόντων οὐδὲν
5 εὐπαθές.

ἡ δὲ τῶν ἀνθῶν σύμμυσις καὶ δίοιξις ἐλαφροτέρα καὶ ῥᾴων [8] ἰδεῖν· ὑπὸ γὰρ τοῦ ψύχους καὶ τῆς ἀλέας γίνεται, ψυχρῶν ὄντων καὶ ἀσθενῶν· συμμύει μὲν γὰρ ξυνιόντος καὶ οἷον πηγνυμένου τοῦ ὑγροῦ,
10 συναπολείπει γὰρ καὶ τὸ θερμόν, ἀνοίγεται δὲ πάλιν διαχεομένου καὶ ἀνιέντος, ὅπερ ὁ ἥλιος ποιεῖ.

19. 4 τὰ δὲ πλέον καταδυόμενα καὶ ὑπερίσχοντα [9] δῆλον ὅτι ψυχρότερα καὶ ἀσθενέστερα, διὸ μᾶλλον συμπάσχει ταῖς μεταβολαῖς. ἡ δὲ αἴσθησις οὕτως ὀξεῖα γινομένη τοῖς καθ' ὕδατος οὐκ ἄλογος,

[1] συμβαίνειν U: -ει N HP.
[2] ⟨γὰρ⟩ Gaza (*enim*), Schneider.
[3] δ' ἀνομοίως ego: δᾶν ὁμοίως U.
[4] λεύκης Dalecampius: πεύκης U.
[5] [τινα] ego.
[6] συμβαίνοι HP: συμβαῖνον U N.
[7] τῶν u: τὸν U.
[8] ῥᾴων u: ῥᾷον U.

among them must dry out and drop whereas the rest merely curl. Now this curling must occur in all trees to a greater or lesser extent; but it is most noticeable in these, because in them the contrast of colour between the two surfaces of the leaf is greatest, the upper surface being a fresh green, the lower having a feeble and washed-out colour. The difference is more prominent in the lime, since its leaves are larger and the underside is paler (and the two sides must contrast) and in the white poplar. In those evergreens where this curling does not take place [a] because at that time there is a certain peak and (as it were) strength of vegetative growth, the reason for its non-occurrence would be the presence of this strength, since when plants are strong nothing in them is easily affected. 19. 3

The closing and opening of the flowers is a less difficult matter and easier to solve, since it is brought about by cold and heat, the flowers being cold and weak. Thus they close up when their fluid condenses and (as it were) freezes (since at this time their heat leaves them too), and open when the fluid dissolves again and thaws, this being done by the sun.

The plants that sink under the water and emerge above it to a greater extent [b] are evidently colder and weaker than the flowers, and for this reason more affected by the changes. That a plant under water should be so keenly sensitive is not unreasonable, 19. 4

[a] That is, all evergreens but the olive.
[b] Than the lotus of the Nile: *cf. HP* 4 8. 9, translated in note *b* on *CP* 2 19. 1.

---

9 ὑπερίσχοντα Scaliger: ὑπερϊσχύοντα U.

5 ἄλλως τε καὶ ἐν τόποις θερμοῖς καὶ ἐμπύροις.[1] ἐπεὶ καὶ ἐν τοῖς μὴ τοιούτοις αἱ διαδόσεις ταχεῖαι πάντων ἀπὸ τοῦ ἡλίου καὶ τῶν ἄστρων· φαίνεται γοῦν συμπάσχειν οὐ μόνον τὰ ἐπὶ γῆς, ἀλλὰ καὶ τὰ ὑπὸ γῆς ὕδατα τροπαῖς τε καὶ ἐπιτολαῖς· ἐπὶ 10 ἐνίων δὲ ἄστρων καὶ αὐτὴ ἡ γῆ καὶ ἡ θάλαττα μεταβάλλει.

19. 5 πάσχει δέ τι παραπλήσιον τούτῳ καὶ τῶν ἀνθῶν πολλὰ καθ' ἡμέραν· ἀεὶ γὰρ συμπεριφέρεται τῷ ἡλίῳ νεύοντα καὶ ἐγκλίνοντα πρὸς αὐτόν· μᾶλλον δ' ἐστὶν τοῦτο καταμαθεῖν ἐν τοῖς ἐλάτ- 5 τοσιν. ἐνίων δὲ καὶ τὰ φύλλα πάσχει[2] ταὐτό, καθάπερ τῆς μαλάχης καὶ τῶν τοιούτων. αἴτιον δὲ τοῦ ὑγροῦ ἡ ἀφαίρεσις· ᾗ[3] γὰρ ἂν ἐξάγῃ θερμαίνων ὁ ἥλιος, ἐν τούτῳ ἡ ἔγκλισις· ἐξάγει δὲ καθ' αὐτόν, ὡς καί, τοῦτο[4] ποιοῦντος, ἀποστρέ- 10 φεταί[5] τε[6] καὶ περιάγεται.

τοῦτο μὲν οὖν ὡς πίστεως χάριν εἰρήσθω πρὸς τὸ πρότερον ῥηθέν.

1–3. *Cf.* Varro, *R.R.* i. 46: nec minus admirandum quod fit in floribus quos vocant heliotropia ab eo quod ad solis ortum mane spectant et eius iter ita secuntur ad occasum, ut ad eum semper spectent.

[1] ἐμπύροις Schneider: ἐκπύροις U.
[2] πάσχει U^r: πάσχειν U^ar; συμπάσχει N HP.
[3] ᾗ u: ἡ U.
[4] ὡς καὶ τοῦτο U: τοῦτο δὲ Dalecampius; ὥστ' ἀεὶ (or ὥστε) τοῦτο Schneider.
[5] ἀποστρέφεται U: ἐπιστρέφεται Schneider.
[6] τε ego (Dalecampius and Schneider delete): γὰρ U.

especially in a torrid region of fiery heat. Indeed even in temperate regions the transmission to water of all effects arising from the position of the sun and the heavenly bodies is rapid. At any rate not only surface waters but also waters under ground appear to be influenced by the solstices and the risings of the stars,[a] and with some stars land itself and sea are changed.[b]

19. 5 Many flowers too are affected by day in a way similar to this, since they keep moving around with the sun, nodding and bending in its direction; one can see this better in the smaller ones. In some plants the leaves as well are affected in the same way, as in the mallow and the like. The cause is the removal of their fluid, since at whatever point the sun warms and removes it, in that part the bending occurs; but the sun removes it on its own side, so that, as the sun keeps this up, the plant slants from its position and moves around with it.

This point is to be taken as added in support of the preceding explanation.[c]

[a] *Cf. CP* 1 13. 6 (waters burst forth from the earth in the dog days).

[b] *Cf.* Aristotle, *On the Generation of Animals*, iv. 9 (777 b 30–35): "For just as we observe not only the sea but the whole nature of fluids to come to rest and to change with the motion and rest of the winds, and the air and the winds with the revolution of the sun and moon, so too the things that have their growth from these and in these must follow suit . . ."

Cf. also A. Rehm, article Episemasiai in *RE* Supplementband vii (1950), coll. 175. 63–185. 19, and especially the tables (coll. 187–188).

[c] *CP* 2 19. 4.

19. 6 ὑπὲρ δὲ τῶν ἄλλων ὅσα συμβαίνει τοῖς δένδροις ἢ φυτοῖς πειρατέον ἐκ τῶν δένδρων μετιέναι καὶ θεωρεῖν, τὴν ἰδίαν οὐσίαν ἑκάστου λαμβάνοντας [1] καὶ τὴν τῆς χώρας φύσιν· ἐκ τούτων γὰρ τὰ κοινὰ 5 πάθη καὶ αἱ κατὰ γένη παραλλαγαὶ καὶ τὸ ἁρμόττον καὶ τὸ οἰκεῖον ἑκάστοις γίνεται φανερόν. δεῖ δὲ καὶ τὸ ὅμοιον καὶ τὸ ταὐτὸ [2] δύνασθαι θεωρεῖν, πολλὰ γὰρ δοκεῖ διαφέροντα οὐ διαφέρειν, ὥσπερ καὶ ἐπὶ τῶν ἄλλων.

10 καὶ ταῦτα μὲν ἐνταῦθα ἐχέτω τὸ πέρας· ὅσα δὲ κατὰ τὰς γεωργίας συμβαίνει, καὶ ὧν χάριν ἕκαστα ποιοῦσιν, λεκτέον ὁμοίως.[3]

[1] λαμβάνοντας u HP: -ος U N.
[2] ταὐτὸ ego: αὐτὸ U.
[3] U subscribes θεοφραστου περι φυτῶν αἰτιων τὸ β̄.

Conclusion to the Discussion of Natural Occurrences

In dealing with all other occurrences in trees or plants we must endeavour to start with the study of the trees, taking as our bases the distinctive nature of each and the nature of the country, since from these bases the common affections and differences between kinds and what comports with and is appropriate to[a] each kind become evident.[b] We must also be able to discern what is merely similar and what is identical, since many occurrences that differ are considered not to do so, as in other matters.[c] 19. 6

Let the present subject[d] find its conclusion here. We must now discuss on the same lines what occurs in the procedures of husbandry, and the purposes for which the different procedures are undertaken.

[a] For the distinction *cf*. *CP* 2 16. 8.
[b] *Cf*. Plato, *Phaedrus*, 270 E 2–271 B 5.
[c] *Cf*. Plato, *Phaedrus*, 261 D 10–262 C 3.
[d] *Cf*. *CP* 2 1. 1; the discussion of this subject covers *CP* 2 1. 2–2 19. 6.

Printed in Great Britain
by Richard Clay (The Chaucer Press), Ltd,
Bungay, Suffolk

THE LOEB CLASSICAL LIBRARY

VOLUMES ALREADY PUBLISHED

Latin Authors

Ammianus Marcellinus. Translated by J. C. Rolfe. 3 Vols.

Apuleius: The Golden Ass (Metamorphoses). W. Adlington (1966). Revised by S. Gaselee.

St. Augustine: City of God. 7 Vols. Vol. I. G. E. McCracken. Vol. II and VII. W. M. Green. Vol. III. D. Wiesen. Vol. IV. P. Levine. Vol. V. E. M. Sanford and W. M. Green. Vol. VI. W. C. Greene.

St. Augustine, Confessions of. W. Watts (1631). 2 Vols.

St. Augustine, Select Letters. J. H. Baxter.

Ausonius. H. G. Evelyn White. 2 Vols.

Bede. J. E. King. 2 Vols.

Boethius: Tracts and De Consolatione Philosophiae. Rev. H. F. Stewart and E. K. Rand. Revised by S. J. Tester.

Caesar: Alexandrian, African and Spanish Wars. A. G. Way.

Caesar: Civil Wars. A. G. Peskett.

Caesar: Gallic War. H. J. Edwards.

Cato: De Re Rustica. Varro: De Re Rustica. H. B. Ash and W. D. Hooper.

Catullus. F. W. Cornish. Tibullus. J. B. Postgate. Pervigilium Veneris. J. W. Mackail.

Celsus: De Medicina. W. G. Spencer. 3 Vols.

Cicero: Brutus, and Orator. G. L. Hendrickson and H. M. Hubbell.

[Cicero]: Ad Herennium. H. Caplan.

Cicero: De Oratore, etc. 2 Vols. Vol. I. De Oratore, Books I and II. E. W. Sutton and H. Rackham. Vol. II. De Oratore, Book III. De Fato; Paradoxa Stoicorum; De Partitione Oratoria. H. Rackham.

Cicero: De Finibus. H. Rackham.

Cicero: De Inventione, etc. H. M. Hubbell.

Cicero: De Natura Deorum and Academica. H. Rackham.

Cicero: De Officiis. Walter Miller.

Cicero: De Republica and De Legibus: Somnium Scipionis. Clinton W. Keyes.

Cicero: De Senectute, De Amicitia, De Divinatione. W. A. Falconer.

Cicero: In Catilinam, Pro Flacco, Pro Murena, Pro Sulla. New version by C. Macdonald.

Cicero: Letters to Atticus. E. O. Winstedt. 3 Vols.

Cicero: Letters to His Friends. W. Glynn Williams, M. Cary, M. Henderson. 4 Vols.

Cicero: Philippics. W. C. A. Ker.

Cicero: Pro Archia, Post Reditum, De Domo, De Haruspicum Responsis, Pro Plancio. N. H. Watts.

Cicero: Pro Caecina, Pro Lege Manilia, Pro Cluentio, Pro Rabirio. H. Grose Hodge.

Cicero: Pro Caelio, De Provinciis Consularibus, Pro Balbo. R. Gardner.

Cicero: Pro Milone, In Pisonem, Pro Scauro, Pro Fonteio, Pro Rabirio Postumo, Pro Marcello, Pro Ligario, Pro Rege Deiotaro. N. H. Watts.

Cicero: Pro Quinctio, Pro Roscio Amerino, Pro Roscio Comoedo, Contra Rullum. J. H. Freese.

Cicero: Pro Sestio, In Vatinium. R. Gardner.

Cicero: Tusculan Disputations. J. E. King.

Cicero: Verrine Orations. L. H. G. Greenwood. 2 Vols.

Claudian. M. Platnauer. 2 Vols.

Columella: De Re Rustica. De Arboribus. H. B. Ash, E. S. Forster and E. Heffner. 3 Vols.

Curtius, Q.: History of Alexander. J. C. Rolfe. 2 Vols.

Florus. E. S. Forster. Cornelius Nepos. J. C. Rolfe.

Frontinus: Stratagems and Aqueducts. C. E. Bennett and M. B. McElwain.

Fronto: Correspondence. C. R. Haines. 2 Vols.

Gellius. J. C. Rolfe. 3 Vols.

Horace: Odes and Epodes. C. E. Bennett.

Horace: Satires, Epistles, Ars Poetica. H. R. Fairclough.

Jerome: Selected Letters. F. A. Wright.

Juvenal and Persius. G. G. Ramsay.

Livy. B. O. Foster, F. G. Moore, Evan T. Sage, and A. C. Schlesinger and R. M. Geer (General Index). 14 Vols.

Lucan. J. D. Duff.

Lucretius. W. H. D. Rouse. Revised by M. F. Smith.

Martial. W. C. A. Ker. 2 Vols.

Minor Latin Poets: from Publilius Syrus to Rutilius Namatianus, including Grattius, Calpurnius Siculus, Nemesianus, Avianus and others, with "Aetna" and the "Phoenix." J. Wight Duff and Arnold M. Duff.

Ovid: The Art of Love and Other Poems. J. H. Mozley.

Ovid: Fasti. Sir James G. Frazer.

Ovid: Heroides and Amores. Grant Showerman.
Ovid: Metamorphoses. F. J. Miller. 2 Vols.
Ovid: Tristia and Ex Ponto. A. L. Wheeler.
Persius. Cf. Juvenal.
Petronius. M. Heseltine. Seneca: Apocolocyntosis. W. H. D. Rouse.
Phaedrus and Babrius (Greek). B. E. Perry.
Plautus. Paul Nixon. 5 Vols.
Pliny: Letters, Panegyricus. Betty Radice. 2 Vols.
Pliny: Natural History. Vols. I.–V. and IX. H. Rackham. VI.–VIII. W. H. S. Jones. X. D. E. Eichholz. 10 Vols.
Propertius. H. E. Butler.
Prudentius. H. J. Thomson. 2 Vols.
Quintilian. H. E. Butler. 4 Vols.
Remains of Old Latin. E. H. Warmington. 4 Vols. Vol. I. (Ennius and Caecilius.) Vol. II. (Livius, Naevius, Pacuvius, Accius.) Vol. III. (Lucilius and Laws of XII Tables.) Vol. IV. (Archaic Inscriptions.)
Sallust. J. C. Rolfe.
Scriptores Historiae Augustae. D. Magie. 3 Vols.
Seneca, The Elder: Controversiae, Suasoriae. M. Winterbottom. 2 Vols.
Seneca: Apocolocyntosis. Cf. Petronius.
Seneca: Epistulae Morales. R. M. Gummere. 3 Vols.
Seneca: Moral Essays. J. W. Basore. 3 Vols.
Seneca: Tragedies. F. J. Miller. 2 Vols.
Seneca: Naturales Quaestiones. T. H. Corcoran. 2 Vols.
Sidonius: Poems and Letters. W. B. Anderson. 2 Vols.
Silius Italicus. J. D. Duff. 2 Vols.
Statius. J. H. Mozley. 2 Vols.
Suetonius. J. C. Rolfe. 2 Vols.
Tacitus: Dialogus. Sir Wm. Peterson. Agricola and Germania. Maurice Hutton. Revised by M. Winterbottom, R. M. Ogilvie, E. H. Warmington.
Tacitus: Histories and Annals. C. H. Moore and J. Jackson. 4 Vols.
Terence. John Sargeaunt. 2 Vols.
Tertullian: Apologia and De Spectaculis. T. R. Glover. Minucius Felix. G. H. Rendall.
Valerius Flaccus. J. H. Mozley.
Varro: De Lingua Latina. R. G. Kent. 2 Vols.
Velleius Paterculus and Res Gestae Divi Augusti. F. W. Shipley.
Virgil. H. R. Fairclough. 2 Vols.
Vitruvius: De Architectura. F. Granger. 2 Vols.

Greek Authors

Achilles Tatius. S. Gaselee.

Aelian: On the Nature of Animals. A. F. Scholfield. 3 Vols.

Aeneas Tacticus. Asclepiodotus and Onasander. The Illinois Greek Club.

Aeschines. C. D. Adams.

Aeschylus. H. Weir Smyth. 2 Vols.

Alciphron, Aelian, Philostratus: Letters. A. R. Benner and F. H. Fobes.

Andocides, Antiphon. Cf. Minor Attic Orators.

Apollodorus. Sir James G. Frazer. 2 Vols.

Apollonius Rhodius. R. C. Seaton.

The Apostolic Fathers. Kirsopp Lake. 2 Vols.

Appian: Roman History. Horace White. 4 Vols.

Aratus. Cf. Callimachus.

Aristides: Orations. C. A. Behr. Vol. I.

Aristophanes. Benjamin Bickley Rogers. 3 Vols. Verse trans.

Aristotle: Art of Rhetoric. J. H. Freese.

Aristotle: Athenian Constitution, Eudemian Ethics, Vices and Virtues. H. Rackham.

Aristotle: Generation of Animals. A. L. Peck.

Aristotle: Historia Animalium. A. L. Peck. Vols. I.–II.

Aristotle: Metaphysics. H. Tredennick. 2 Vols.

Aristotle: Meteorologica. H. D. P. Lee.

Aristotle: Minor Works. W. S. Hett. On Colours, On Things Heard, On Physiognomies, On Plants, On Marvellous Things Heard, Mechanical Problems, On Indivisible Lines, On Situations and Names of Winds, On Melissus, Xenophanes, and Gorgias.

Aristotle: Nicomachean Ethics. H. Rackham.

Aristotle: Oeconomica and Magna Moralia. G. C. Armstrong (with Metaphysics, Vol. II).

Aristotle: On the Heavens. W. K. C. Guthrie.

Aristotle: On the Soul, Parva Naturalia, On Breath. W. S. Hett.

Aristotle: Categories, On Interpretation, Prior Analytics. H. P. Cooke and H. Tredennick.

Aristotle: Posterior Analytics, Topics. H. Tredennick and E. S. Forster.

Aristotle: On Sophistical Refutations.
On Coming to be and Passing Away, On the Cosmos. E. S. Forster and D. J. Furley.

Aristotle: Parts of Animals. A. L. Peck; Motion and Progression of Animals. E. S. Forster.

Aristotle: Physics. Rev. P. Wicksteed and F. M. Cornford. 2 Vols.

Aristotle: Poetics and Longinus. W. Hamilton Fyfe; Demetrius on Style. W. Rhys Roberts.

Aristotle: Politics. H. Rackham.

Aristotle: Problems. W. S. Hett. 2 Vols.

Aristotle: Rhetorica Ad Alexandrum (with Problems. Vol. II). H. Rackham.

Arrian: History of Alexander and Indica. Rev. E. Iliffe Robson. 2 Vols.

Athenaeus: Deipnosophistae. C. B. Gulick. 7 Vols.

Babrius and Phaedrus (Latin). B. E. Perry.

St. Basil: Letters. R. J. Deferrari. 4 Vols.

Callimachus: Fragments. C. A. Trypanis. Musaeus: Hero and Leander. T. Gelzer and C. Whitman.

Callimachus: Hymns and Epigrams. Lycophron. A. W. Mair. Aratus. G. R. Mair

Clement of Alexandria. Rev. G. W. Butterworth.

Colluthus. Cf. Oppian.

Daphnis and Chloe. Thornley's Translation revised by J. M. Edmonds. Parthenius. S. Gaselee.

Demosthenes I: Olynthiacs, Philippics and Minor Orations I–XVII and XX. J. H. Vince.

Demosthenes II: De Corona and De Falsa Legatione. C. A. Vince and J. H. Vince.

Demosthenes III: Meidias, Androtion, Aristocrates, Timocrates and Aristogeiton I and II. J. H. Vince.

Demosthenes IV–VI: Private Orations and In Neaeram. A. T. Murray.

Demosthenes VII: Funeral Speech, Erotic Essay, Exordia and Letters. N. W. and N. J. DeWitt.

Dio Cassius: Roman History. E. Cary. 9 Vols.

Dio Chrysostom. J. W. Cohoon and H. Lamar Crosby. 5 Vols.

Diodorus Siculus. 12 Vols. Vols. I–VI. C. H. Oldfather. Vol. VII. C. L. Sherman. Vol. VIII. C. B. Welles. Vols. IX and X. R. M. Geer. Vol. XI. F. Walton. Vol. XII. F. Walton. General Index. R. M. Geer.

Diogenes Laertius. R. D. Hicks. 2 Vols. New Introduction by H. S. Long.

Dionysius of Halicarnassus: Roman Antiquities. Spelman's translation revised by E. Cary. 7 Vols.

Dionysius of Halicarnassus: Critical Essays. S. Usher. 2 Vols.

Epictetus. W. A. Oldfather. 2 Vols.

Euripides. A. S. Way. 4 Vols. Verse trans.

Eusebius: Ecclesiastical History. Kirsopp Lake and J. E. L. Oulton. 2 Vols.

Galen: On the Natural Faculties. A. J. Brock.

The Greek Anthology. W. R. Paton. 5 Vols.

Greek Elegy and Iambus with the Anacreontea. J. M. Edmonds. 2 Vols.

The Greek Bucolic Poets (Theocritus, Bion, Moschus). J. M. Edmonds.

Greek Mathematical Works. Ivor Thomas. 2 Vols.

Herodes. Cf. Theophrastus: Characters.

Herodian. C. R. Whittaker. 2 Vols.

Herodotus. A. D. Godley. 4 Vols.

Hesiod and The Homeric Hymns. H. G. Evelyn White.

Hippocrates and the Fragments of Heracleitus. W. H. S. Jones and E. T. Withington. 4 Vols.

Homer: Iliad. A. T. Murray. 2 Vols.

Homer: Odyssey. A. T. Murray. 2 Vols.

Isaeus. E. W. Forster.

Isocrates. George Norlin and LaRue Van Hook. 3 Vols.

[St. John Damascene]: Barlaam and Ioasaph. Rev. G. R. Woodward, Harold Mattingly and D. M. Lang.

Josephus. 9 Vols. Vols. I–IV. H. Thackeray. Vol. V. H. Thackeray and R. Marcus. Vols. VI–VII. R. Marcus. Vol. VIII. R. Marcus and Allen Wikgren. Vol. IX. L. H. Feldman.

Julian. Wilmer Cave Wright. 3 Vols.

Libanius. A. F. Norman. Vol. I.

Lucian. 8 Vols. Vols. I–V. A. M. Harmon. Vol. VI. K. Kilburn. Vols. VII–VIII. M. D. Macleod.

Lycophron. Cf. Callimachus.

Lyra Graeca. J. M. Edmonds. 3 Vols.

Lysias. W. R. M. Lamb.

Manetho. W. G. Waddell. Ptolemy: Tetrabiblos. F. E. Robbins.

Marcus Aurelius. C. R. Haines.

Menander. F. G. Allison.

Minor Attic Orators (Antiphon, Andocides, Lycurgus, Demades, Dinarchus, Hyperides). K. J. Maidment and J. O. Burtt. 2 Vols.

Musaeus: Hero and Leander. Cf. Callimachus.

Nonnos: Dionysiaca. W. H. D. Rouse. 3 Vols.

Oppian, Colluthus, Tryphiodorus. A. W. Mair.

Papyri. Non-Literary Selections. A. S. Hunt and C. C. Edgar. 2 Vols. Literary Selections (Poetry). D. L. Page

Parthenius. Cf. Daphnis and Chloe.

Pausanias: Description of Greece. W. H. S. Jones. 4 Vols. and Companion Vol. arranged by R. E. Wycherley.

PHILO. 10 Vols. Vols. I–V. F. H. Colson and Rev. G. H. Whitaker. Vols. VI–IX. F. H. Colson. Vol. X. F. H. Colson and the Rev. J. W. Earp.

PHILO: two supplementary Vols. (*Translation only.*) Ralph Marcus.

PHILOSTRATUS: THE LIFE OF APOLLONIUS OF TYANA. F. C. Conybeare. 2 Vols.

PHILOSTRATUS: IMAGINES. CALLISTRATUS: DESCRIPTIONS. A. Fairbanks.

PHILOSTRATUS and EUNAPIUS: LIVES OF THE SOPHISTS. Wilmer Cave Wright.

PINDAR. Sir J. E. Sandys.

PLATO: CHARMIDES, ALCIBIADES, HIPPARCHUS, THE LOVERS, THEAGES, MINOS and EPINOMIS. W. R. M. Lamb.

PLATO: CRATYLUS, PARMENIDES, GREATER HIPPIAS, LESSER HIPPIAS. H. N. Fowler.

PLATO: EUTHYPHRO, APOLOGY, CRITO, PHAEDO, PHAEDRUS. H. N. Fowler.

PLATO: LACHES, PROTAGORAS, MENO, EUTHYDEMUS. W. R. M. Lamb.

PLATO: LAWS. Rev. R. G. Bury. 2 Vols.

PLATO: LYSIS, SYMPOSIUM, GORGIAS. W. R. M. Lamb.

PLATO: REPUBLIC. Paul Shorey. 2 Vols.

PLATO: STATESMAN, PHILEBUS. H. N. Fowler. ION. W. R. M. Lamb.

PLATO: THEAETETUS and SOPHIST. H. N. Fowler.

PLATO: TIMAEUS, CRITIAS, CLITOPHO, MENEXENUS, EPISTULAE. Rev. R. G. Bury.

PLOTINUS. A. H. Armstrong. Vols. I–III.

PLUTARCH: MORALIA. 17 Vols. Vols. I–V. F. C. Babbitt. Vol. VI. W. C. Helmbold. Vols. VII and XIV. P. H. De Lacy and B. Einarson. Vol. VIII. P. A. Clement and H. B. Hoffleit. Vol. IX. E. L. Minar, Jr., F. H. Sandbach, W. C. Helmbold. Vol. X. H. N. Fowler. Vol. XI. L. Pearson and F. H. Sandbach. Vol. XII. H. Cherniss and W. C. Helmbold. Vol. XV. F. H. Sandbach.

PLUTARCH: THE PARALLEL LIVES. B. Perrin. 11 Vols.

POLYBIUS. W. R. Paton. 6 Vols.

PROCOPIUS: HISTORY OF THE WARS. H. B. Dewing. 7 Vols.

PTOLEMY: TETRABIBLOS. Cf. MANETHO.

QUINTUS SMYRNAEUS. A. S. Way. Verse trans.

SEXTUS EMPIRICUS. Rev. R. G. Bury. 4 Vols.

SOPHOCLES. F. Storr. 2 Vols. Verse trans.

STRABO: GEOGRAPHY. Horace L. Jones. 8 Vols.

THEOPHRASTUS: CHARACTERS. J. M. Edmonds. HERODES, etc. A. D. Knox.

Theophrastus: De Causis Plantarum. B. Einarson and G. K. K. Link. Vol. I.
Theophrastus: Enquiry into Plants. Sir Arthur Hort, Bart. 2 Vols.
Thucydides. C. F. Smith. 4 Vols.
Tryphiodorus. Cf. Oppian.
Xenophon: Cyropaedia. Walter Miller. 2 Vols.
Xenophon: Hellenica. C. L. Brownson. 2 Vols.
Xenophon: Anabasis. C. L. Brownson.
Xenophon: Memorabilia and Oeconomicus. E. C. Marchant. Symposium and Apology. O. J. Todd.
Xenophon: Scripta Minora. E. C. Marchant and G. W. Bowersock.

IN PREPARATION

Greek Authors

Arrian I. New version by P. A. Brunt.
Plutarch: Moralia XIII 1–2. H. Cherniss.

Latin Authors

Manilius. G. P. Goold.

DESCRIPTIVE PROSPECTUS ON APPLICATION

CAMBRIDGE, MASS. **HARVARD UNIVERSITY PRESS**
LONDON **WILLIAM HEINEMANN LTD**

Theophrastus.
De causis
plantarum.English & Greek.

QK
41
T23
1976
v.1

CARLOW COLLEGE
Grace Library
Pittsburgh, Pa. 15213